Sustainable Energy Systems: From Primary to End-Use

Special Issue Editors

João Carlos de Oliveira Matias
Radu Godina
Leonel Jorge Ribeiro Nunes

MDPI • Basel • Beijing • Wuhan • Barcelona • Belgrade

Special Issue Editors

João Carlos de Oliveira Matias
University of Aveiro
Portugal

Radu Godina
Universidade NOVA de Lisboa
Portugal

Leonel Jorge Ribeiro Nunes
University of Aveiro
Portugal

Editorial Office
MDPI
St. Alban-Anlage 66
4052 Basel, Switzerland

This is a reprint of articles from the Special Issue published online in the open access journal *Sustainability* (ISSN 2071-1050) from 2018 to 2019 (available at: https://www.mdpi.com/journal/sustainability/special_issues/Sustainable_Energy_Systems)

For citation purposes, cite each article independently as indicated on the article page online and as indicated below:

LastName, A.A.; LastName, B.B.; LastName, C.C. Article Title. *Journal Name* **Year**, *Article Number*, Page Range.

ISBN 978-3-03921-096-1 (Pbk)
ISBN 978-3-03921-097-8 (PDF)

Sustainable Energy Systems

Contents

About the Special Issue Editors

João Carlos de Oliveira Matias holds a B.Sc. in Mechanical Engineering (1994), a Ph.D. degree in Production Engineering (2003), and Habilitation for Full Professor "Agregação" (2014) in Industrial Engineering and management. He is a Full Professor of the Department of Economics, Management, Industrial Engineering, and Tourism (DEGEIT), University of Aveiro—Portugal and member of the Industrial Engineering and Management Research Group. He is the Coordinator of the Scientific Area of Industrial Engineering and Management. His areas of research focus on industrial management in general and in sustainable energy systems and management systems in particular. He has been involved in several research projects (Local Coordinator of Projects: SINGULAR, Grant Agreement No: 309048, FP7-EU, from December 2012 to November 2015. Funding: 5,259,445.00 and PTDC/EEA-EEL/118519/2010, FCT-Portugal, from February 2012 to January 2015. Funding: 133,000.00). Additionally, he is the author or co-author of more than 200 articles published in several international journals and congress proceedings.

Radu Godina holds a B.Sc. in Electromechanical Engineering (2011), as well as an M.Sc. and Ph.D. degree in Industrial Engineering and Management, obtained in 2013 and 2016, respectively. Radu Godina is currently a researcher and an Invited Assistant Professor at the NOVA University Lisbon. He is an author and co-author of more than 90 journal, book chapter and conference proceedings papers in topics closely related to industrial engineering and energy management.

Leonel Jorge Ribeiro Nunes holds a Ph.D. in Industrial Engineering and Management from the University of Beira Interior, a M.Sc. in Hydraulics and Water Resources from the University of Lisbon, a M.Sc. in Geological Engineering from the University NOVA of Lisbon, a degree in Geology from the University of Coimbra, a degree in Geology from the University of Minho, and a degree in Mining and Geo-Environment Engineering from the Faculty of Engineering of the University of Oporto. He is an Invited Assistant Professor in the University of Aveiro, at the Department of Economy, Management, Industrial Engineering, and Tourism (DEGEIT), in the Agrarian Higher School of the Polytechnic Institute of Viana do Castelo and in the University Institute of Maia (ISMAI). He lectures undergraduate and graduate courses on natural resources management, sustainability, production efficiency, biomass energy and biomass conversion technologies. He is a researcher on governance, competitiveness and public policy (GOVCOPP) and has had more than 50 scientific works published in books, chapters, articles, and conference proceedings. He is a reviewer in several journals as well as international conferences. Has more than 20 years of experience in industry and is presently a Member of the Board of Directors at AFS—Advanced Fuel Solutions SA, a large-scale biomass torrefaction plant.

Preface to "Sustainable Energy Systems: From Primary to End-Use"

The long-term transformation of the energy industry, from fossil to sustainable energy sources, is one of the central challenges of the 21st century. Such a transition requires many changes on a technical and organizational level. The successful implementation of the energy transition requires not only sufficient generation capacity from renewable energies, but also an efficient, intelligent, decentralized and secure infrastructure for the distribution, storage and use of electricity and heat. The resulting change in the energy industry poses major challenges for small and medium-sized enterprises (SMEs) in particular. On the one hand, this creates great opportunities for the development of new business models, products and processes, on the other hand, a stronger networking of decentralized system components also entails risks, which are due in particular to the increased use of information and communication technology (ICT) and the automation of processes. The goal must therefore be to create solutions for a sustainable energy system that is economically, environmentally and socially viable, while meeting high security requirements.

This book focuses on sustainable energy systems. On the one hand, several innovative and alternative concepts are presented, but also the topics of energy policy, life cycle assessment, thermal energy, and renewable energy play a major role. Models on various temporal and geographical scales are developed to understand the conditions of technical as well as organizational change. New methods of modeling, which can fulfil technical and physical boundary conditions and nevertheless consider economic environmental and social aspects, are also developed.

João Carlos de Oliveira Matias, Radu Godina, Leonel Jorge Ribeiro Nunes

Special Issue Editors

Article

Research on Increasing the Performance of Wind Power Plants for Sustainable Development

Adriana Florescu [1,*], Sorin Barabas [1] and Tiberiu Dobrescu [2]

[1] Faculty of Technological Engineering and Industrial Management; Transilvania University of Brasov, 500222 Brasov, Romania; sorin.barabas@unitbv.ro
[2] Faculty on Engineering and Management of Technological Systems; University Polytechnic of Bucharest, 060040 Bucharest, Romania; tibidobrescu@yahoo.com
* Correspondence: fota.a@unitbv.ro; Tel.: +40-735-788-292

Received: 20 November 2018; Accepted: 11 February 2019; Published: 27 February 2019

Abstract: A topical issue globally is the development and implementation of renewable energy sources for sustainable development. To meet current requirements, the research in this paper is directed towards finding solutions to increase the performance and efficiency of wind power plants by implementing innovative solutions for hollow roller bearings developed through the use of sustainable growth programs in the field of green energy. Another solution that has the effect of increasing wind power performance consists of the implementation of a new large-size lubrication system for large-size bearings in wind energy units, which will increase their durability by developing maintenance capabilities. In this research, we will explore the possibility of introducing an innovative automated lubrication system in hollow roller bearings. The main results of the research, the innovative constructive solutions, will lead to important savings by lowering wind farm maintenance costs, increasing the durability of large bearings, and increasing the energy efficiency and yield of the whole system. The expected impact of implementing the solutions found will mainly be in the field of sustainable growth and environmental development.

Keywords: renewable energies; wind power plants; hollow rollers; large bearings

---

1. Introduction

Wind power is currently one of the most important alternative sources of renewable energy, offering the major advantage of zero emissions of greenhouse gases by reducing and even eliminating the use of fossil fuels. By developing the wind energy sector as the energy of the future, environmental and sustainable development requirements will be met while achieving energy security objectives. Implementation of wind power projects aiming at building wind turbines that produce more energy, thus reducing costs in the context of sustainable development, has been a major concern in recent years.

Over the years, researchers have consistently sought new ways of making wind turbines more and more eco-friendly [1]. Research in recent years has provided an overview of different solutions to develop a new generation of wind turbines, and feasible and technological trends in wind power generation that could reduce costs. The need to use renewable energy, as well as a review of various wind technologies in connection with their applications and operating devices, is detailed in this paper [2]. It is demonstrated that the cost of generating electricity, as well as the current economic and environmental policies, supports the installation and development of wind power systems. This paper provides arguments on the usefulness of this research in the development of wind assemblies. In [3], some technological solutions regarding the power electronics in the wind turbine systems are presented. The paper provides an overview of the state of the technology and discusses the technological trends in the field. There are also some difficulties that arise in the operation of the control part, generators and gearboxes, where the main components are known to be bearings. It is necessary to find solutions to improve the operation and reliability of wind turbines, and the solutions proposed in the present paper follow this direction. Another technical solution, consisting of the implementation of a new concept, "the adaptive-blade concept in wind-power applications", is presented in [4]. One of the technological challenges in the wind energy field is the development of a new generation of feasible, improved turbines, which further reduce production costs. As the size of the typical turbine increases, weight reductions, as well as complexity in the design of rotors and their auxiliary mechanisms, are becoming increasingly important. The study proposed for reducing inertial masses by implementing large bearings with hollow rollers is of great interest. After some research on the reliability of wind turbines [5], analysis of the failure modes, and causes of defects, it was found that the highest failure rate occurs in some key components of the wind turbine assembly, including the generator, gearbox and blades. Transmission and generator faults are mainly caused by bearings. Ref. [6] discusses the analysis requirements for the design and operation of the main bearings in modern multi-megawatt wind turbines, with a view to finding technical solutions for bearings with high reliability and profitability.

Wearing processes considerably reduce the precision and durability of bearings [7]. In this paper, an analysis is carried out of bearing failures caused by wear. The presence of different types of wear and the detection of bearing defects are studied. The bearing elements (inner and outer rings, rollers) are put into operation with complex loads, such as: high strains and alternating stretch and compression, rolling friction, abrasive friction and corrosion.

Worldwide, in the field of large-sized bearings, studies only began relatively late (1970–1975), compared with studies in the field of small- and medium-sized bearings. In countries with bearing manufacturing traditions (USA, Japan and Germany), a great deal of research on large bearings has been seen in the last decade. It has been found that wind turbine generator bearings have a surprisingly high failure rate, with failures happening too early due to classical rolling contact fatigue [8]. Reference [9] studies the behavior of the bearings in the gearbox of a wind turbine and the impact on its reliability. Inertial forces on small- and medium-sized bearings do not have the negative effects that they have on large-sized bearings.

Today, all the construction solutions for wind turbines use large roller bearings with solid rollers, with expensive logarithmic profiles to reduce contact pressures and increase resistance under variable load conditions [10,11]. Based on all of this research, this paper proposes new technical solutions for improving the lifespan of wind turbines, especially the large-sized bearings inside the gearing system, by implementing an automated lubrication system in the hollow rollers. Following previous studies by the authors [12–14], a solution to using hollow roller large bearings in the construction of wind power plant assemblies is proposed.

The proposed solution using hollow rollers requires only a cylindrical profile (easy to obtain) with a similar behavior to the solid rollers with a logarithmic profile. Due to the weight of the rollers, inertial masses have been widely investigated, with the use of hybrid or ceramic bearings being proposed [15]. Their high cost requires new research on the reduction of inertial masses of large-size bearings. Attempts have been made to reduce the negative influence of centrifugal loads by decreasing the density of the rolling elements, in turn achieved by using lower density materials.

There are constructive solutions that have adopted drilled or hollow steel rolling elements [16,17], but these are not intended for the construction of large-size bearings for wind turbines. A solution for determining the size of the hollow roller was described in [18]. The analysis has shown that the durability of hollow cylindrical roller bearings operating at the maximum allowable stress level can be substantially higher than the durability of solid roller bearings.

After analyzing the specialty literature, it was found that there is no standard method for calculating the optimal hollowness of hollow roller bearings. This depends mainly on the applied load, and the size and the material of the roller. The use of hollow rollers in large bearings has some advantages over the solid rollers. These include reducing the material used in roller making, reducing the weight of the roller bearing, and attaining the preloading capability of the hollow cylindrical element, thus generating more stability, and less noise and vibrations. Hollow roller bearings are single- or double-row radial bearings with an inner ring, outer ring, and cylindrical or thin wall rollers. The thin wall of the rollers allows preloading, as opposed to cylindrical bearings with solid rollers. Preloading rollers on large bearings with hollow rollers increases radial stiffness and reduces vibrations [19].

On the other hand, major problems in the operation of a wind turbine occur in the lubrication of the bearings. Appropriate lubrication is essential for the correct operation and lifespan of the bearing. Existing lubrication system automation involves expensive and complex solutions [20], or solutions that require highly qualified personnel and maintenance stops for wind farms. Operation and maintenance costs are estimated to reach up to 30% of the total lifecycle expenditure [21]. The proposed lubrication system is inexpensive, and easy to carry out and maintain.

Reducing subsidies in this area transforms this topic into a mainstream one, as it leads to important savings by lowering the cost of maintenance of wind farms by increasing the durability of large bearings, as well as by increasing the energy efficiency and yield of the whole system. The expected impact of the implementation of hollow roller bearings and the automated lubrication system is important in times of economic crisis, mainly in the area of sustainable growth.

1.1. Wind Energy, Strategies and Directions of Development

The wind power industry has grown greatly around the world, and is among the green energy resources. The use of wind energy has increased nearly four times between 2004 and 2015 and currently accounts for about one third of renewable electricity. The use of terrestrial wind energy is quite close to the anticipated trajectory over the years [22]. The Global Wind Energy estimation [23] explores the future of the wind energy industry by 2020, 2030 and by 2050. Several scenarios have been developed by the International Energy Agency [24]. Based on the advanced scenario, GWEC estimates that by 2050, global wind power will reach 5806 GW (Table 1).

Table 1. Estimated global wind power (Source: GWEC 2017).

	Global Total						
Total Capacity in MW	**2013**	**2014**	**2015**	**2020**	**2030**	**2040**	**2050**
New Policies Scenario	318,354	369,596	432,656	639,478	1.259,974	2.052,528	2.869,611
450 Scenario	318,354	369,596	432,656	658,009	1.454,395	2.458,757	3.545,595
Moderate Scenario	318,354	369,596	432,656	797,028	1.675,624	2.767,351	3.983,995
Advanced Scenario	318,354	369,596	432,656	878,446	2.110,161	3.720,919	5.805,882

At the end of 2016, in the Global Wind Energy Council Report [23], 341.320 wind turbines were catalogued as being in operation, of which 104.934 were in China, 52.343 were in the USA, and 3589 were offshore wind turbines in Europe at the end of 2016. The evolution of the total capacity of all wind turbines installed worldwide during the period 2013–2017 is presented in Figure 1, according to preliminary World Energy Association (WWEA) statistical data [25]. Thus, in 2017, a capacity of 6145 MW was installed, with a growth rate of 12.30% compared to the previous year (Figure 2), representing a revival.

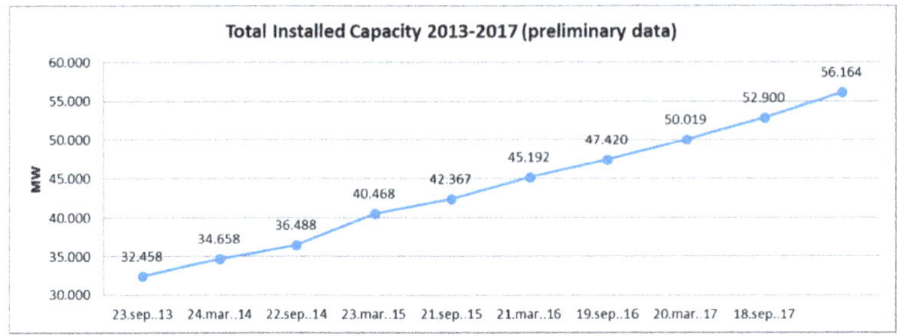

Figure 1. Total installed capacity 2013–2017 (Source: WWEA).

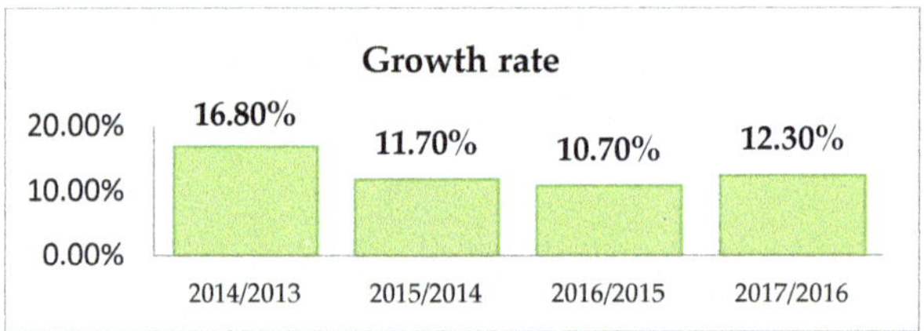

Figure 2. Growth rate of total installed capacities (Source: WWEA).

Many countries have based their strategies of phasing out fossil and nuclear energies on the development of wind power energy sources. A recent study in the specialty literature [26] presents the current state of lifespan extension of offshore wind turbines in Germany, Spain, Denmark and UK. According to [23,27], among the top 20 countries in total wind power capacity in 2017, the fastest growth of new facilities was found in the United Kingdom (+4.3 GW/+29.2%), Brazil (+2.0 GW/+18.8%), Ireland (+0.4 GW/15.8%), India (+4.1 GW/+14.5%) and France (+1.7 GW/+14.0%). Year 2017 was a spectacular year in the field of wind energy generation, with Denmark setting a new world record, with 43% of its power coming from wind. Countries like Germany, Ireland, Portugal, Spain, Sweden or Uruguay have reached a double-digit electricity share, as shown in the WWEA Report [25].

The Global Wind Energy Council (GWEC) considers, based on market statistics for February 2018, that wind energy has a major role to play in the sustainable development, "wind is the most competitively priced technology in many if not most markets, and the emergence of wind/solar hybrids, the more sophisticated grid management and the increasingly affordable storage, begin to paint a picture of what a fully commercial fossil-free power sector will look like [23]." Cumulative total installations are expected to reach 840 GW by the end of 2022 (Figure 3, [23]).

Figure 3. Global cumulative wind capacity at the end of 2017 (Source: GWEC).

For the European Union, combating climate change is an important objective to ensure sustainable development. The measures taken by the EU to meet this main target, as well as the necessary activities in this area, both for the next period up to 2020 and for the period after 2020, are detailed in the GWEC Report [23]. Thus, it is considered that "Europe is expected to align with its 2020 targets, and current discussions within the EU indicate that overall renewable targets could be raised to 35% by 2030, putting the industry in a position stronger for the post-2020 market. In Europe, it is expected to install 76 GW of new wind power energy by the end of 2022, reaching a cumulative total of 254 GW".

1.2. Wind Energy Potential in Romania

To make the EU a truly smart and sustainable low-carbon economy, the European Commission and the member states need to work together to use renewable energies. The strategy of the European Union and its member states on wind energy development consists of "stepping up urgent efforts to use wind energy as part of a global strategy for renewable energy and to develop a roadmap for a future 100% renewable energy", as highlighted in the (WWEA Report, 2018) [25]. The adoption of the "Clean Energy for All Europeans" package of measures aims to maintain the EU's competitiveness as the transition to clean energy changes the world's energy markets [28,29].

Currently, in Romania, wind energy has priority development through national strategies for sustainable development by 2030 and 2050 [30,31]. Recent Eurostat statistics for 2017 [32] highlight the efficiency of wind energy implementation in the EU Member States. Thus, at the end of 2016, EU wind turbines generated 315.00 GWh of electric energy (equivalent to more than 10% of the 3.1 million GWh produced in the EU). Denmark ranks first (43%), and is followed by other countries: Ireland (21%), Portugal (20%), Spain (18%) and the United Kingdom (14%). Romania, with a rate of about 10%, ranked tenth (Figure 4).

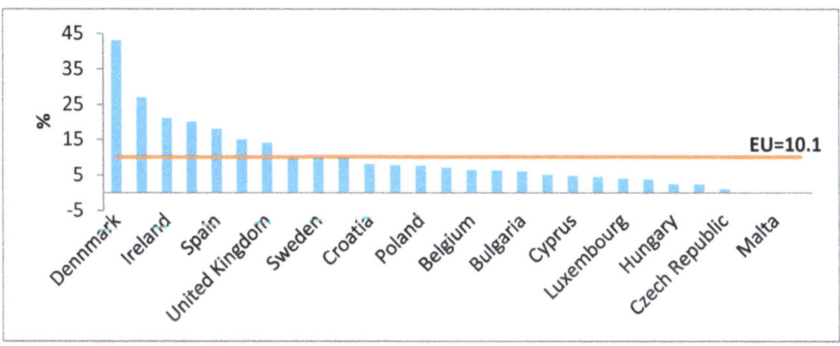

Figure 4. Share of wind in total gross electricity generation in the EU Member States, 2016 (Source: Eurostat 2017).

Regarding Romania's wind energy, five wind zones were identified, depending on the environmental, topographical and geographical conditions, taking into account the level of energy potential of such resources at an average height of 50 m and over. The results of recorded measurements show that Romania is in a temperate continental climate with a high energy potential, especially in the coast and coastal areas (mild climate), as well as in alpine areas with mountain plateaus and peaks (severe climate). Based on the evaluation and interpretation of recorded data, it is possible to conclude that in Romania, the wind energy potential is most favorable on the Black Sea coast, in the mountain areas and plateaus in Moldova or Dobrogea.

Measures performed in our country demonstrate a high wind capacity, confirming that Dobrogea is, along with northern Scotland, the most promising wind exploiting region in Europe, according to the Romanian Wind Energy Association (RWEA) specialists. In the RWEA Report [33], "Wind Energy and Other Renewable Energy Sources in Romania", there are areas (Scotland, Iceland, Denmark, Costa Rica, Tasmania, Australia) that have planned different deadlines for full reduction of carbon emissions and the development of programs to ensure green energy independence for the period 2020–2030. On 1 January 2017, Romania recorded 3025 MW in investments of over 5 billion EUR. Today, our country numbers 20 large wind farms, ranging from 70 MW up to 600 MW of installed power. Romania sources 12.3% of its electricity consumption from wind power [34].

Analyzing the widespread use of wind energy in Romania, the proposal to implement a completely new automated lubrication system positioned in the hollow rollers of large bearings is, from an economic and technical point of view, one of major interest with respect to the sustainable development of the green energy field.

2. Materials and Methods

Renewable energy sources are playing a major role in turning the EU into a world leader in innovation, with the EU owning 30% of all world-wide patents on renewable energy [22]. In the field of wind energy generation, a significant part of the cost reductions can be achieved through technological improvements and the implementation of innovative solutions that lead to sustainable development.

This research was conducted in two stages. In the first stage, the design of the hollow rollers was carried out using the finite element analysis method applied by established software, Nastran and Catia, and validation was done through direct measurement. The results consisted of studying the distribution of stresses and deformations in the rollers and bearings. The second stage was the designing, developing and testing of a prototype lubrication system implemented in the hollow rollers of large bearings.

Large-size bearing modeling has become a mandatory issue, precisely because of its size. Reducing inertial masses on these bearings would be a leap forward across the whole bearing industry. Referring to the product catalogues of the leading bearings companies (SKF, TIMKEN [35,36], INA, FAG [37,38]), cylindrical, conical or barrel roller bearings were identified in sizes up to 7 m and weighting several tons. All of these bearings have massive rollers—solid, large masses, with high inertial moments, which are disadvantages. The implementation of hollow roller bearings in wind power plants responds to the two major problems encountered in the operation of wind farms: increasing lifespan and increasing energy yield.

At present, the lifespan of a turbine is approximately 20 years, with 1/3 of this time being scheduled for maintenance. Maintenance costs increase with the age of the wind farm. According to WMI—Wind Measurement International [39]—a first-generation turbine has a maintenance cost of about 3% of its initial value; currently, this cost is falling to about 1.5%. Reduction of maintenance time can be done by increasing turbine durability to external factors: wind turbulence, variable air density, humidity, salinity, temperature. It has been found that the resulting defects lead to a 10% reduction in produced energy [40], and half of these are due to defects in bearings. All wind turbines have their own wind measuring devices and, based on the information gathered by them, the computer itself makes the adjustments necessary for optimal operation without the need for a human operator to be present at all times. The wind turbines include remote means of verification and remote control, thus making management of production units easier. Since this is an installation with moving parts, wear and maintenance costs will occur [35,36].

2.1. Operating Principle of a Wind Turbine

The system describing the operation of a turbine is based on a simple principle. The wind moves the blades that, in turn, actuate the electric generator. The mechanical system [41] includes a speed multiplier that directly operates the central shaft of the electric generator.

Figure 5 shows the main rotation movements of a horizontal wind turbine; the pivot system can be seen to feature a large bearing.

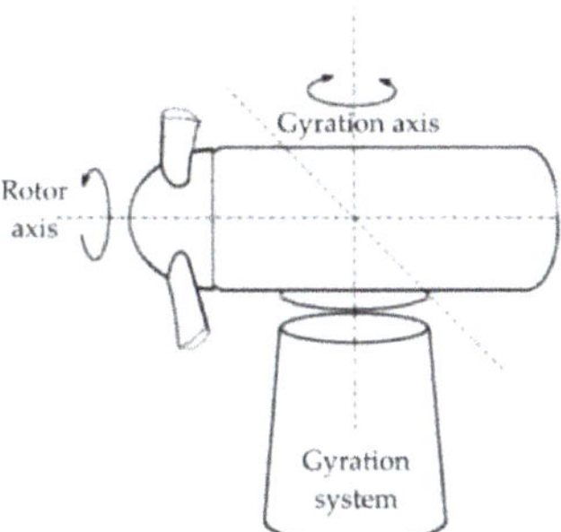

Figure 5. Rotary systems: rotor, blade, pivot, provided with large bearings.

The electrical current obtained is either transmitted for storage in batteries and then used by an inverter for low-capacity turbines or delivered directly to the AC network for distribution. The turbine that drives the electric generator is also driven by the wind pressure. The amount of electricity produced by a wind farm depends on the type and size of the turbine and the location of the farm. At low speeds, it does not generate electricity. From Beaufort 2 (about 3 m/s) onward, the turbine delivers its maximum power. At a wind speed of over 25 m/s, the turbines are designed to lock and brake in a controlled way to avoid overloading and damaging the turbine installation or construction. The latest developments are equipped with a tilt angle control device that changes the angle of the rotor blade under unfavorable weather conditions [42].

Wind turbines are equipped with a robust safety system including an aerodynamic locking system. In case of danger, or for shutdowns needed in maintenance, a locking disc is used. The rotation systems that actuate the electric generator have roller bearings provided with logarithmic profile rollers on which carburetor thermal treatment is made at high depths of 7–10 mm [12].

Reducing roller masses directly reduces inertial moments, increasing start-up speed, offering easier handling of the plant, reducing static load in the bearing, and reducing vibrations. The influence of inertial masses in the wind turbine construction, implicitly in the construction of large-size bearings, is studied by Song, Dhinakaran and Bao [43].

2.2. Logical Scheme for Choosing the Optimal Construction Solution for Large Bearings

Figure 6 illustrates the logical scheme of the interaction of influence factors in the operation of large-size bearings. Each category of factors interacts and influences the end point of the logic scheme, namely the duration of operation. Both the static load created by the bearing construction and the dynamic load due to the external wind conditions directly determine the contact stresses, hence the deformations. The roller construction, material conditions and thermal treatment are designed to counteract the negative effects of variable loads, and the temperature generated by additional frictional forces.

An appropriate lubrication system, reducing loads by reducing inertial masses, as well as faster and clearer system response to external disturbances, is a result of using hollow roller bearings in the construction of large-size bearings. Generally, when studying a bearing, the lifespan of the bearing and the system in which it is mounted are studied. In the analysis of a bearing, dynamic loading is an essential element in its definition and mathematical modeling. Dynamic loading occurs between rolling elements and rolling treads due to the movement of the rolling elements, both the movement around the bearing axis and the movement around its own axis. These movements at higher or lower speeds, depending on the use of the bearing, generate forces that interact and produce stresses between rolling elements and rolling treads.

Controlling the stresses that appear on the surface of the rolling elements leads to increased lifespan and improved bearing efficiency. In the case of large cylindrical roller bearings, due to their mass, the stresses are higher and generate significant defects and high maintenance costs.

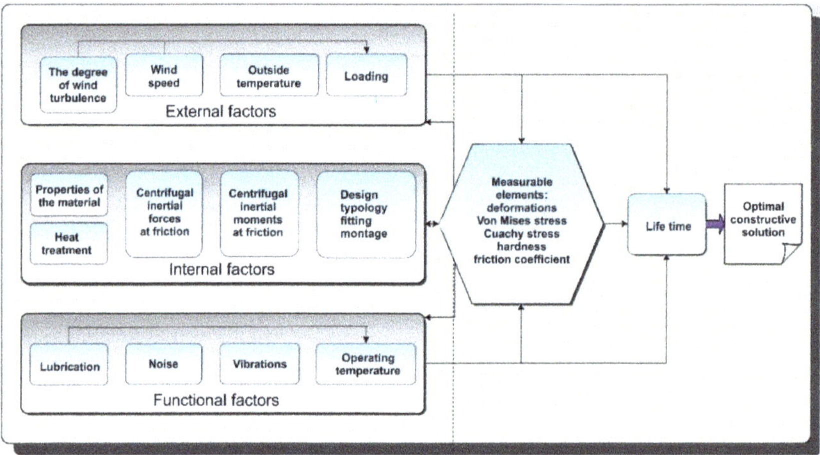

Figure 6. Logical scheme for choosing the optimal construction solution.

3. Results

The aim of this research is to implement the constructive solution of hollow rollers in the construction of large-size bearings for wind power plants. Hollowness, size and shape are features that are optimized, starting from the basic necessity of complying with the initial functionality. Studies in [12] on the behavior of hollow rollers confirmed the advantages of their use in the construction of large-size bearings, mainly for use in the wind energy industry.

3.1. Finite Element Analysis Model of Hollow Rollers

Choosing an optimal constructive variation is influenced by internal, external and operating factors, as previously presented (Figure 7). Their influence is quantified and measured by using Finite Element Analysis—FEA (Von Mises stresses and Cauchy stresses), as well as by direct measurements (hardness, friction coefficients, microscopy and macroscopy, tensometry, roughness).

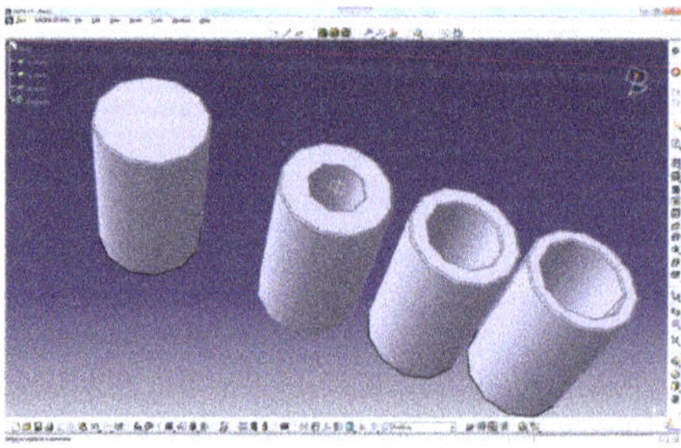

Figure 7. Cylindrical hollow roller models proposed for implementation.

3.1.1. Developing Hollow Roller Virtual Models

CAD/CAM software packages (Catia and Solid Works) were used to develop virtual models of hollow rollers (Figure 7). Many models have been developed based on the real situation of large solid roller bearings. The rollers' cavities were the main element in their design. Three hollow roller 3D models with an outer diameter of 120 mm and internal diameters of 60 mm, 80 mm and 90 mm were developed.

To obtain a generalized model for the study of stresses that appear in the bearing designed in CATIA, the Geometrical Set instrument was used. Thus, it was possible to create links between the roller types in such a way that if the inner diameter of the rollers is changed, all rollers would change. The possibility of installing covers for encapsulating a lubricant inside them to improve the lubrication of the bearings was also considered (Figure 8).

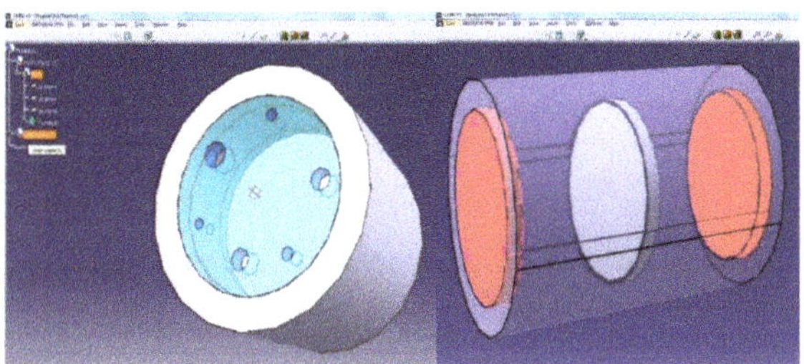

Figure 8. Hollow roller bearing with lubrication covers and extra support in the middle.

Three other virtual models of hollow rollers with 2 mm, 4 mm covers were developed, the third model having, in addition to the two covers, a third inner support in the middle of the roller. Two virtual solid roller models were developed with an outer diameter of 120 mm, one cylindrical and the other logarithmic. Assembly of the roller covers is done by welding, MAG process, or hoop wrapping, on one of the rollers.

Hollow rollers designed and manufactured in this way (Figure 7) were then subjected to finite element analysis. The purpose of this analysis was to compare the contact stresses and deformations encountered during contact with the roller types proposed in this research.

3.1.2. Results of the FEA for the Hollow Rollers Bearings

The finite element analysis (FEA) was performed on a wide roller typology, taking into account the use of two specialized programs, ABAQUS and NASTRAN, and taking into account variable loads, making the study of contact stresses and deformations a safe tool.

The analysis of stresses by finite elements consists of replacing the assembly studied with a structural system made up of sub-regions, called finite elements and which, in fact, are parts of that assembly. The method uses meshing of the analyzed elements, based on which the nodes are created and their status evaluated. The computational algorithm used to make the inner-ring roller contact has the following steps [44]: it identifies from all master segments and all slave nodes those that can form potential pairs for contact, then it determines the nodal pairs that form contact, as well as the size and direction in which the contact took place, and applies to them an interaction force.

In the analysis applied to the hollow rollers, the kinematic iteration method was used, a method based on the assumption that the nodal forces are known to be uniformly distributed, and the contact forces are still to be calculated. The value of these forces is determined by successive iterations, resulting in the moment when the node's movement speed becomes null.

Several different-sized rollers were analyzed. After applying the constraints required to run the computer program, the values of the deformations of the hollow roller bearings (Figure 9) were obtained.

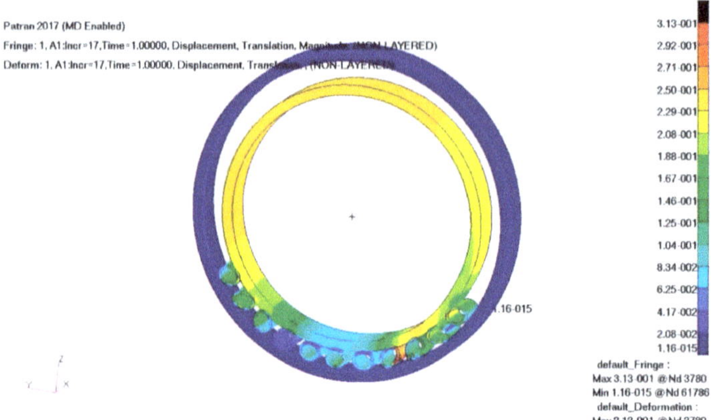

Figure 9. Deformations of the hollow cylindrical roller bearing (mm).

The deformation analysis clearly shows that the deformations at the bottom of the bearing (the simulated and constrained part) are extremely small, at 1.16 (μm). This result shows that the maximum deformation is in the red area, and has a value of 3.13 (μm). By applying the FEA method, it was found that the specific deformations and stresses described by discrete quantities are distributed throughout the entire structure of the hollow roller bearing within the specified limits. The validation of the results obtained with the finite element analysis and the tensometric measurements leads to the conclusion that implementation of the hollow rollers in the large bearings is advantageous.

The bearing stress distribution after the application of a radial force equal to Q_R = 275 KN. The previously designed hollow rollers (Figure 7) are shown in Figure 10.

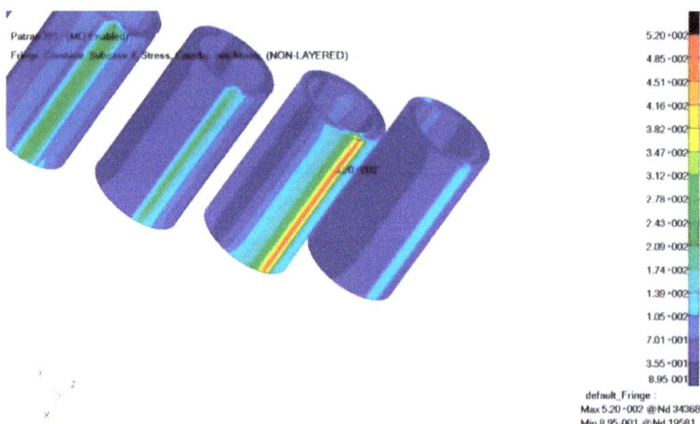

Figure 10. Cauchy and Von Mises stresses occurring in hollow rollers (MPa).

By using the Patran-Nastran software package, the Von Mises graph of stresses can be plotted by the roller type and the loading force (N/mm^2); Figure 11.

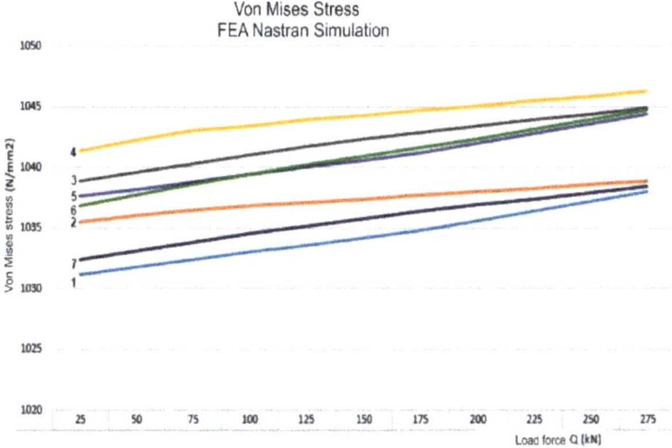

Figure 11. Von Mises stress graph according to roller type and load force (N/mm^2).

The analysis was carried out on 7 roller types and sizes: 1—solid roll; 2—hollow roller (Di = 60 mm); 3—hollow roller (Di = 80 mm); 4—hollow roller (Di = 90 mm); 5—hollow roller with 2 mm covers (Di = 80 mm); 6—hollow roller with 4 mm covers (Di = 80 mm); 7—hollow roller with 2 mm caps and support in the middle (Di = 80 mm). The loading forces to which roller bearings were subjected varied between 25 (kN) and 275 (kN). By implementing the previously designed solution (Figure 9, Section 3.1.1) consisting of the mounting of caps and a bearing support in the middle of the roller, an improvement is achieved by decreasing the stresses in the roll due to the stiffening of the system. Thus, analysis of roller 7 demonstrates that stresses decrease relative to the solid roller's stresses if the system is supported in the middle of the roller.

3.2. Dynamic Analysis of the Rotor Assembly with Hollow Roller Bearings

The elements taken into account in the dynamic analysis of the rotor assembly are the main shaft, the bearings and the mounting part in the housing [12]. Dynamic analysis is done by taking into account the rotation movement around a given axis and the forces generated on the axes by the forces and moments induced by the inertial masses, both from the blades, which take the wind force and transmit it to the main axis of the turbine, and from the reaction forces generated by the axle-bearing contact, as well as the housing bearing assembly.

Reducing inertial masses by using hollow roller bearings is the purpose of this research, and leads to an increase in the wind power system lifespan. The dynamic study of the forces acting on the main shaft and their effect—the reduction of deformations, the reduction of stresses and the reduction of temperature—are subject of this chapter.

3.2.1. Modeling Pattern

The design of the turbine shaft was executed taking into account the existing constructive variants [45]. The aim of the research is to demonstrate the advantages of introducing hollow roller bearings while preserving the current design of the other power plant components.

The dynamic analysis of the rotor assembly is based on the comparison between a solid roller bearing assembly and a hollow roller bearing assembly. The scheme used (Figure 12) took into account: two large bearings (pos. 1 and pos. 3), the main shaft (pos. 2) and the blade clamping system (pos. 4), which was considered to be removable in the finite element analysis. The generation of the virtual model (Figure 13) was done using Dassault Catia V5 and Solid Works.

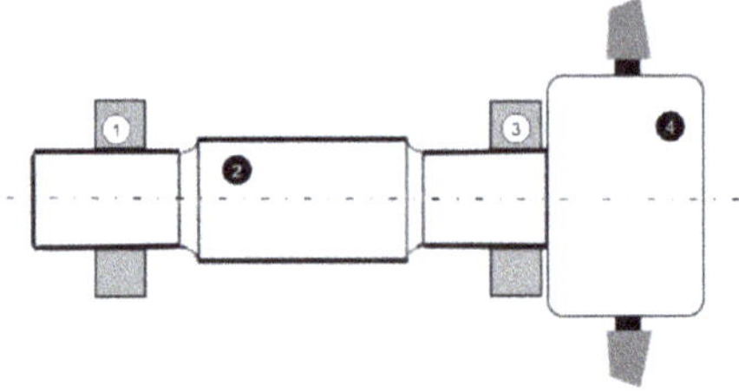

Figure 12. Study scheme of the main shaft.

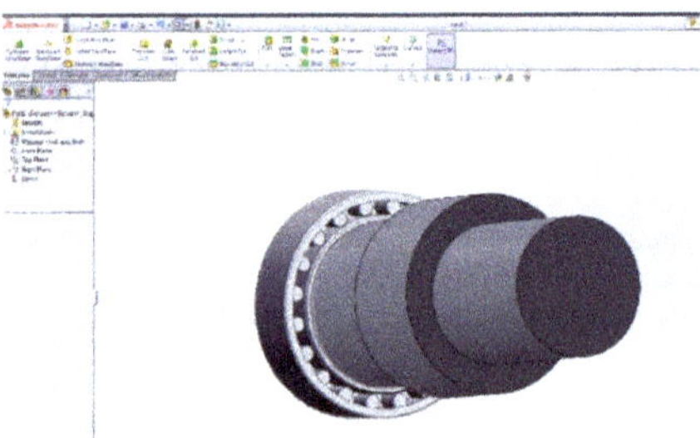

Figure 13. Main shaft and bearing.

3.2.2. Finite Element Analysis. Results and Conclusions

After creating the 3D model shown in Figure 13, it was subjected to finite element analysis. The model was imported into the MD Nastran program and meshed (Figure 14) by running the mesh module. The association of material properties, the creation of deformable components and the application of constraints lead to the possibility of triggering the finite element analysis operation. For the examined case study, the modeling is based on the reduction of inertial masses on a bearing with a hollow of 80 mm at a roller diameter of 120 mm.

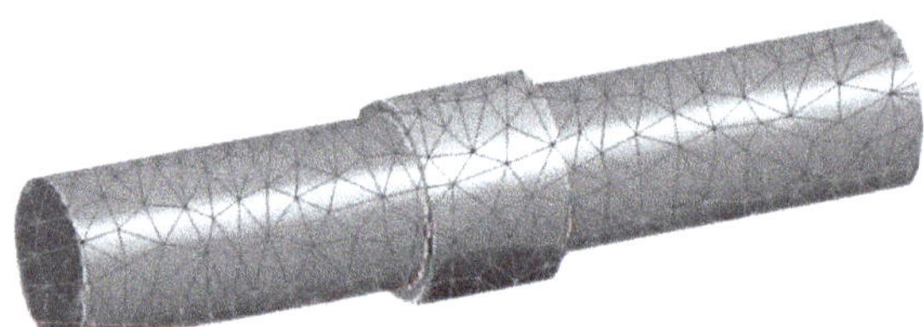

Figure 14. Meshing of the main shaft in MD Nastran software.

The results from the finite element analysis are shown in Figure 15. Computational elements were considered throughout the research. Variable forces (25 kN to 275 kN) were applied. The maximum pushing force Q_{max} = 250 kN is the pushing force of the moving blades. The reaction force in the first bearing is Q_{React} = 275 kN, and the force in the opposite bearing is Q_{bearer} = 150 kN. Then, the stress in the rotor assembly was measured, both for the solid roller and for the hollow roller. The Von Mises stresses on the analyzed rotor assembly are shown in Figure 15, both for solid roller bearings and for hollow roller assemblies. The results show that as the inertial masses decrease, the loading forces decrease and, with them, the effort across the system decreases. Using hollow rollers brings long-lasting benefits to all wind farms. It was observed that the maximum stress occurs in the blade section mounting area. These are the elements that set the entire assembly in motion, creating a momentum of maximum torsion. The support bearing response induces strong stresses. The decrease of inertial masses results in decreased startup resistance, forming an interesting phenomenon in the direction of decreasing stresses in the blade mounting area. In the area of the first bearing, the stress also decreases compared to the solid roller version.

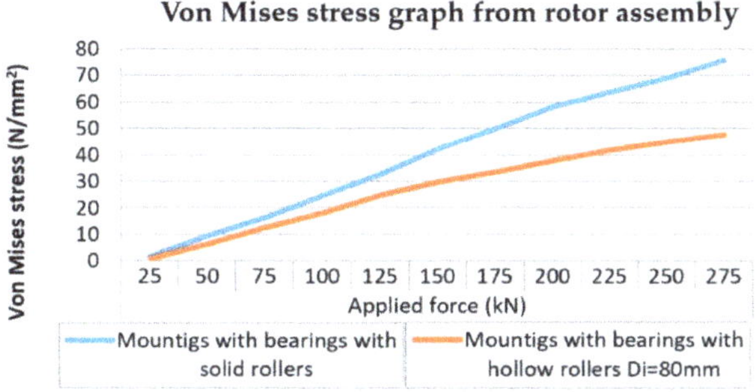

Figure 15. Von Mises stress graph from rotor assembly.

The implementation of hollow rollers in large-size bearings is a clear solution that may revive the wind energy industry, which has had a slower growth over recent years. Theoretical analyses made with dedicated software tools have proven the real advantages of using hollow rollers. Increasing internal stresses in rollers leads to very small deformations that do not influence the behavior of the bearings. The three hollow rollers solution used by the company where these were tested for slippage has proven to be a viable solution that does not jeopardize the durability of the bearing, and the decrease in inertial mass causes a decrease in the effort across the system.

3.3. Automated Lubrication System Model for Large Bearings

The purpose of the applied research is to design and test a functional hollow roller model provided with an automated lubrication system for large bearings. Based on the research performed in the literature and on the Internet, the use of hollow rollers as an area for the implementation of an automated lubrication system is an entirely new approach.

There are automatic lubrication systems (SKF, Lincoln, ATS), there are large hollow roller bearings (INA Schaeffler, EDAC Technology) and there are patents related to hollow roller bearings (US 5033877, US 5071265, US 6682226), which have claims for the mounting mode or other claims different from those in this proposal. There is also the Invention Patent [13], which proposes the use of hollow rollers, containing a spongy foam impregnated with lubricant on the inside.

The development of an automated innovative lubrication system, implemented in the hollow roller, which equips the large wind turbine bearings, together with the design of a test bench for testing are among the results of this research. The implementation of such an automated system in the cylindrical hollow roller bearing has major patenting potential, and the research has an industrially applicable character.

Automated Lubrication System for Bearings. Proposed Solution

The concept of an automated lubrication system implemented in the hollow rollers that fit the large-size bearings installed in wind farms presents fundamental novelties. The innovative elements consist of the positioning of an automated lubrication system inside the hollow roller. The system would be equipped with a control unit, temperature sensors, closing/opening module for the lubrication area, wireless transmitter to the computer, power supply, and software for monitoring, command and control of the bearing's function.

Figure 16 shows the scheme of the proposed lubrication system to be installed inside the hollow roller. Thus, the opening–closing system may be comprised of electrically actuated valves or thermal valves or magnetically actuated locking systems. The source could be a durable battery or brush collectors. The control unit could be programmable or programmed. The assembly may take place inside or outside the roller [12]. The operation of the system is managed by a control unit that unlocks the valves, leaving the lubricant to enter the runways when the sensors announce an increase in the operating temperature.

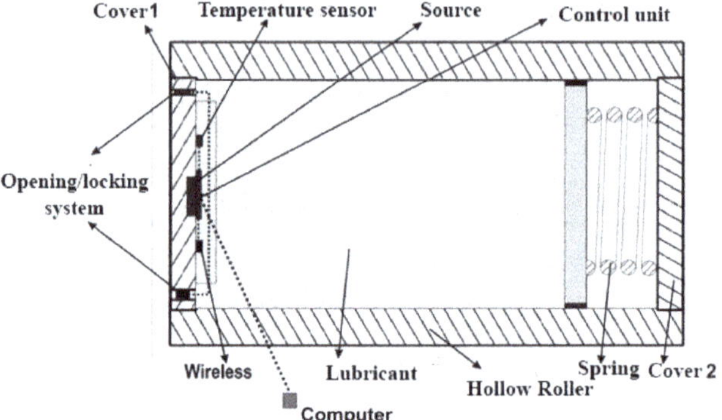

Figure 16. Diagram of the proposed automated lubrication system.

To determine all design factors, use is made of robust design by insensitivity of the chosen lubrication system to disturbing factors. With increasing temperature when the valves are open, the lubricating fluid is pushed into the tread by the piston actuated by the pressure spring. A movement of the center of gravity occurs in the roller, which can lead to additional stresses in the contact areas.

The dynamic behavior of hollow rollers provided with an automated lubrication system will be tested on a sample bench, depending on the variables acting on the system, namely:

- variables with predictable behavior that have permanent real time control (lubricant working temperature, etc.);
- variables with unpredictable behavior such as sudden burdens due to changes in the rotation speed or resonance vibration with resonant amplification.

The main problems are the identification of the optimal design solution for the discharge of the lubricant from the hollow roller and the dynamic behavior of such a roller whose center of gravity changes with the removal of the lubricant from the roller to the lubrication pathways, where previous studies have demonstrated the advantages of using hollow rollers in large-size bearings fitted in the wind power plants.

To implement the designed system, several steps will be followed:

- Develop an innovative automated lubrication system for large hollow roller bearings in wind energy assemblies.
- Design SCADA-like software to monitor, control, and lead the automated lubrication process.
- Dynamically model hollow roller bearings provided with the proposed lubrication system under variable load conditions and roller centers of gravity, which are variable over time due to lubricant removal to the tread.
- The development of a hollow roller test stand equipped with the automated lubrication system, based on previous research.
- Development of at least 3 models, based on different constructive solutions, regarding the way to introduce the lubricant from the hollow roller on the treads of the bearing.
- Elaborate the functional model of the automated lubrication system by adapting the monitoring, command and control program to the executed models.

- Test, verify and validate the functional models. Compare the experimental results with the theoretical results.
- Determine the parameters of the new computerized architecture with which the lubrication system is equipped. Optimize the system functionality.
- Assess the economic impact of the proposed solutions by making estimations of the introduction on the market of the automated lubrication for large bearings.

The research topic is broad and proposes that the future research direction will validate a new type of automated and computer-monitored bearing lubrication technology, starting from the formulation of the product concept "Hollow Roller with Built-in Bearing Lubrication".

4. Assessment of the Economic Impact of the Proposed Solution

4.1. Estimate of the Marketing the Hollow Roller Bearings

Marketing the product is closely related to the technical innovation activity. In the launch phase, sales must be fostered by smart marketing. Since the product has a high degree of novelty, and the number of competitors is small, the distribution can reach high values. Launching can be done through intense, planned distribution. The economic analysis for marketing the hollow roller bearings is based on the current product, with estimates being made for the new product. Factors influencing the analysis include changes made to rollers: additional cost changes, or cost reductions. The two categories were quantified by estimating according to values in Figure 17.

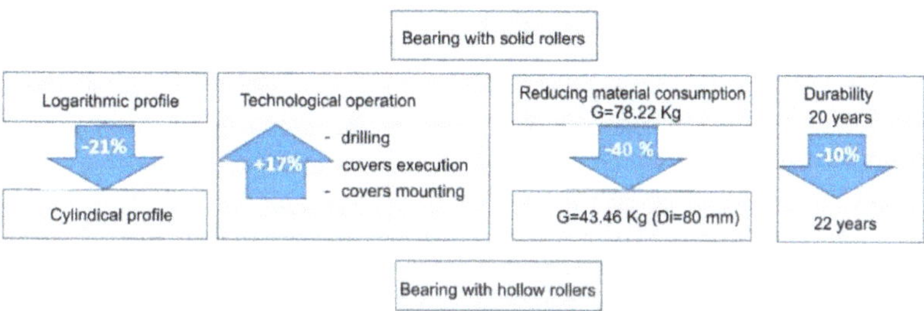

Figure 17. Diagram of the economic impact assessment of marketing hollow roller bearings.

The following priorities have been considered: operational safety, reliability, environmental impact, technological change, effort to implement the solution, and benefits achieved. Priorities have been assimilated with variables within a tolerable range. In determining the benefit, two potential outlets were considered:

- the existing market, due to demand for new products;
- the market generated by the need to provide maintenance to old products.

Price discrimination in the two markets was achieved by identifying the specificities and characteristics of each market in order to maximize profit.

4.2. Estimate of the Benefit Gained by Implementing Hollow Roller Bearings in Wind Energy Assemblies

The study of energy gains was achieved with the help of dynamic control. The optimal setting of influence parameters in the power system was done using optimization techniques and the probabilistic approach. The quantification of energy savings showed significant increases with respect to the proposed objective: maximization of energy yield, and hence the cost/kW.

Reducing system inertia is one of the most important effects of using hollow roller bearings. Fast start, easy stop and easy control determine an increase in energy efficiency. At very low wind speeds, there is not enough torque exerted by the wind on the turbine blades to rotate them. The speed at which the turbine starts to rotate to generate energy is called the cut-in speed, and this is usually between 3 and 4 m/s. When the speed increases to values that are too high, the forces on the structure increase, generating a risk of damaging the rotor. As a result, a braking system is used to bring the rotor to a dead stop. This is called cut-off speed and is usually around 25–30 m/s. The direct relationship between wind speed and inertia leads to a decrease in the starting speed in the case of a decrease in the rotor inertial mass (Figure 18) [46].

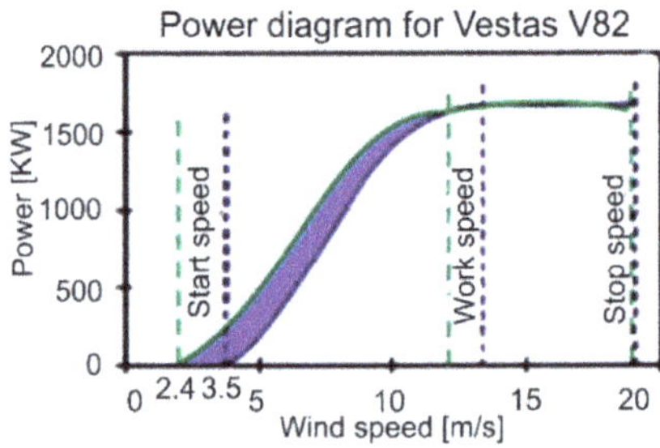

Figure 18. Vestas V82 turbine power diagram.

Figure 18 shows the blue curve for the power plant equipped with solid roller bearings and green for the center with hollow roller bearings. For the Vestas V82, a reduction in speed from 3.5 m/s to 2.4 m/s was observed in the experimental observation, leading to an actual energy gain at a single start of 250 kWh. For 15 starts/month, the energy gained for a turbine is approx. 45 MWh in one year, representing 3% of the total turbine capacity.

The dynamic analysis performed by using the finite element method as well as the results of the analytical calculation leads to the conclusion that hollow cylindrical rollers can replace the more expensive and heavy logarithmic profile rollers, while increasing bearing lifespan by reducing the uneven wear of the rolling elements. Uniform stress, combined with vibration reduction and operation at a lower temperature (additional lubricant can be stored in the hollow roller) leads to an increase in energy of approx. 7% for a turbine. The expected lifespan could be around 21–22 years compared to 20 years for the traditional bearings. For a 20-roller bearing, the difference is about 700 kg, 30% lighter in weight. For a product, the energy gain for 1 year is approx. 1.5% if we approximate the weight gained with the energy accumulated. The technology energy savings for a product for one year is based on cutting the consumption of tools and requires a complex calculation. It is approximately 2% and 6%, respectively. The total energy saving is approx. 26% [14].

For a power plant of 1.65 MW/h, the energy gain is estimated to be 0.43 MW/h (1.65MW/h × 26%). The energy performance increase by implementing a hollow roller bearing in wind plants leads to increased sustainability by enhancing maintenance capabilities. Decreasing subsidies in this area make the implementation of hollow roller bearings of great interest because it leads to significant savings by reducing the maintenance cost of wind power plants through increased sustainability and energy efficiency.

5. Conclusions

The purpose of this applied research was to design and test a functional hollow roller model provided with an internal automated lubrication system for the large bearing as a sustainable solution in the field of wind energy. The optimal solution of the automated roller lubrication system was identified and the dynamic behavior of such a roller was studied. The center of gravity of the roller changes with the removal of the lubricant from the roller to the lubrication pathways. The prototype was made, and the test stand will be made as the next research step. The stand will validate or invalidate, based on actual measurements, the behavior of the lubrication system using different geometries as a result of the lubricant discharge from the rollers.

Several hollow roller patterns with different hollowness were modeled and developed, starting from the actual situation of large bearings with solid rollers. CAD/CAM programs were used, the hollow rollers being the main element in their design. To encapsulate a lubricant that improves the lubrication of the bearings, the covers were also designed and assembled. For the study of stresses by finite element analysis, the contact stresses and deformations encountered during contact with the roller types proposed in this research were compared.

Based on the dynamic analysis of the rotor assembly, a comparison was made between a solid roller bearing assembly and a hollow roller bearing assembly. By performing Von Mises analysis, the results demonstrate that with the decrease of inertial masses, the loading forces decrease and, at the same time, the effort across the system decreases.

The solution of using hollow rollers in large bearings has important effects in reducing material consumption, increasing energy yield (in the field of wind farms), increasing the durability of large bearing systems, reducing operating noise, increasing resistance to vibrations, decreasing losses through friction and lowering the working temperature. Performing studies on the behavior of these bearings under extreme vibrational conditions and variable forces would be one of the future research directions. The benefits obtained by introducing large bearings with hollow rollers into wind power assemblies are demonstrated by presenting the estimated economic calculation of product implementation. The economic impact is achieved by reducing the cost price of the whole energy package, which could be the starting point for the revitalization of the wind energy industry.

Author Contributions: All authors contributed equally to this work and all authors have read and approved the final manuscript.

Funding: This research received no external funding.

Conflicts of Interest: The authors declare no conflict of interest.

References

1. Oxygen. Energy evolution on the globe. *Engie Rom.* **2016**, *3*, 03–06.
2. Kumar, Y.; Ringenberg, J.; Depuru, S.S.; Devabhaktuni, V.K.; Lee, J.W.; Nikolaidis, E.; Andersen, B.; Afjeh, A. Wind energy: Trends and enabling technologies. *Renew. Sust. Energ. Rev.* **2016**, *53*, 209–224. [CrossRef]
3. Blaabjerg, F.; Ke, M. Wind Energy Systems. *Proc. IEEE* **2017**, *105*, 11.
4. Ponta, F.; Otero, A.; Rajan, A. The adaptive-blade concept in wind-power applications. *Energy. Sustain. Dev.* **2014**, *22*, 3–21. [CrossRef]
5. Li, Y.; Caichao, Z.; Chaosheng, S.; Jianjun, T. Research and Development of the Wind Turbine Reliability. *Int. J. Mech. Eng. Applic.* **2018**, *6*, 35–45. [CrossRef]
6. Torsvik, J.; Nejad, A.R.; Pedersen, E. Main bearings in large offshore wind turbines: Development trends, design and analysis requirements. *Iop Conf. Ser. J. Phys.* **2018**. [CrossRef]
7. Tazi, N.; Châtelet, E.; Bouzidi, Y. Wear Analysis of Wind Turbine Bearings. *Int. J. Renew. Energy* **2017**, *7*, 2120–2129.
8. Whittle, M.; Trevelvan, J.; Tavner, P.J. Bearing currents in wind turbine generators. *J. Renew. Sustain. Energy* **2013**, *5*, 053128. [CrossRef]
9. Gallego-Calderon, J.; Natarajan, A.; Dimitrov, N. Effects of bearing configuration in wind turbine gearbox reliability. *Energy Procedia* **2015**, *80*, 392–400. [CrossRef]
10. Fujiwara, H.; Kawase, T. Logarithmic Profile of Rollers in Roller Bearing and Optimization of the Profile. *Trans. Jpn. Soc. Mech. Eng.* **2009**, *18*, 3022–3029.
11. Barnsby, R.; Duchoski, J.; Harris, T.; Ioannides, E.; Losche, T.; Nixon, H.; Webster, M. Life ratings for modern rolling bearings. A design guide for the application of International Standard ISO 281/2. ASME, New York. *TRIB.* **2013**, *14*, 90.
12. Barabaş, S.A.; Florescu, A. Analysis of bearing behavior with cylindrical rollers with variable center of gravity. *Matec Web Conf.* **2017**, *94*, 02001. [CrossRef]

13. Luca, V.; Şerban, C.E.; Barabaş, S.A. Hollow roller bearing. In *Patent*; No. 125038; Transilvania Univesity of Brasov: Brasov, Romania, 2013.

14. Barabas, B.; Florescu, A.; Barabas, S.A. Saving energy estimation for use of hollow rollers in bearings utilized in wind energy turbines. *Sci. Res. Ed. Air Force Afases* **2015**, *2*, 377–382.

15. NSK Premium Technology for the Wind Industry. 2009. Available online: https://www.pkservis.com/data/web/upload/nsk/en-wind-industry-bearings.pdf (accessed on 3 July 2018).

16. Bowen, W.L.; Bhateja, C.P. The Hollow Roller Bearing. *ASME Trans. J. Lubr. Technol.* **1980**, *102*, 222–228. [CrossRef]

17. Pasdari, M.; Gentle, C.R. Computer Modelling of a Deep Groove Ball Bearing with Hollow Balls. *WEAR* **1986**, *111*, 101–114. [CrossRef]

18. Darji, P.H.; Vakharia, D.P. Development of Graphical Solution to Determine Optimum Hollowness of Hollow Cylindrical Roller Bearing Using Elastic Finite Element Analysis. In *Finite Element Analysis-Applications in Mechanical Engineering*; Chapter 11; InTech: Vienna, Austria, 2012. [CrossRef]

19. Abu-Jadayil, W.M. Relative fatigue life estimation of cylindrical hollow rollers in general pure rolling contact. *Tribotest* **2008**, *14*, 27–42. [CrossRef]

20. SKF Evolution. 2011. Available online: http://evolution.skf.com/centralized-lubrication-system-for-wind-turbines-offers-improved-efficiency/ (accessed on 27 August 2018).

21. Martin, R.; Lazakis, I.; Barbouchi, S.; Johanning, L. Sensitivity analysis of offshore wind farm operation and maintenance cost and availability. *Renew. Energy* **2016**, *85*, 1226–1236. [CrossRef]

22. European Commission. *COM. Renew. Energy Progress Report*; European Commission: Brussels, Belgium, 2017; Available online: https://ec.europa.eu/commission/sites/beta-political/files/report-renewable-energy_en.pdf (accessed on 7 October 2018).

23. GWEC (Global Wind Energy Council). Glob. Wind Rep. 2017. Available online: www.gwec.net (accessed on 10 October 2018).

24. International Energy Agency. IEA. 2017. Available online: https://www.iea.org/ (accessed on 9 October 2018).

25. WEEA Report. 2018. Available online: https://wwindea.org/ (accessed on 9 October 2018).

26. Ziegler, L.; Gonzalez, E.; Rubert, T.; Smolka, U.; Malero, J. Lifetime extension of onshore wind turbines: A review covering Germany, Spain, Denmark, and the UK. *Renew. Sust. Energ. Rev.* **2018**, *82*, 1261–1271. [CrossRef]

27. PowerWeb. Renew. Energy. 2018. Available online: http://www.fi-powerweb.com/Renewable-Energy.html (accessed on 2 September 2018).

28. European Comision. Clean Energy for All Europeans—Unlocking Europe's Growth Potential. 2016. Available online: http://europa.eu/rapid/press-release_IP-16-4009_en.htm (accessed on 7 October 2018).

29. Calanter, P. The role of renewable energy in action to combat climate change in the European Union. *EUROINFO* **2018**, *2*, 19–31.

30. Government of Romania; Programul Natiunilor Unite pentru Dezvoltare. *Nat. Strategy for Sustain. Dev. Romania. Horizon 2013-2020-2030*; Government of Romania: Romania, Bucharest, 2008.

31. Government of Romania. *Energy Strategy of Romania 2016–2030, with the Perspective of 2050*; Government of Romania: Bucharest, Romania, 2016.

32. Eurostat. 2018. Available online: http://ec.europa.eu/eurostat/statisticsexplained/index.php?title=Renewable_energy_statistics/ro (accessed on 15 August 2018).

33. RWEA (Romanian Wind Eolian Association). 2017. Available online: http://rwea.ro (accessed on 7 June 2018).

34. ANRE. 2017. Available online: https://www.anre.ro/ro/energie-electrica/rapoarte/rapoarte-indicatori-performanta (accessed on 8 June 2018).

35. SKF. *Rolling bearings*; SKF Group: Gothenburg, Sweden, 2018.

36. TIMKEN. *Engineering Lubrication and Seals*; The Timken Company: North Canton, OH, USA, 2009; pp. 146–162.

37. INA Schaeffler. *Tapered Roller Bearing Catalogue*; Schaeffler Technologies AG & Co.: Schweinfurt, Germany, 2017.

38. FAG; OEM; Handel AG. *The Design of Rolling Bearing Mountings*; Publ. No. WL 00 200/6 EA; Schaeffler Technologies AG & Co.: Schweinfurt, Germany, 2012.

39. Wind Measurement International. WMI. Available online: http://www.windmeasurementinternational.com/ (accessed on 11 October 2018).

40. Turi, M.B.; Marks, C.S. *Analysis Helps Wind Turbine Designers Find Their Bearings*; Timken, Co.: Canton, OH, USA, 2010.
41. Burton, T. *Wind Energy Handbook*, 2nd ed.; Wiley: Chichester, UK; New York, NY, USA, 2011.
42. Lyatkher, V.M. *Wind Power: Turbine Design, Selection, and Optimization*; Scrivener Publishing, Wiley: New Jersey, NJ, USA, 2014.
43. Song, Y.D.; Dhinakaran, B.; Bao, X.Y. Industrial Aerodynamics. *J. Wind Eng.* **2000**, *85*, 293–308.
44. Ebert, F.J. Fundamentals of Design and Technology of Rolling Element Bearing. *Chin. J. Aeronautics* **2014**, *23*, 123–136. [CrossRef]
45. Smith, E.H. *Mechanical Engineer's Reference Book*; Linacre House: Jordan Hill, Oxford, UK, 2000; pp. 9–56. ISBN 0-7506-4218-1.
46. Kakuta, K. High Speed Rolling Bearings for Gas Turbines. *Jpn. J. Tribol.* **1990**, *35*, 877–889.

Technical Note

Day-Ahead Forecasting of Hourly Photovoltaic Power Based on Robust Multilayer Perception

Chao Huang [1,2,3], Longpeng Cao [1,2,3], Nanxin Peng [4], Sijia Li [5], Jing Zhang [1,2,3], Long Wang [1,2,3,*], Xiong Luo [1,3] and Jenq-Haur Wang [6]

[1] School of Computer and Communication Engineering, University of Science and Technology Beijing (USTB), Beijing 100083, China; chahuang3-c@my.cityu.edu.hk (C.H.); s20170672@xs.ustb.edu.cn (L.C.); g20178747@xs.ustb.edu.cn (J.Z.); xluo@ustb.edu.cn (X.L.)

[2] Key Laboratory of Wind Energy and Solar Energy Technology (Inner Mongolia University of Technology), Ministry of Education, Hohhot 010051, China

[3] Beijing Key Laboratory of Knowledge Engineering for Materials Science, Beijing 100083, China

[4] School of International Business, Southwestern University of Finance and Economics, Chengdu 611130, China; nancypeng.sib@2015.swufe.edu.cn

[5] National Internet Finance Association of China, Beijing 100080, China; lsijia@nifa.org.cn

[6] Department of Computer Science and Information Engineering, National Taipei University of Technology, Taipei City 106, Taiwan; jhwang@csie.ntut.edu.tw

* Correspondence: lwang@ustb.edu.cn

Received: 13 November 2018; Accepted: 14 December 2018; Published: 19 December 2018

Abstract: Photovoltaic (PV) modules convert renewable and sustainable solar energy into electricity. However, the uncertainty of PV power production brings challenges for the grid operation. To facilitate the management and scheduling of PV power plants, forecasting is an essential technique. In this paper, a robust multilayer perception (MLP) neural network was developed for day-ahead forecasting of hourly PV power. A generic MLP is usually trained by minimizing the mean squared loss. The mean squared error is sensitive to a few particularly large errors that can lead to a poor estimator. To tackle the problem, the pseudo-Huber loss function, which combines the best properties of squared loss and absolute loss, was adopted in this paper. The effectiveness and efficiency of the proposed method was verified by benchmarking against a generic MLP network with real PV data. Numerical experiments illustrated that the proposed method performed better than the generic MLP network in terms of root mean squared error (RMSE) and mean absolute error (MAE).

Keywords: forecasting; multilayer perception; photovoltaic; sustainable energy; pseudo-Huber loss

---

1. Introduction

Solar energy is considered to be one of the most renewable and sustainable energy resources. Photovoltaics (PV), which convert solar energy into electricity, is the most widely used technique to make use of solar energy. The increasing penetration of PV power, however, brings challenges for the planning and scheduling of the power grid due to the uncertainty of PV power production [1]. Forecasting is an essential technique to alleviate the negative impacts on the grid operation [2] and to facilitate the management of grids including renewable energies [3–5].

PV power forecasting methods can be categorized into direct methods and indirect methods. The former produces PV power production as model outputs; while the latter first generates forecasts of solar irradiance, then PV performance models are applied to derive the PV power production based on solar irradiance [6,7]. The presented modeling approaches include numerical weather prediction (NWP) models, statistical models, and artificial intelligence. NWP-based approaches dynamically model the atmospheric states and project their impacts on solar irradiance and PV power production [8]. These approaches are computationally expensive, making them difficult to use for short-term forecasting. In the early stages of PV power forecasting, statistical models such as autoregressive moving average (ARMA) [9] and its variants [10] were frequently used. However, these models are linear, which cannot capture the nonlinear characteristics in PV power production. Recently, artificial intelligence including artificial neural network (ANN) [11,12] and its variants such as extreme learning machine [13], support vector regression (SVR) [14], and Gaussian process regression [15] have been widely applied to renewable energy forecasting. In [16], a wavelet recurrent neural network (WRNN) was proposed for the prediction of energy production in a PV park. As for the day-ahead forecasting of hourly solar irradiance and PV power, ANN topped the methodologies for multi-input multi-output forecasting [17–19].

In the aforementioned studies on ANN based forecasting of PV power production, the learnable parameters such as weights and biases of an ANN were usually derived by minimizing the mean squared error on the training dataset. However, the mean squared error can be dominated by a few particularly large errors due to the sudden change in weather patterns that are difficult to predict. This can lead to poor asymptotic relative efficiency of the mean squared error based estimator in terms of estimation theory.

This paper contributes to developing a robust ANN model for day-ahead hourly forecasting of PV power based on a robust loss function, the pseudo-Huber loss, which combines the best properties of squared loss and absolute loss. The pseudo-Huber loss is less sensitive to large errors to train a more robust ANN model. The efficacy of the proposed method was validated on real PV power data by benchmarking against the generic ANN trained on the squared loss and the persistence model. Numerical experimental results showed that the proposed method outperformed the generic ANN model and persistence model in terms of root mean squared error and mean absolute error.

2. Methodology

In this section, the fundamentals of ANN will first be introduced, followed by the training of neural network based on the pseudo-Huber loss function.

2.1. Multilayer Perception Network

The proposed ANN model is based on a typical network structure, the multilayer perception (MLP), as shown in Figure 1. The MLP consists of the input layer, hidden layers, and output layer, and the model can be mathematically expressed as Equations (1) and (2):

$$\mathbf{H} = f_h\left(\omega_h \mathbf{X}^T + \mathbf{b}_h\right) \tag{1}$$

$$\mathbf{Y} = f_y\left(\omega_y \mathbf{H}^T + \mathbf{b}_y\right) \tag{2}$$

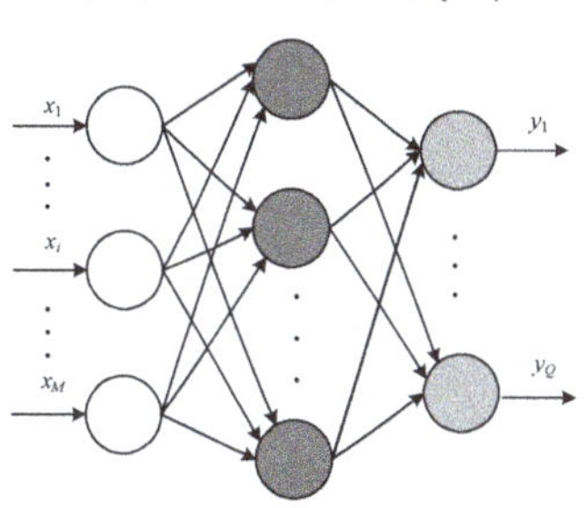

Figure 1. The structure of an MLP [20].

In Equations (1) and (2), $\mathbf{X} = (x_1, x_2, \ldots, x_M)$ and $\mathbf{Y} = (y_1, y_2, \ldots, y_Q)$ denote the M-dimensional and Q-dimensional model input and output, respectively; $\boldsymbol{\omega}$ and $\mathbf{b}$ represent the weights and biases of the network, respectively; f denotes the activation function; and the subscripts h and y stand for the hidden layer and output layer, respectively.

The learnable parameters including $\boldsymbol{\omega}$ and $\mathbf{b}$ are usually obtained by minimizing the mean squared error with optimizing algorithms such as the Adam optimization [21]. The mean squared error is sensitive to a few particularly large errors and can lead to a poor estimator. In this paper, the pseudo-Huber loss was applied to the training of the MLP.

2.2. Pseudo-Huber Loss

The pseudo-Huber loss function is defined in Equation (3) [22]:

$$L_\delta(e) = \delta^2 \left(\sqrt{1 + (e/\delta)^2} - 1 \right) \tag{3}$$

In Equation (3), δ is a controlling parameter. The pseudo-Huber loss function combines the best properties of squared loss and absolute loss that with small errors e, $L_\delta(e)$ approximates $e^2/2$, which is strongly convex, and with extremely large e, $L_\delta(e)$ approximates a straight line with a slope of δ, which is less steep than the squared loss. This property of the pseudo-Huber loss makes it less sensitive to large errors. This paper took advantage of this property of the pseudo-Huber loss to train the MLP as the sudden change in weather patterns can result in large modeling errors. The objective function based on the pseudo-Huber loss for the training of the MLP is expressed in Equation (4):

$$L = \sum_{i=1}^{N} \sum_{q=1}^{Q} \delta^2 \left(\sqrt{1 + (e_{i,q}/\delta)^2} - 1 \right) \tag{4}$$

In Equation (4), $e_{i,q} = \hat{y}_{i,q} - y_{i,q}$ for $i = 1, \ldots, N$ and $q = 1, \ldots, Q$ where i denotes the number of training data point and q indexes the output element, respectively, and y and $\hat{y}$ denote the observed and modeling PV power, respectively. To improve the forecasting performance, swarm intelligence-based optimization algorithms such as the Jaya algorithm [23–26] can be adopted to optimize δ.

3. Case Study

3.1. Data Description

The hourly PV power production between 1 January 2012 and 31 December 2017 of a PV plant installed at the Andre Agassi Preparatory Academy Building B (36.19N, 115.16W, elevation of 620 m) in the USA was used to verify the efficacy of the proposed MLP for day-ahead hourly PV power forecasting. The specifications of the PV power plant is shown in Table 1, and the data are available at https://maps.nrel.gov/pvdaq/. The histogram on the hourly power production is shown in Figure 2 where it can be observed that the hourly PV power production showed high variability, which induces great challenges for the forecasting model.

The hourly PV power produced over 2017 is illustrated in Figure 3 (data on some days were not available). It can be seen that the hourly PV power production showed strong seasonality and diurnal cycle. In the case study, day time data from 6:00 am to 19:00 pm including 14 h a day were considered. The input of the MLP model was composed of the PV power production of the last seven days for the hourly forecasting of the upcoming day. Hence, the dimensions of input vector and output vector were $M = 7 \times 14$ and $Q = 14$, respectively. The total data were divided into two groups, the training dataset consisted of observations from 1 January 2012 to 31 December 2016 and the test dataset included data over the whole year of 2017. The data were not always available for the PV plant, so the real training and test data sizes were 1742 and 309, respectively. To simplify the training of the MLP model, all data were normalized to [0, 1] by dividing the nominal power. The operation of normalization also facilitated the comparison of forecasting performance between the different datasets.

Table 1. Specifications of the PV power plant.

Nominal DC power	68.48 kW
PV module type	NU-U240F1
PV module manufacturer	Sharp
Inverter type	50 kW
Inverter manufacturer	SatCon Technology

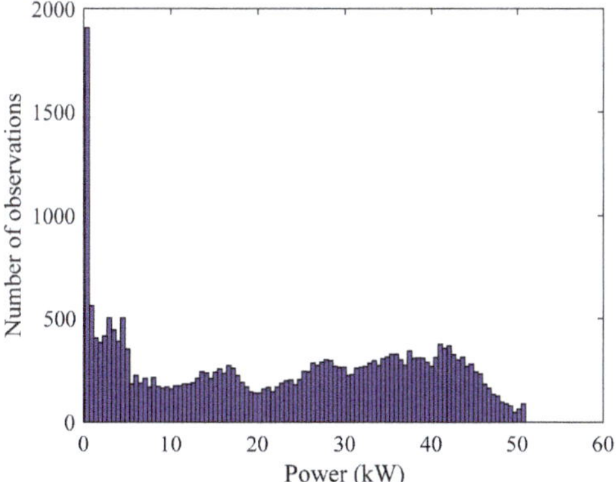

Figure 2. Histogram on the hourly power production.

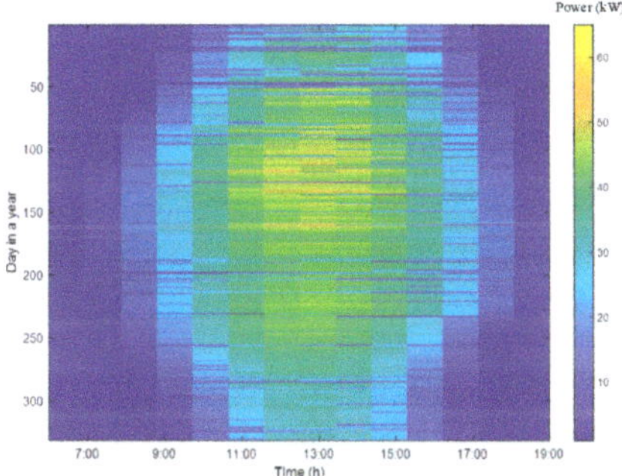

Figure 3. PV power production over 2017.

3.2. Performance Metrics

To compare the forecasting performance by different methods, root mean squared error (RMSE) and mean absolute error (MAE) as defined in Equations (5) and (6) were employed.

$$RMSE = \sqrt{\frac{1}{K}\sum_{k=1}^{K}(\hat{p}_k - p_k)} \tag{5}$$

$$MAE = \frac{1}{K}\sum_{k=1}^{K}|\hat{p}_k - p_k| \tag{6}$$

In Equations (5) and (6), $\hat{p}_k$ and p_k for $k = 1, \dots, K$ are the normalized forecasts and observations of PV power, respectively.

3.3. Numerical Results and Analysis

The forecasting performance of the proposed method based on robust loss function was compared with the generic MLP network trained by minimizing the mean squared error. The persistence model, $\hat{p}_{d,h} = p_{d-1,h}$ where the power production at hour h on the targeted day d was assumed to be equal to the observation of power production at hour h on the day indexed by $d - 1$, was also considered as a benchmark. The RMSE and MAE computed based on the whole test dataset for all of the forecasting methods are provided in Table 2. It can be observed that the proposed method performed better than the generic MLP model and the persistence model in terms of both RMSE and MAE.

Table 2. RMSE and MAE on the test dataset.

Method	RMSE	MAE
Robust-MLP	0.0775	0.0439
Generic-MLP	0.0788	0.0459
Persistence	0.0978	0.0474

To provide insight into the forecasting performance for each hour, Table 3 gives the RMSE and MAE indexed by the hour. From Table 3, the proposed forecasting method outperformed the generic MLP model for each hour with a lower RMSE and MAE. However, the RMSE and MAE were greater in the middle of the day. At early and late hours (6:00 am, 18:00 pm, and 19:00 pm) in the day, the performance of the persistence model was better than the proposed method and the generic MLP. This is because PV power production at early and late hours are usually very small (in winter the power production at these hours can be zero) and there is little difference each day, which fits the persistence model well.

Table 3. RMSE and MAE for each hour in the test dataset.

Hour	RMSE			MAE		
	Robust-MLP	Generic-MLP	Persistence	Robust-MLP	Generic-MLP	Persistence
6:00	0.0089	0.0110	0.0086	0.0053	0.0071	0.0032
7:00	0.0277	0.0297	0.0319	0.0167	0.0191	0.0139
8:00	0.0566	0.0572	0.0630	0.0391	0.0398	0.0339
9:00	0.0790	0.0808	0.0961	0.0530	0.0551	0.0538
10:00	0.0995	0.1013	0.1221	0.0649	0.0670	0.0686
11:00	0.0990	0.1012	0.1245	0.0657	0.0687	0.0725
12:00	0.1072	0.1086	0.1407	0.0724	0.0754	0.0849
13:00	0.1178	0.1197	0.1546	0.0768	0.0793	0.0937
14:00	0.1112	0.1125	0.1446	0.0729	0.0753	0.0879
15:00	0.1020	0.1038	0.1245	0.0690	0.0719	0.0775
16:00	0.0703	0.0711	0.0881	0.0520	0.0539	0.0492
17:00	0.0289	0.0299	0.0351	0.0200	0.0207	0.0193
18:00	0.0099	0.0118	0.0093	0.0060	0.0074	0.0042
19:00	0.0025	0.0029	0.0019	0.0013	0.0016	0.0007

The daily forecasting performance was also investigated as shown in Table 4. The daily forecasting of power production was computed by summing the forecasts of each hour in a day. It was also observable that the proposed method beat the generic MLP-based forecasting method and the persistence model for daily performance.

Table 4. Daily forecasting performance on the test dataset.

Method	RMSE	MAE
Robust-MLP	0.6508	0.4370
Generic-MLP	0.6635	0.4511
Persistence	0.7988	0.4990

The forecasting of PV power production by the proposed method compared with the observations and forecasts by the generic MLP model on consecutive days is depicted in Figure 4. It can be observed that the proposed method well captured the evolution of hourly PV power production and the forecasting errors by the proposed method were generally smaller than the generic MLP network.

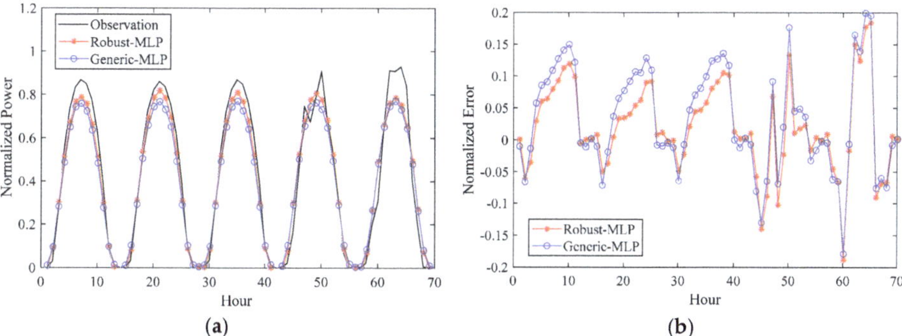

Figure 4. Forecasting performance comparison: (**a**) forecasting values; (**b**) forecasting errors.

4. Conclusions

In this paper, a robust MLP network was proposed for day-ahead hourly forecasting of PV power production. The generic MLP network is usually trained by minimizing the mean squared error; however, the mean squared error loss is sensitive to a few particularly large errors, which can result in an estimator with poor asymptotic relative efficiency. To tackle this problem, this paper adopted the pseudo-Huber loss, which combines the best properties of squared loss and absolute loss.

The efficacy of the proposed method was validated by testing on real PV data. The numerical experiments demonstrated that the proposed method outperformed the generic MLP network in terms of both RMSE and MAE for various time scales. This study suggests that the pseudo-Huber loss can be employed to optimize artificial intelligence-based approaches in handling solar data.

Author Contributions: Conceptualization and writing, C.H.; methodology, L.W.; software, L.C. and J.Z.; validation, N.P. and S.L.; funding acquisition, J.-H.W. and X.L.

Funding: This research was supported in part by the Foundation of Key Laboratory of Wind Energy and Solar Energy Technology, Ministry of Education under Grant 2018ZD02, in part by the Fundamental Research Funds for the Central Universities under Grants 06500103 and 06500078, in part by the University of Science and Technology Beijing–National Taipei University of Technology Joint Research Program under Grant TW2018008, in part by the National Natural Science Foundation of China under Grant 71473155, and in part by the National Key Research and Development Program of China under Grant 2018YFC0808306.

Conflicts of Interest: The authors declare no conflict of interest.

References

1. Huang, C.; Wang, L.; Lai, L. Data-driven short-term solar irradiance forecasting based on information of neighboring sites. *IEEE Trans. Ind. Electron.* **2018**. [CrossRef]
2. Sobri, S.; Koohi-Kamali, S.; Rahim, N.A. Solar photovoltaic generation forecasting methods: A review. *Energy Convers. Manag.* **2018**, *156*, 459–497. [CrossRef]
3. Zhang, Y.; Gatsis, N.; Giannakis, G.B. Robust Energy Management for Microgrids With High-Penetration Renewables. *IEEE Trans. Sustain. Energy* **2013**, *4*, 944–953. [CrossRef]
4. Hosseinzadeh, M.; Salmasi, F.R. Robust Optimal Power Management System for a Hybrid AC/DC Micro-Grid. *IEEE Trans. Sustain. Energy* **2015**, *6*, 675–687. [CrossRef]
5. Sardou, I.G.; Zare, M.; Azad-Farsani, E. Robust energy management of a microgrid with photovoltaic inverters in VAR compensation mode. *Int. J. Electr. Power Energy Syst.* **2018**, *98*, 118–132. [CrossRef]
6. Wang, L.; Huang, C. A novel Elite Opposition-based Jaya algorithm for parameter estimation of photovoltaic cell models. *Optik* **2018**, *155*, 351–356. [CrossRef]
7. Huang, C.; Wang, L. Simulation study on the degradation process of photovoltaic modules. *Energy Convers. Manag.* **2018**, *165*, 236–243. [CrossRef]
8. Andrade, J.R.; Bessa, R.J. Improving Renewable Energy Forecasting With a Grid of Numerical Weather Predictions. *IEEE Trans. Sustain. Energy* **2017**, *8*, 1571–1580. [CrossRef]
9. Wu, J.; Chan, C.K. Prediction of hourly solar radiation using a novel hybrid model of ARMA and TDNN. *Sol. Energy* **2011**, *85*, 808–817.
10. Li, Y.T.; Su, Y.; Shu, L.J. An ARMAX model for forecasting the power output of a grid connected photovoltaic system. *Renew. Energy* **2014**, *66*, 78–89. [CrossRef]
11. Leva, S.; Dolara, A.; Grimaccia, F.; Mussetta, M.; Ogliari, E. Analysis and validation of 24 hours ahead neural network forecasting of photovoltaic output power. *Math. Comput. Simul.* **2017**, *131*, 88–100. [CrossRef]
12. Capizzi, G.; Napoli, C.; Bonanno, F. Innovative Second-Generation Wavelets Construction With Recurrent Neural Networks for Solar Radiation Forecasting. *IEEE Trans. Neural Netw. Learn. Syst.* **2012**, *23*, 1805–1815. [CrossRef] [PubMed]
13. Luo, X.; Sun, J.; Wang, L.; Wang, W.; Zhao, W.; Wu, J.; Wang, J.-H.; Zhang, Z. Short-Term Wind Speed Forecasting via Stacked Extreme Learning Machine With Generalized Correntropy. *IEEE Trans. Ind. Inform.* **2018**, *14*, 4963–4971. [CrossRef]
14. Zendehboudi, A.; Baseer, M.A.; Saidur, R. Application of support vector machine models for forecasting solar and wind energy resources: A review. *J. Clean. Prod.* **2018**, *199*, 272–285. [CrossRef]
15. Huang, C.; Zhang, Z.J.; Bensoussan, A. Forecasting of daily global solar radiation using wavelet transform-coupled Gaussian process regression: Case study in Spain. In Proceedings of the 2016 IEEE Innovative Smart Grid Technologies-Asia (ISGT-Asia), Melbourne, Australia, 28 November–1 December 2016; pp. 799–804.
16. Capizzi, G.; Sciuto, G.L.; Napoli, C.; Tramontana, E. Advanced and Adaptive Dispatch for Smart Grids by means of Predictive Models. *IEEE Trans. Smart Grid* **2017**, *9*, 6684–6691. [CrossRef]
17. Mellit, A.; Pavan, A.M. A 24-h forecast of solar irradiance using artificial neural network: Application for performance prediction of a grid-connected PV plant at Trieste, Italy. *Sol. Energy* **2010**, *84*, 807–821. [CrossRef]
18. Ehsan, R.M.; Simon, S.P.; Venkateswaran, P. Day-ahead forecasting of solar photovoltaic output power using multilayer perceptron. *Neural Comput. Appl.* **2017**, *28*, 3981–3992. [CrossRef]
19. Gigoni, L.; Betti, A.; Crisostomi, E.; Franco, A.; Tucci, M.; Bizzarri, F.; Mucci, D. Day-Ahead Hourly Forecasting of Power Generation From Photovoltaic Plants. *IEEE Trans. Sustain. Energy* **2018**, *9*, 831–842. [CrossRef]
20. Huang, C.; Bensoussan, A.; Edesess, M.; Tsui, K.L. Improvement in artificial neural network-based estimation of grid connected photovoltaic power output. *Renew. Energy* **2016**, *97*, 838–848. [CrossRef]
21. Kingma, D.P.; Ba, J. Adam: A method for stochastic optimization. *arXiv*, 2014; arXiv:1412.6980.
22. Barron, J.T. A more general robust loss function. *arXiv*, 2017; arXiv:1701.03077.

23. Wang, L.; Zhang, Z.J.; Huang, C.; Tsui, K.L. A GPU-accelerated parallel Jaya algorithm for efficiently estimating Li-ion battery model parameters. *Appl. Soft Comput.* **2018**, *65*, 12–20. [CrossRef]
24. Huang, C.; Wang, L.; Yeung, R.S.C.; Zhang, Z.J.; Chung, H.S.H.; Bensoussan, A. A Prediction Model-Guided Jaya Algorithm for the PV System Maximum Power Point Tracking. *IEEE Trans. Sustain. Energy* **2018**, *9*, 45–55. [CrossRef]
25. Huang, C.; Wang, L.; Long, H.; Luo, X.; Wang, J.-H. A Hybrid Global Maximum Power Point Tracking Method for Photovoltaic Arrays under Partial Shading Conditions. *Optik* **2018**, *180*, 665–674. [CrossRef]
26. Wang, L.; Huang, C.; Huang, L.M. Parameter estimation of the soil water retention curve model with Jaya algorithm. *Comput. Electron. Agric.* **2018**, *151*, 349–353. [CrossRef]

 *sustainability*

Article

Integrated Model of Economic Generation System Expansion Plan for the Stable Operation of a Power Plant and the Response of Future Electricity Power Demand

Jang-yeop Kim [1] and Kyung Sup Kim [2,*

[1]   Institute of Defense Acquisition Program, Kwangwoon University, Seoul 01897, Korea; jykim1670@kw.ac.kr
[2]   Department of Industrial Engineering, Yonsei University, Seoul 03722, Korea
*   Correspondence: kyungkim@yonsei.ac.kr; Tel.: +82-2-2123-4012

Received: 6 June 2018; Accepted: 3 July 2018; Published: 11 July 2018

Abstract: The current study aims to establish an optimal Generation System Expansion Plan that can satisfy the increasing electricity demand while maintaining operational elements and the stability of the energy supply. The architecture is composed of plan-level and operation-level models, which are basically based on optimization. In the first step, we estimated future power demand data through time series analysis. In addition, power plant data were defined and verified data were collected. In the next step, the previous Generation System Expansion Plan methodology was used to deduce a feasible solution and construction costs that satisfy the reserve rate. In the third step, mixed integer programming (MIP)-based power generation system operation plan methodology was used to deduce numbers on the operation of power generation system. In addition, power plants with similar characteristics were grouped to reduce the calculation complexity of unit commitment. In the last step, a feasible solution for the duration of the plan (deduced in Stage II) and operations and maintenance cost information were combined to produce the optimal solution that minimizes the total cost. Experiments were conducted to demonstrate the proposed integrated generation system expansion planning architecture for establishing the optimal generation system expansion planning. This study has academic implications for the establishment of optimal power plant expansion plans to meet future increasing power demand while maintaining operational considerations and supply stability. The effectiveness of the proposed methodology is also illustrated through comparison and verification with the National Plan for Electricity Supply and Demand.

Keywords: generation system scheduling; integrated model; basic plan for long-term electricity supply and demand; forecasting model for electricity demand

---

1. Introduction

In order to satisfy the power demands that have been increasing since 2002, the Korean government establishes 15-year Generation System Expansion Plans biannually, of which there are seven to date. These plans are based on the National Energy Master Plan, which is the governing document for the Basic Plan for Electricity Demand and Supply. As its name indicates, the Generation System Expansion Plan minimizes the aspects of the operation of the power generation system, focusing more on the expansion plan for the generation system.

In establishing the Basic Plan for Electricity Demand and Supply, which contains construction plans for power generators, the dynamic optimal energy mix planning model WASP (Wein Automatic System Planning) is used, but it only utilizes reliability criteria determined by the reserve rate, which causes it to provide only the most limited index to secure a stable power supply. However, as the

Basic Plan for Electricity Demand and Supply is completed after consideration of a comprehensive set of data regarding the intent to construct and further modification measures are taken, the results of the energy mix calculated by the basic plan is adequate. If such background conditions are ignored and only WASP is used to calculate the energy mix, the model will produce high ratios of nuclear power, which has an absolute superiority over other power sources in terms of cost, making it difficult to find a truly adequate energy mix enabling a stable power supply.

As such, the current paper seeks to find a method for calculating the optimal energy mix that minimizes the total cost (including fixed cost determined by whether power generators are constructed, fuel costs related to power output, and environmental cost) while making it possible to operate a stable power generation system responsive to changes to load. In order to achieve this end, this paper formalizes the question of optimal energy mix and selects an optimization tool that can linearize the non-linear characteristics of the question to suggest the results of the analysis. Therefore, this study presents the algorithm of the process from the methodology for analyzing the demand data of the future to the establishment of the generation system expansion and the operation planning of the power generation facility, and the composition of an integrated model for deriving an optimal solution. The following are important elements to be presented through this study.

(1) Establishment of a methodology for power generation extension at the planning level
(2) Development of a model for calculating the realized operation and maintenance costs including the operating conditions of power utilities.
(3) Development of an integrated model for connecting planning and operation levels
(4) Estimation of the verification of the proposed model based on the domestic electricity demand supply program

This research focuses on model development for the extension of power plans in the domestic electricity supply and demand program. The electricity that is the object of this research has a particular characteristic in that all kinds of power plants generate identical goods, with the same quality but in different ways.

The product costs per utilities are different and the price predominance among those is determined in advance. Electricity is a volatile product that has no conception of stock and becomes extinct just after generation. If supply is lacking, it can cause a national crisis, thus it is necessary to retain a margin of power, even though it is not efficient. The criteria for the extension of power utilities include not just average demand but peak demand, in order to prevent the worst outcome.

To establish the extension planning of a generation system, the next issue of concern is to have a margin of power. This research designs the model with two types of reserved margin, such as a utility reserved margin and an instant reserved margin. The utility reserved margin is to prevent an uncertain situation by retaining a utility capacity that is greater than the expected demand of the peak times of year based on the concept of safe stock. The instant margin is additional reserved electricity that enables a stable supply at the operational level, even if the demand of electricity increases suddenly.

In reality, the model of the electricity system can be classified into six stages. Figure 1 shows a cascading chain of these models working across a range of timescales from milliseconds to years with an associated trade-off in the level of engineering detail captured [1].

Among the six stages, this research proposes an integration model considering four stages such as capacity planning, production cost, unit commitment and economic dispatching. At the stage of capacity planning, feasible solutions are generated through the setting of a utility reserved ratio. Next, the unit commitment stage establishes the optimization model, reflecting the operational constraints that should be considered in the activation of a realistic generation system. In the economic dispatch stage, we derive the power generation amount per utilities and other expenses that satisfy time-based demand based on the optimization model in the unit commitment stage. Lastly, the production cost stage generates the construction, operation and maintenance costs through cost-minimizing.

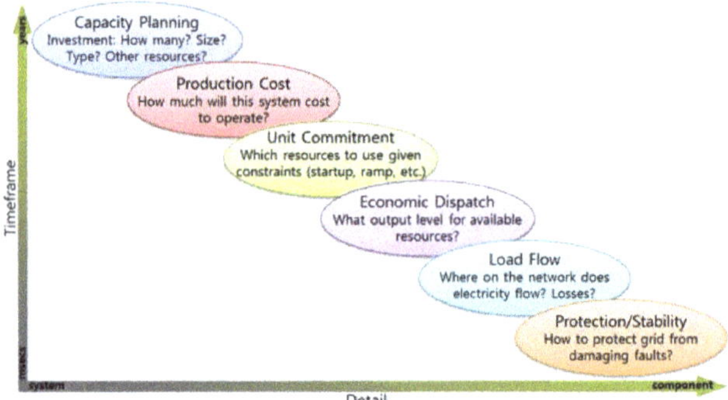

Figure 1. Electricity Modeling Types. Source: Palmintier [1].

We finally design the optimal extension planning of power generation utilities to maintain both the operational constraints and supply stabilization, as well as to satisfy the future increasing electricity demand, based on the concept of such electricity systems.

In this respect, this study has academic implications for the establishment of optimal power plant expansion plans to meet future increasing power demand while maintaining operational considerations and supply stability.

The structure of the current thesis is as follows: Section 2 summarizes previous studies on Generation System Expansion Plan, and how the current study differs from them. Section 3 explains the research model and experiment environment for the optimal Generation System Expansion Planning presented in this study. In Section 4, we first generate future electricity demand data to be used in the statistical experiments, and defines detailed data related to the operation of power plants. We propose a generation system scheduling model that determines which system will be turned on and off when, in order to satisfy hourly electricity demands. Section 5 discusses an Integrated Generation System Expansion Planning methodology that can cater to operational elements and maintain supply stability while satisfying increasing electricity demands. Lastly, Section 6 makes conclusions based on the experiment results and discusses implications of the current research as well as recommendations for future studies in the area.

2. Literature Review

2.1. Generation System Expansion Planning

Recently, research on trading in the electric power market has been actively conducted. Bahrami and Amini [2] proposed an energy trading algorithm that can optimize the cost of load aggregators and profit of the generators in a power market environment where a distributed power source is expanding such as a smart grid. The future smart grid aims to empower utility companies and users to make more informed energy management decisions [3]. In addition, Demand Response is becoming more important as the smart grid is spread and spread. In this regard, much research has been conducted to manage the load on the consumer side and to minimize the electricity cost [4,5]. Although these studies are important, as mentioned above, we have studied the construction of power generation sources that focus on the demand and supply of electricity considering the scale of the research model.

In general, the power generation system plan refers to the Generation System Expansion Plans that power companies set to predict and satisfy the future electricity demand of their jurisdiction. Investment costs are returned through electricity fees, and in case the electricity market is competitive, the power generation operator forecasts future market conditions to review collecting the investment and make decisions on the construction of power generators [6].

The Generation System Expansion Plan is an act of determining the construction of various types of power generators to adequately satisfy electricity demands. Construction of power plants requires immense financial and time investment, and the generators operate for long terms of more than thirty years in general. Because power generators have varied construction costs, construction periods, plant capacities, variable costs, operational characteristics, and operational periods, various studies have been conducted to determine how to select power generators for construction.

Several methods have been developed to solve the problem, such as stochastic dynamic programming [7], non-linear programming (NLP) [8], mixed-integer linear programming (MILP) [9], multi-objective programming [10], evolutionary programming (e.g., GAs) [11–18], and other heuristics and mathematical approaches [19–21]. The formulation of the problem objective and constraints varies in each implementation, incorporating emissions costs and other environmental constraints (NOx, SOx), transmission constraints, reliability criteria, demand-side management programs, reserve margins, location and financial constraints.

The following are papers that use optimization methodology: Majumdar and Chattopadhyay [9] generated solutions using mixed-integer linear programming (MILP) methodology, and executed sensitivity analysis based on the application of financial planning and changes to various options. Meza et al. [10] reflected construction costs, operation and maintenance costs, transmission costs, environmental effects, and changes in the fuel process, as limiting factors to solve the problems via multi-objective linear programming. Tekiner et al. [21] utilized Monte-Carlo simulation to calculate the power plant operation rate and power generation output, and proposed a multi-objective optimization model that considers environmental effects in addition to limitations already applied to the previous Generation System Expansion Plans. Ahmed et al. [22] generated solutions using a mixed-integer linear programming (MILP) model that applies a levelized cost, but was only able to calculate the capacity ratio structure of various power generation system types in the final year, rather than a yearly construction plan. Cheng et al. [23] developed the Chinese Generation System Expansion Plan methodology, which considers geographical conditions in China to propose a model that takes into account geographical structures and inter-regional transmission conditions. Flores et al. [24] mathematically suggested a typical Generation System Expansion Plan model that takes into account power generation conditions in Argentina, but it cannot calculate short-term operation because of characteristics of the long-term expansion plan model.

The following are representative studies using the stochastic methodology. Mo et al. [7], Botterud and Ilic [25] proposed solutions to the Generation System Expansion Plan problem through a stochastic dynamic programming model that takes into account the uncertainties in demand and fuel prices. In addition, most expansion plan research using the stochastic methodology apply probability values through the Markov chain process in order to define uncertainty factors as yearly variability of demand and fuel prices and reflect them in a probabilistic manner.

In addition, there are multiple studies that use heuristic methodologies. Park et al. [12], Kannan et al. [16], Firmo and Legey [13], Sirikum et al. [14], Sepasian et al. [20] are examples of this. Most heuristics-related studies solve the problem through GA (Genetic Algorithm), hybrid GA and develop customized models that fit the situations of each nation, and deduce a cost function based on the models.

Integrated software packages have also been developed for the solution of the centralized Generation System Expansion Planning (GSEP). WASP, which is the most general method for modifying the dynamic planning method, began to be developed in 1973 by TVA (Tennessee Valley Authority) and ORNL (Oak Ridge National Laboratory). Afterwards, in order to use WASP more effectively, International Atomic Energy Agency (IAEA) set out to complement Model for Analysis of Energy Demand (MAND) and hydro-thermal power system, developing further versions up to WASP-IV. WASP-IV is a computing model to establish long-term energy-mix development plan, which aims to minimize the system cost by generating the optimal construction plan for each year, from the beginning to the end of the plan period. The model uses a probabilistic simulation and dynamic planning method to select candidate plans that minimize total power generation costs, taking into account investment costs, operation costs, fuel costs, and remaining values. The WASP model was used widely in Korea as the mathematical planning model for selecting the minimum cost alternative for an electricity supply and demand plan [26].

2.2. Generation System Operation Planning

A power generation system operation plan is a scheduling problem to determine which power plant to start-up and shut-down at what times in order to satisfy hourly electricity demands. Through this plan, an hourly power generation plan by plant and realistic operation and maintenance costs are calculated. Many previous unit commitment studies solve highly complex problems that consider various limiting factors, but because of this, a power generation schedule that can be calculated through the models are limited to the short-term, such as daily. Depending on their purpose, unit commitments do not have to be long-term. However, due to the strength in which unit commitment shows similar results to actual power generation conditions, the model can have much meaning if the problem is expanded to long-term scheduling. In such terms, the following shows the categorization of previous unit commitment studies into short-term and long-term scheduling problems. First, Delarue et al. [27] conducted a study on short-term unit commitment, analyzed the Belgian electricity system through a typical model that takes into account the operational state of power plants, such as start-up and shut-down. Delarue et al. [28] studied the daily operation costs of introducing wind power generation and the resulting environmental influences. Andrianesis et al. [29] suggested a power generation system operation plan model for the Greek electricity system, which reflected a condition in which a monopoly was transitioning to a competitive market system. Simoglou et al. [30] generated the optimal operation scheduling for the stable operation of a power plant based on the previous day's electricity demand data. Li and Shahidehpour [31] proposed an operation planning model aiming to maximize the revenue of power generation companies, rather than establishing a plan from the perspective of supply stability that previous models took.

For long-term SCUC, various methods have also been proposed [32–34]. In Vemuri and Lemonidis [32], the Lagrangian relaxation (LR) method is used to decompose the fuel-constrained unit commitment problem into a linear fuel dispatch problem and a unit commitment problem. LR is also employed in Fu et al. [34] to divide the long-term SCUC into tractable short-term subproblems without fuel or emission constraints. In addition, studies such as Thorin et al. [35] and Seki et al. [36] apply the Lagrangian relaxation (LR) method. Handschin and Slomsmi [33] use a two-stage method to solve the unit commitment problem with long-term energy constraints. The optimal daily energy for each unit is calculated in the first stage, and then the unit commitment problem is solved in the second stage. Wang et al. [37] presents a fast bounding technique to improve the traditional branch-and-cut algorithm. Based on the work in Wang et al. [37], an inducing-objective-function-based method is proposed in Bai et al. [38]. In addition, a study by Chen et al. [39] based on the stochastic method was proposed, and Voorspools and D'haeseleer [40] generated operation plan patterns for each power generation system, designated priority for the patterns, and conducted probabilistic scenario analysis on them.

3. Research Methodology

The current study proposes architecture and a methodology for an Integrated Generation System Expansion Planning Model as shown in Figure 2. The architecture is composed of a plan-level model that satisfies both electricity demands that will increase in the future as well as the reserve rate, and an operation-level model to consider operational elements and maintain the stability of the electricity supply. The methodology consists of Stages I–IV. In Stage I, future electricity demand data is generated and power plant data is collected. In Stage II, the previous Generation System Expansion Plan methodology is used to deduce a feasible solution and construction costs that satisfy the reserve rate. In Stage III, mixed integer programming (MIP)-based power generation system operation plan methodology is used to deduce numbers on the operation of power generation system, such as year-round operation and maintenance costs, power generation output, and a maintenance plan. In Stage IV, a feasible solution for the duration of the plan (deduced in Stage II) and operation and maintenance cost information are combined to produce the optimal solution that minimizes the total cost.

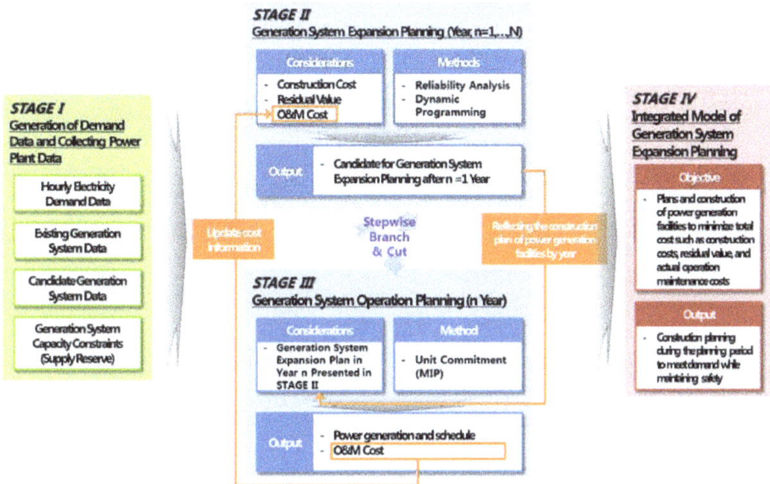

Figure 2. Research Model.

Evaluation in Stage IV is executed in order to simultaneously satisfy the yearly maximum electricity demands in the long term and electricity supply stability in the short term. It is noted that the hybrid integrated models cannot be directly used in stability analysis or optimization due to (1) the varied and often conflicting objectives for the different levels (planning and scheduling); (2) computational complexity of the optimization model, especially the operation planning (unit commitment) model that should consider all power plants and the time horizon.

Prior to configuring the model, the data on future electricity demands were generated to establish an expansion plan. The hourly electricity demand pattern shows that the demands have seasonal characteristics by week or year. In order to utilize such characteristics, dual-seasonality Holt-Winters time series model [41] was used to generate future electricity demand data. The details of the models can be found in Section 3. In addition, data for power plants used in the current study and other related data were defined and collected.

A model similar to the Generation System Expansion Plan model used in Korea was optimized and reconfigured based on a probability model as part of activities in Stage II (refer to Section 4). Through this, the current study deduced the significance and limitations of the current Korean Generation System Expansion Plan model, and the model was verified using its results and the results of the (Sixth) Basic Plan for Electricity Demand and Supply [42].

In Stage III, a power generation system operation plan model is an optimized model expressing the schedules of each power plant that can satisfy the electricity demands. The general methodology of unit commitment is a question of deducing a schedule that can minimize the costs by reflecting detailed elements related to the operation and outage of power plants as limitations. In fact, because different nations and power generation companies have different types and number of power plants, questions adequate for individual situations are defined so that solutions can be deduced when dealing with unit commitment problems. One issue that may arise is that the complexity of calculations may increase. Because of this, most unit commitment problems have a goal of establishing a short-term power plant schedule. The current study, however, sets out to propose a long-term power plant schedule model based on MIP, which can be applied to the new Generation System Expansion Plan methodology to compute the power generation output and operation and maintenance cost of power plant.

Previously in Stage II, we were able to compute the power plant construction plan and construction cost by year through the Expansion Plan, and Stage II enabled the calculation of power generation output and the operation and maintenance cost of each power plant, through the operation plan. In particular, the schedule optimization (Stage III) is described by presenting the decision variable, objective functions and the optimization methodology. The specifications for interactions of the expansion and operation models for use in Stage IV of the proposed architecture are detailed. In order to eliminate the infinitely many feasible solutions that are generated by the Expansion Plan in Stage II, branch-and-cut method for the total cost is applied in a forward direction. Section 6 details other connections of models and their integration.

Statistical experiments were configured with IBM ILOG OPL-CPLEX 6.3, the optimal commercial package used in the industry. The software was executed on a PC (Core(TM) i5-2500 CPU 3.3 GHz).

4. Data Generation and Model Composition

4.1. Generation of Demand Data

Unlike regular time series, electricity demand data take the multiple pattern (classified by time, day of the week, and season), which makes it important to consider such multiple patterns in order to improve the predictability of the model. According to the recent very-short-term power demand forecast research, the predictability of a model can be improved by applying patterns by the day or by the hour, rather than applying the same daily pattern throughout, when considering the seasonal differences in electricity demand. For example, Gould et al. [43] compared the electricity demand curves for each day of the week to categorize data into four patterns (Mondays to Thursdays, Fridays, Saturdays, and Sundays), and applied appropriate patterns for each instance of estimation to increase the predictive power of the model. In addition, Taylor and Snyder [44] segmented the pattern further, by the hour (Monday until 8:30 a.m., Tuesdays to Thursdays, Friday after 11:00, weekends) to construct a general model with even better predictive power. The application of the latest electricity demand forecast model must be preceded by the analysis of electricity demand by the hour and day of the week.

The business day pattern displayed in Figure 3 shows a very similar pattern to the yearly average pattern. Lowest load is reached at around 04:00 in the morning, after which the load continues to increase to surpass daily average at around 09:00, reaching the daily peak at around 12:00. At around 13:00, which is lunchtime, the load decreases to match the daily average, and increases again at around 15:00 to a point near the daily peak.

The non-working day (weekends and holidays) load pattern in Figure 4 shows a large difference from the business day load pattern. From 02:00 to 10:00, the load is lower than the respective daily average, but it increases beginning at 19:00 due to lighting demands, forming daily average at around 23:00.

The time series model considered for the electricity demand forecast model in the current study is the Holt-Winters seasonal method.

Generally, the power demand data does not follow the basic premise of the model in which the same pattern is repeated every cycle. In order to minimize the variation of these unstable patterns, we used the sliding window method in this study. The sliding window method is a method for estimating the coefficient of the model newly at each time point while sequentially moving the starting point while keeping the size of the sample period in the sample fixed. The size of the sliding window is fixed at 1820 (7 days × 52 weeks × 5 years of daily data) days, which can be considered both daytime and year-to-day, and the daily load on the next day of each sliding window is predicted Based on the model coefficients estimated in each of the sliding windows, the next day's forecast was derived over 364 days from 1 January 2015 to 30 December 2015.

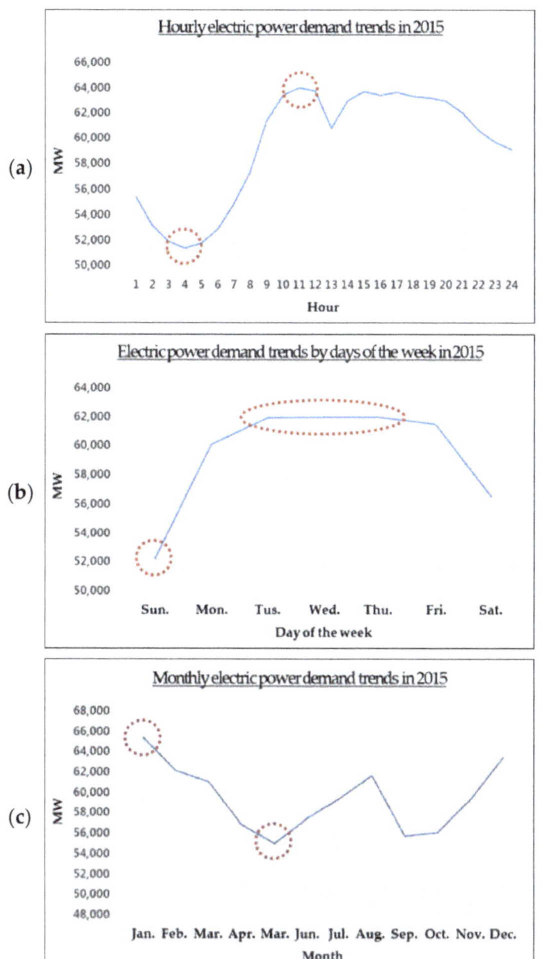

Figure 3. Analysis of Hourly, Daily, Monthly Electricity Demand Trends in 2015.

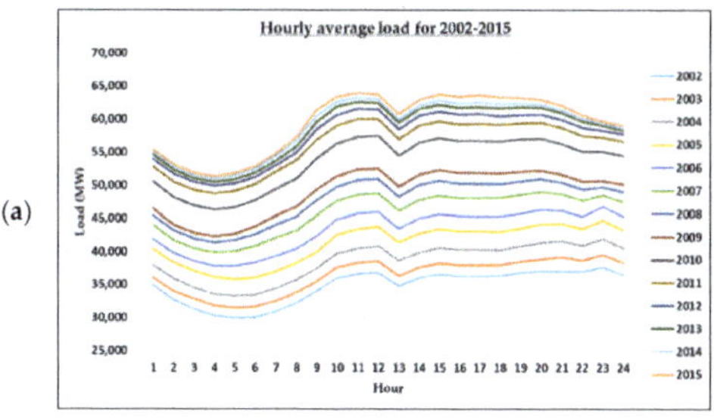

Figure 4. *Cont.*

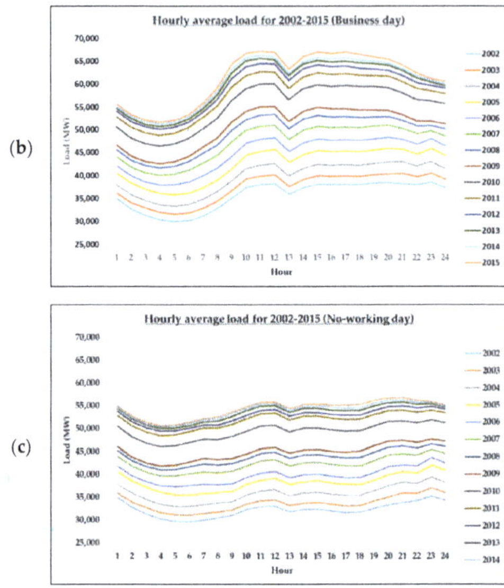

Figure 4. Time-dependent load pattern of annual average, working day and holiday.

The daily power demand data has a periodicity by every week (7 days) and year (364 days), so we consider a double seasonal Holt-Winters. Table 1 shows the results of the parameter estimation. The comparison of the prediction performance of the model is based on the root mean square error (RMSE) and the mean absolute percentage error (MAPE), as shown in Equations (1) and (2). In addition, comparisons were done by month as shown in Table 2.

$$RMSE = \sqrt{\frac{1}{n}\sum_{t=1}^{n}\left(y_t - \widehat{y_t}\right)^2} \tag{1}$$

$$MAPE = \frac{1}{n}\sum_{t=1}^{n}\left|\frac{y_t - \widehat{y_t}}{y_t}\right| \tag{2}$$

Table 1. Estimates of double seasonal Holt-Winters model.

Parameter	Value
α	0.53
β	0.01
γ	0.26
δ	0.42

Table 2. Monthly RMSE and MAPE during test periods.

Month	RMSE	MAPE	Month	RMSE	MAPE
January	1155.7	0.020	July	946.5	0.020
February	1752.3	0.030	August	990.9	0.021
March	1012.3	0.019	September	757.3	0.016
April	954.3	0.019	October	785.4	0.016
May	1236.8	0.027	November	847.6	0.016
June	722.4	0.016	December	1529.1	0.026
			Total	1053.8	0.021

Figure 5 shows the results comparing actual demands and forecasting demand derived from the model based on 2015, through a time series analysis. Data with a deviation greater than the reference value are excluded from the analysis as outliers.

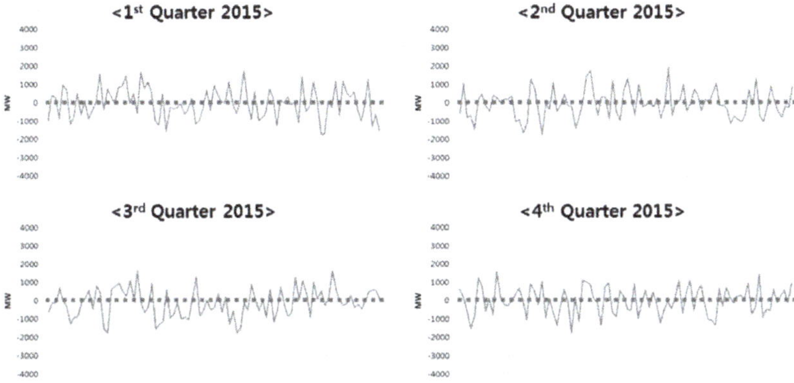

Figure 5. Quarterly Bias (first, second, third, fourth).

In the same way, forecasting demand data from 2016 to 2030 was generated, reflecting the annual growth rate of electricity consumption and peak demand in the national power supply and demand plan. For generating realistic data, the power consumption pattern of past holidays and the pattern of unexpected increase in power consumption due to weather characteristics were utilized in the model.

4.2. Data Collection of Power Plants

Key input variables relating to the generation system include economic variables such as predicted electricity demand, construction unit price, fuel cost, discount rate, inflation, as well as technical variables and reliability indices (LOLP: loss of load probability) such as power generator capacity, forced outage rate, days of preventive maintenance, operating and maintenance costs, and heat consumption of power generators. Such input data are variables that affect the optimization process, whose changes can cause large differences in the yearly power generator mix calculated by the model. As such, in order to generate a reasonable plan, the input variables demanded by the model must be correctly interpreted and the data must undergo sensitivity analyses. For example, the construction unit price for each candidate power generator indicates direct investment cost; the Generation System Expansion Plan uses the yearly investment cost, expressed as "Fixed Prices from Initial Year of the Plan," is converted to the cost at the time of construction completion using the real discount rate, ultimately using the sum of the latter cost as the total construction cost. This method is used to utilize the dynamic planning method and expresses the yearly construction cost in terms of the price at the time of construction completion. Table 3 is a summary of major input variables related to the generation system.

Table 3. Main Input Parameters Related to Generation System.

Parameter and Variable	Unit		Remarks
Demand forecast	MW, GWh	•	The shape of load duration curve for each quarter of each year, maximum demands, electricity usage
Discount Rate	%	•	This number is for comparing the current value of Generation System Expansion Plan alternatives
Generation Capacity	MW	•	Maximum power generation capacity of power plants
Construction Cost	$/kW	•	Cost at the time of completion

Table 3. *Cont.*

Parameter and Variable	Unit	Remarks
Fuel Cost	¢/Gcal	• Nuclear fuel cycle cost, as well as costs of B.C. oil, bituminous coal, and LNG (including transportation, insurance, other utility fees)
Operation and Maintenance Cost	\$/kW-month, \$/MWh	• Categorized into fixed and variable operation and maintenance costs
Heat Rate	Kcal/KWh	• Calories required to produce 1 kwh
Forced Outage Rate	%	• Probability that the power plant will be off due to failures during some time in the future • The rate influences power generation rate, LOLP calculation, operation cost, and the size of adequate reserve rate
Scheduled maintenance days	Day	• Period of planned maintenance to maintain power generator performance levels • Influences power generation output and LOLP calculation, and in turn influences the size of operation cost and adequate reserve rate

4.3. Generation System Expansion Planning Model Composition

The basic goal of the Generation System Expansion Plan is to find the energy mix that minimizes the objective function, which is the sum of current-value yearly investment cost and operating cost, under the restriction of satisfying the supply of reliability. Because generation systems are operated for two to five decades once constructed, and the generation mix of a year impacts the future generation mix, the optimal system mix for one year cannot be deemed optimal for the entire planning period. As such, the issue of power supply development planning takes on dynamic features, in which the generation mix among years are interrelated.

In order to maintain consistency with the ongoing energy policy of South Korea, we refer to the 7th Electricity Supply and Demand Plan to estimate the baseline of the GHG emission of the year 2030 for the generation sector, the major features of which focus on the low carbon energy mix; firstly, the four pre-arranged coal-fired power plants are excluded due to their high GHG emissions, the result of which eventually decreases the ratio of the coal-fired power plants. Policy planners reached an agreement that these excluded plants will be replaced with two new 1500 MW nuclear power plants. Secondly, the facility capacities of renewable energy are expected to be up to 33,890 MW and its portion will be 11% in 2035. Finally, the active demand management plan is fortified in connection with the energy efficiency and information technology.

For the purpose of establishing the baseline of the 2030 GHG emission volume, we make use of the WASP model which the IAEA has distributed to estimate electricity generation quantities. We anticipate the emission volume using the standard emission coefficient issued by the IAEA in Table 4.

Table 4. Emission coefficient. (Unit: g CO_2-eq/kWh).

Classification	Coal	LNG	Oil	Nuclear	Renewables *
Coefficient	1025	492	782	15	49

* solar photovoltaic; Source: IAEA (2006), IAEA (2016).

4.4. Generation System Operation Planning

4.4.1. Model Description

In the 1960s and 70s, mixed integer programming (MIP) was suggested to solve the unit commitment plan problem in earnest [45], and in the 70s and 80s, dynamic programming, as an algorithm for unit commitment plan for energy management system (EMS), was developed [46,47].

Dynamic programming, because it searches for all possible power generator combinations, can come up with the optimal solution, but if the number of power generators is too large, it takes too much time for calculation, effectively disabling the search for the optimal solution. In order to solve these problems, variations of DP, such as DP-sequential combination (DP-SC) and DP-truncated combination (DP-TC) were developed. In recent years, new variations of DP, such as Fuzzy-DP [48,49] were created as well. Since then, from the early 1980s to today, methods such as Lagrange's Method of Undetermined Coefficients, expert system, purge theory, neural network, genetic algorithm, and evolution programming appeared in the field. Lagrange's Method of Undetermined Coefficients, however, is adequate for optimizing large-scale systems, while the duality of the algorithm itself prevents it from reflecting the feasibility of the solution and the limiting conditions on power generators [50,51]. Most recently, a deterministic unit commitment (DUC) algorithm was developed [52], which is able to rapidly obtain solutions that are overwhelmingly economical compared to those yielded by other methods, but the performance of this new algorithm cannot be guaranteed for a multiplicity of limiting conditions. As such, no algorithm currently in existence can find the optimal solution in a short amount of time, in the face of all limiting conditions.

The problem of basic unit commitment planning is to find the optimal power generator outage/operation combination that satisfies multiple limiting conditions while minimizing the total power generation cost. The types of power generating system include steam power (anthracite coal, bituminous coal, heavy oil, LNG), combined thermal power (combined cycle: general, thermal cogeneration), hydroelectric power, internal combustion power, and nuclear power. In the case of nuclear power, the power generation output is almost fixed, because of which it is generally excluded from the unit commitment planning or is a part of a special unit commitment plan, as in the case of France. Hydroelectric power is excluded from the unit commitment plan due to rainfall and social and environmental limitations, or as in the case of Canada and Scandinavian countries, hydroelectric power plants are constructed in mass to operate a separate hydrothermal unit commitment plan [53]. As such, most of the power generators used in the unit commitment plan are thermal ones, which can be started in a matter of hours, making normal unit commitment plans to determine the combination of power generators by the hour.

The combination of operation/outage states during a time period is called the state during time period, and the number of possible states is shown in Equation (3), where the total number of input power generators is [54].

$$C(N,1) + C(N,2) + \cdots + C(N,N-1) + C(N,N) = 2^N - 1 \tag{3}$$

where,

$$C(N,j) = \left[\frac{N!}{(N-1)!j!} \right]$$

That is, the number of possible operation/outage combinations possible during the time period is $(2^N - 1)^M$, which makes it that the number of possible operation/outage combinations for ten power generators for 24 h in a day is $(2^{10} - 1)^{24} = 1.7259 \times 10^{72}$. Furthermore, the limiting conditions, system limiting conditions, and power generator limiting conditions generated from the relationship with previous time periods must be satisfied for every state of every hour, the curse of dimensionality frequently occurs, in which the dimensions are too large to calculate the solution to the optimal unit commitment problem. In other words, the problem of unit commitment planning is an optimization problem in the form of large-scale combined non-convex, which includes numerous equation/inequation limiting conditions and integer/real number variables [52,55].

In solving the unit commitment problem for each power plant, as suggested in previous sections, a high number of cases amounting to NP-hard exist to solve a long-term problem of calculating actual operating and maintenance costs over a year. That is, while it would be possible to solve the problem for a small number of power generation systems, the number of variables and cases will radically increase when there are more power generation systems under consideration. Such a problem can be solved by grouping power generation systems with similar characteristics.

Table 5 displays information on unit commitment for major power plants in Korea. There are 140 power plants of three types (eight types if categorized by capacity), and the number goes up to 200 if various other power plants are considered. In this case, there are many equations to solve the unit commitment problem, as shown in Figure 6, and because a total of 8760 h must be considered in order to calculate a year's worth of total operating and maintenance costs, the number of equations exponentially increase as shown in Figure 6. In order to solve this problem, the power plants are grouped by their characteristics to create less than ten types, thereby simplifying power plant data. Power generation systems cannot be randomly built by power generation companies; rather, there are established sizes and types of power plants that can be constructed, and power generation systems within the same category share many constraints, which makes it acceptable to group seemingly similar generators. A significance test is conducted in the next Section, by comparing with other grouping problems at the demand level, with which the basic problem can deduce results. The process by which a decision variable is deduced in the unit commitment problem shows that the previous problem of determining the on/off status of all power plants in the group is simplified into an integer programming problem in which the number of power plants participating in the operation among the group of power plants. For example, if the calculations determined that there are twelve power plants being operated during a certain period in power plant group 1, in which there are twenty members, the number of stopped power plants in the said group would automatically be eight.

Table 5. Characteristics data for each generation system.

Division	Nuclear	Coal Fired	LNG
Main capacity (MW)	1400, 1500	500, 800, 1000	400, 800
Construction cost per unit ($/kW)	2360	1419	955
Standard construction period (Month)	66	56	28
Fuel cost per unit (₵/Gcal)	149	2052	6715
Setup time after stop (Hour)	76–339	2.3–32.5	0.37–7.25

Source: Roh [56].

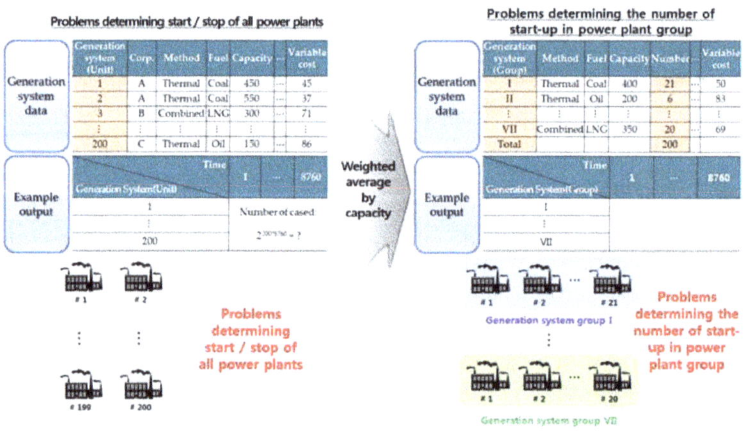

Figure 6. Grouping generation system to solve real problems.

The grouping of power generation systems necessitates partial changes to the previous equation. First, weights were given for the capacity of each group when calculating its characteristic values. Next, since piecewise linear approximation is impossible for heat consumption rate, it was changed to linear expression. This is possible because most of the power plants have a secondary coefficient of fuel usage function that is under 10^{-3}, which has a similar form to the linear form. Lastly, as the major decision variables have been changed to integer, not binary, form, it became impossible to distinguish the three starting conditions for each power plant in the group (hot/warm/cold). As such, these were unified into a single starting condition for application. For this, the frequency of starting conditions in normal situations were considered to apply the average as the weight of the characteristic.

4.4.2. Model Formulation

The GSOP model objective function to be minimized is formulated as follows:

$$C^{total} = \min \sum_{g \in G} \left[\sum_{d \in D} C^{maint}_{g,d} + \sum_{t \in T} \left(C^{Oper}_{g,t} + C^{start}_{g,t} \right) \right] \tag{4}$$

$$C^{maint}_{g,d} = M_{g,d} \cdot \frac{c^{fixO\&M}_g \cdot s^{maintfrac}_g}{s^{maint}_g} \qquad \forall g \in G, d \in D \tag{5}$$

$$C^{Oper}_{g,t} = f_{g,t}\left(P_{g,t}\right) \cdot c^{fuel}_g + P_{g,t} \cdot c^{varO\&M}_g \qquad \forall g \in G, t \in T \tag{6}$$

$$C^{start}_{g,t} = I_{g,t} \cdot \left[f^{start}_g \cdot c^{fuel}_g + c^{fixstart}_g \right] \qquad \forall g \in G, t \in T \tag{7}$$

As shown in Equation (4), objective function includes the maintenance and repair costs, operating costs including fuel cost, and costs for turning the power generation system on and off. The maintenance and repair cost of the system is determined by variable $M_{g,d}$, which determines the time of maintenance and repair for power generation system in each year, as shown in Equation (5). Equation (6) is on the operating cost, and includes the fuel use volume, fuel cost, and variable operating costs. Costs for turning the system on consider the costs, and are elaborated in Equation (7).

$$f_{g,t}\left(P_{g,t}\right) = a_g + b_g \cdot P_{g,t} \qquad \forall g \in G, t \in T \tag{8}$$

Formula (8) expresses the fuel use for power generation volume; fuel use volume function generally takes the form of quadratic expression, that is, a nonlinear form. Nonlinear quadratic expression, however, must be converted to linear function form for MILP solution.

The model constraints are as follows:

◆ Grouping

$$0 \leq U_{g,t} \leq n^{max}_g - M_{g,d} \qquad \forall g \in G, t \in d, d \in D \tag{9}$$

◆ System energy balance:

$$\sum_{g \in G} P_{g,t} = ed_t \qquad \forall t \in T \tag{10}$$

Power Output:

$$P_{g,t} \geq U_{g,t} p^{min}_g + R^{2,down}_{g,t} \qquad \forall g \in G, t \in T \tag{11}$$

$$U_{g,t} p^{max}_g \geq P_{g,t} + R^{1,GF}_{g,t} + R^{1,AGC}_{g,t} + R^{2,up}_{g,t} \qquad \forall g \in G, t \in T \tag{12}$$

◆ Inter-Period Ramping Limits:

$$P_{g,t-1} - P_{g,t} \leq U_{g,t} \cdot \Delta p^{downmax}_g + \max\left(p^{min}_g, \Delta p^{downmax}_g \right) \cdot D_{g,t} \qquad \forall g \in G, t \in T \tag{13}$$

$$P_{g,t} - P_{g,t-1} \leq U_{g,t} \cdot \Delta p^{upmax}_g + \max\left(p^{min}_g, \Delta p^{upmax}_g \right) \cdot S_{g,t} \qquad \forall g \in G, t \in T \tag{14}$$

◆ Minimum Up and Down Times:

$$U_{g,t} \geq \sum_{\tau=t-a_g^{minup}}^{t-1} I_{g,\tau} \quad \forall g \in G, t \in T \tag{15}$$

$$1 - U_{g,t} \geq \sum_{\tau=t-a_g^{mindown}}^{t-1} O_{g,\tau} \quad \forall g \in G, t \in T \tag{16}$$

◆ Unit Minimum and Maximum Output

$$U_{g,t} \cdot p_g^{min} \leq P_{g,t} \leq U_{g,t} \cdot p_g^{max} \quad \forall g \in G, t \in T \tag{17}$$

◆ frequency regulation Reserve Requirements:

$$\sum_{g \in G^1} R_{g,t}^{1,GF} \geq r^{1,GF} \quad \forall t \in T \tag{18}$$

$$\sum_{g \in G^1} R_{g,t}^{1,AGC} \geq r^{1,AGC} \quad \forall t \in T \tag{19}$$

$$\sum_{g \in G^1} \left[R_{g,t}^{1,GF} + R_{g,t}^{1,AGC} \right] \geq r^1 \quad \forall t \in T \tag{20}$$

◆ Contingency (Non-spinning), Supplementary (on-demand) Reserve Requirements:

$$R_{g,t}^{1,GF} \leq s_g^{1,GF} \cdot p_g^{max} \quad \forall g \in G^1, t \in T \tag{21}$$

$$R_{g,t}^{1AGC} \leq s_g^{1,AGC} \cdot p_g^{max} \quad \forall g \in G^1, t \in T \tag{22}$$

$$R_{g,t}^{2,up} \leq s_g^{2,up} \cdot p_g^{max} \quad \forall g \in G^2, t \in T \tag{23}$$

$$R_{g,t}^{2,down} \leq s_g^{2,down} \cdot p_g^{max} \quad \forall g \in G^2, t \in T \tag{24}$$

◆ Reserve Upper Bound Vector Constraints:

$$R_{g,t}^{1,GF} \leq s_g^{1,GF} \cdot n_g^{max} \cdot p_g^{max} \quad \forall g \in G^1, t \in T \tag{25}$$

$$R_{g,t}^{1AGC} \leq s_g^{1,AGC} \cdot n_g^{max} \cdot p_g^{max} \quad \forall g \in G^1, t \in T \tag{26}$$

$$R_{g,t}^{2,up} \leq s_g^{2,up} \cdot n_g^{max} \cdot p_g^{max} \quad \forall g \in G^2, t \in T \tag{27}$$

$$R_{g,t}^{2,down} \leq s_g^{2,down} \cdot n_g^{max} \cdot p_g^{max} \quad \forall g \in G^2, t \in T \tag{28}$$

◆ Commitment State:

$$U_{g,t} = U_{g,t-1} + I_{g,t} - O_{g,t} \quad \forall g \in G, t \in T \tag{29}$$

◆ Peak Period (Winter/Summer) Reserve Rate Lower-bound Constraints:

$$\frac{\sum_g \left[1 - U_{g,t} - M_{g,d} - f_g^{outage} \right] \cdot p_g^{max}}{\sum_g p_g^{max}} \geq r^{oper,winter} \quad \forall t \in T^1 \tag{30}$$

$$\frac{\sum\limits_{g}\left[1 - U_{g,t} - M_{g,d} - f_g^{outage}\right] \cdot p_g^{\max}}{\sum\limits_{g} p_g^{\max}} \geq r^{oper,summer} \qquad \forall t \in T^2 \tag{31}$$

◆   Maintenance Sufficiency:

$$\sum_{d \in D} M_{g,d} \geq s_g^{maint} \cdot n_g^{\max} \qquad \forall g \in G \tag{32}$$

$$M_{g,d} = M_{g,d-1} + M_{g,d}^{in} - M_{g,d}^{out} \qquad \forall g \in G, d \in D \tag{33}$$

$$M_{g,d} \geq \sum_{\delta = d - s_g^{maint}}^{d} M_{g,\delta}^{begin} \qquad \forall g \in G, d \in D \tag{34}$$

$$f_{g,t}^{outage} \cdot M_{g,d} = 0 \qquad \forall g \in G, t \in d, d \in D \tag{35}$$

◆   Simultaneity Constraints:

$$M_{g,d} \leq s_g^{maintfrac} \cdot n_g^{\max} \qquad \forall g \in G, d \in D \tag{36}$$

Equation (9) refers to the range of power plants in which the unit commitment can be performed, excluding the power plants under maintenance in the group. Equation (10) is the balance equation for power generation volume by each power plant. Equations (11) and (12) are equations for deducing the actual power generation output in consideration of operating status and primary and secondary reserve power. In addition, when the actual demand shows a large difference from forecasted demand, power plants may need to radically increase or decrease output in a short amount of time. In such cases, Equations such as (13) and (14) may be required, which are related to the ramping up and down of power generation which also takes into account the characteristics of each power plant. Furthermore, Equations (15) and (16) are about the minimal time required to maintain the respective statuses when the power plants are started or shut down. Equation (17) is the boundary of power generation considering unit commitment. Equations related to reserve power are categorized into primary reserve power, shown in Equations (18)–(20), and secondary reserve power, expressed through Equations (21)–(24). The current study defines primary reserve power as being used to satisfy the power shortage caused by rapid changes in demands, while secondary reserve power is used to respond to electricity shortage caused by output decline from force power plant outages. Primary reserve power was calculated for LNG, petroleum, pumped storage, and hydroelectric power plants, which are more rapidly responsive, and secondary reserve power was calculated for coal power plants that are in states of operation or outage. Equations (25)–(28) are interaction equations that show the possible maximum output of primary and secondary reserve powers, taking the operating statuses and inherent characteristics of each power plant into account. Equation (29) is a balance equation for the change in the number of power plants in operation during this period compared to the previous period. Equations (30) and (31) express the minimum reserve power that must be maintained as per policy during winter and summer times with high electricity demands.

Next, Equations (32)–(35) are related to the maintenance and repair of power plants. They are interaction equations that express the satisfaction of minimum maintenance and repair days required for each power plant, calculation of timing for maintenance and repairs, and minimum maintenance period. Equation (36) is a constraint on the minimum value of maintenance required in a group.

4.5. Results of the Experiment: National Plan for Electricity Supply and Demand

Using the mathematical model of unit commitment for grouped power generation systems that verified statistical significance, this section generated Table 6, which contains unit commitment plans for all Korean power plants, as well as their detailed operation and maintenance costs and power generation outputs in January 2013. For the experiment, major Korean power plants were grouped into five types and ten types. The total number of power plants is 168, which excludes new and renewable energy and collective energy power generation systems.

Table 6. Result of the Operation Plan in January 2013.

Power Plant		Unit Capacity (MW)	No.	Total Capacity (MW)	Cost (K$)					Power Generation Output (GWh)	Operational Rate (%)	Start up	Shutdown	Total Time in Operation	Average Days in Operation
					Total		Variable		Fixed						
					C^{total}	C^{fuel}	C^{var}	C^{start}	C^{maint}						
1	Nuclear	896	24	21,504	232,533	5395	177,226	4086	45,826	13,076	84.5	1	1	15,955	27.7
2	Coal	451	47	21,197	261,968	89,454	97,839	60,257	14,418	13,593	89.1	14	14	32,689	29.0
3	Coal	835	4	3340	33,574	14,112	11,712	6310	1440	2201	91.5	2	2	2670	27.8
4	Oil	260	16	4160	30,689	2495	1019	24,174	3002	129	4.3	5	5	992	2.6
5	Oil	58	4	232	676	227	0	0	449	0	0	0	0	0	0
6	LNG	399	52	20,748	154,678	65,148	33,828	42,750	12,952	5907	39.5	36	20	31,228	25.0
7	LNG	783	4	3132	32,472	19,704	8270	2838	1659	2036	90.3	2	0	2784	29.0
8	Pumped Storage	294	16	4704	12,685	0	10,807	268	1609	3254	96.1	16	16	11,136	29.0
9	Nuclear	1500	1	1500	13,826	475	11,111	215	2025	1004	93.0	1	1	672	28.0
10	Coal	1000	0	0	416	416	0	0	0	0	0	0	0	0	0
	Total		168	80,517	773,516	197,426	351,812	140,898	83,380	41,200	71.1	77	59	98,126	24.3

From the operation and maintenance costs perspective, the price of fossil fuel is causing the fuel to take large amounts of operation and maintenance costs. In contrast, nuclear power plants, which have lower fuel prices and generate much base power, have a high rate of variable costs, which rise and fall depending on the power generation output. Furthermore, the fact that operation/shutdown costs take up 20% of all operation and maintenance costs indicates that precise calculation of operation/shutdown cost is important for computing operation and maintenance costs.

In addition, in order to compare the detailed operation and maintenance costs deduced by the operation plan in this section with operation and maintenance cost deduced from Stage II (Generation System Expansion Plan), Table 7 was set up to comparatively view fuel costs, detailed operation and maintenance costs, capacities, power generation outputs, operation rate, and average days in operation for January 2013.

Table 7. Comparison of the Operation Plan Results for Stage II and III in January 2013.

Category (UC)		Fuel Cost (K$)	Operation and Maintenance Cost (K$)				Capacity (MW)	Power Generation Output (GWh)	Operation Rate (%)	Average Days in Operation
			Total	Variable Cost	Repair Cost	Shut Down Cost				
Stage III Power Generation System Operation Plan	Nuclear	5871	240,489	188,336	4301	47,852	23,004	14,080	85.0	27.7
	Coal	103,982	191,976	109,551	66,566	15,858	24,537	15,794	89.4	28.9
	LNG	84,853	102,297	42,098	45,589	14,611	23,880	7943	46.2	25.3
	Oil	2722	28,642	1019	24,174	3450	4392	129	4.1	2.1
	Pumped Storage	0	12,685	10,808	268	1609	4704	3254	96.1	29.0
	Total	197,428	576,090	351,812	140,898	83,380	80,517	41,200	71.1	24.3
Stage II Power Generation System Expansion Plan	Nuclear	5739	317,809				23,116	16,109	95.5	29.0
	Coal	107,690	104,076				24,534	16,955	94.7	28.8
	LNG	82,352	85,464				23,579	7231	42.0	12.8
	Oil	94	23,452				4781	5	0.1	0.0
	Pumped Storage	0	10,572				4700	3379	98.5	30.0
	Total	195,875	541,373				80,710	43,679	74.1	22.5

Because the results are deduced for the same period, they show similar values for most items, but a comparison of operation and maintenance costs reveals that those calculated in Stage II (Power Generation System Expansion Plan) and Stage III (Power Generation System Operation Plan) differ to a degree. Because the experiment was conducted for a macro-model of a long-term expansion plan, operation and maintenance costs calculated in Stage II were used as a parameter after simplifying operation and maintenance cost history collected in advance. In contrast, operation and maintenance costs calculated in Stage III are a result of the unit commitment plan of all power plants responding to actual hourly demands, which also provides all detailed operation and maintenance cost items. Comparing the two operation and maintenance costs yields similar results, with a slightly higher value to that deduced by the unit commitment plan. This is thought to be the result of a difference in unit commitment cost. The major factor is thought to be the increase in operation and shutdown frequency of fossil fuel generators using coal, oil, and LNG. Excluding these, the operation and maintenance costs for Stage II, 3 during the same experiment time is the same.

The following Figures 7 and 8 compare the monthly operation and maintenance costs (minus fuel cost) in 2013 and yearly operation and maintenance costs. For the monthly operation and maintenance costs, the same in Stage II is used as a parameter-fixing value as mentioned before, but that in Stage III is variable depending on seasonal features, similar to actual power plant operation conditions. Because of this, operation and maintenance costs calculated in Stage II and Stage III become more different towards the end of the planned period, which indicates that the operation and maintenance costs of Stage II may be distorted.

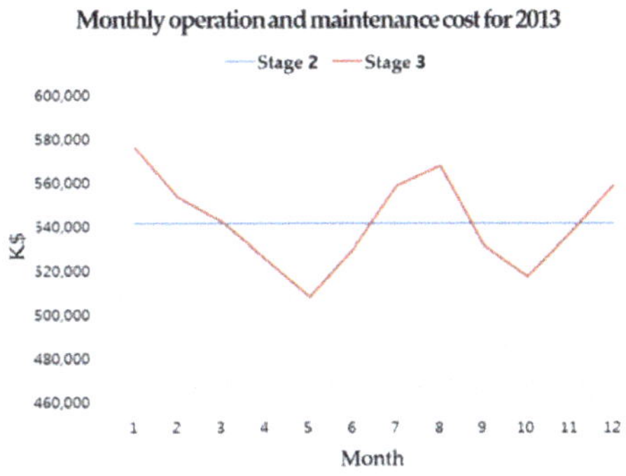

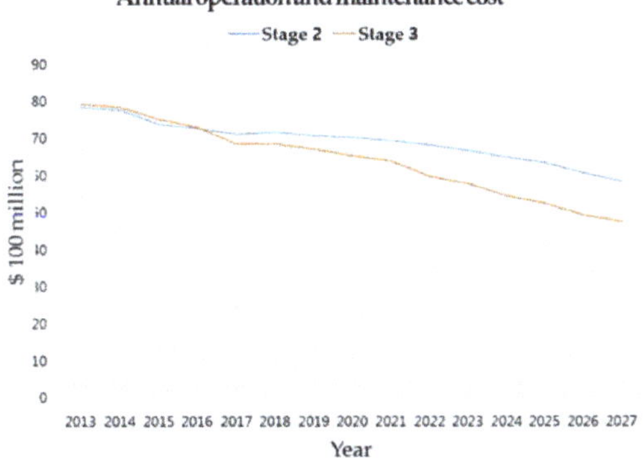

Figure 7. Monthly operation and maintenance costs (minus fuel cost) in 2013.

Figure 8. Yearly operation and maintenance costs.

5. Integrated Model Composition and Case Study

A long-term plan is established using the Generation System Expansion Planning methodology suggested in Stage II. The most problematic thing in this is the algorithm for calculating operating costs. The previous Generation System Expansion Plan used a probability simulation to compute the operating costs, and the precision of the simulation differed depending on whether the results were to be used for long-term or short-term planning. In the short-term plan, such as a power generation plan or a fuel supply plan, the evaluation of operating costs utilizes hourly load data, considering the maintenance plan and power generator unit commitment, in the process of precision simulation. However, the evaluation for long-term plans such as the Generation System Expansion Plan does not take into account a more detailed unit commitment plan, and only simplified versions of repair plans as well as the load duration curve—as opposed to hourly load—are used. This is because the calculation time increases exponentially if mathematical planning techniques such as dynamic programming are used to find the Generation System Expansion Plan that minimizes investment and operating costs, as they consider thousands of simulations for each state in each year. As such, the current study used a probability simulation for a MILP-based and operating cost-based Generation System Expansion Plan despite lower reliability. However, the less-precise operating cost-based Generation System Expansion Plan will yield errors as years advance.

The operating cost evaluation in short-term plans utilizes unit commitment analysis to establish an operating schedule for short periods of one day or one week. The process was discussed in the previous section. The current study proposed a methodology for analyzing unit commitment via grouping power plants, in order to enable the consideration of operating cost evaluation in long-term plans. Under the assumption that this methodology guarantees the precision of demand data, the current study executed statistical verification that it has a similar result to the previous unit commitment methodology. Through this, the current study assesses that it would be possible to realistically and precisely evaluate the operating costs by applying a long-term unit commitment analysis method via grouping of power plants, which can replace the previous system that has the limitation of lower precision.

As such, the current section looks to propose a related model and algorithm that utilize the advantages of generation system expansion planning suggested in Stage II, and unit commitment analysis suggested in Stage III.

5.1. Algorithm of Integrated Model

As show in Figures 9 and 10, using the Generation System Expansion Plan utilizing the DP concept, step 1 generates a feasible solution for the aggregate paths including the number of power generation system constructed in each year. Step 2 deduces the total operating and maintenance costs that reflects the initial years' construction plan for all routes, using a generation system operating plan. Step three deduces the construction costs for each year and the remaining value for all routes in the Generation System Expansion Plan, and calculates the total cost, adding the operating and maintenance cost (converted to the current value) resulting from step 2. In step 4, the second and third steps are repeated in order for each year. In addition, because there are too many cases to deduce all total operating and maintenance cost for each year, we cut the routes that do not require calculation. In step 5, a branch-and-cut method is applied; if the aggregate paths for multiple same construction plans overlap in the same year, the path with the minimum total cost until the previous year is selected and the remaining paths are cut. Here, after the branches are unified into one, one optimal path will be selected for further procedure; the path may not be optimal if the total cost until the previous year is not the minimum among many options. Step 6 is a repetition of step 2 to step 5 until the end of the planned year. At the end, the total cost of paths that are not cut until the completion of the algorithm are compared to select one most optimal path.

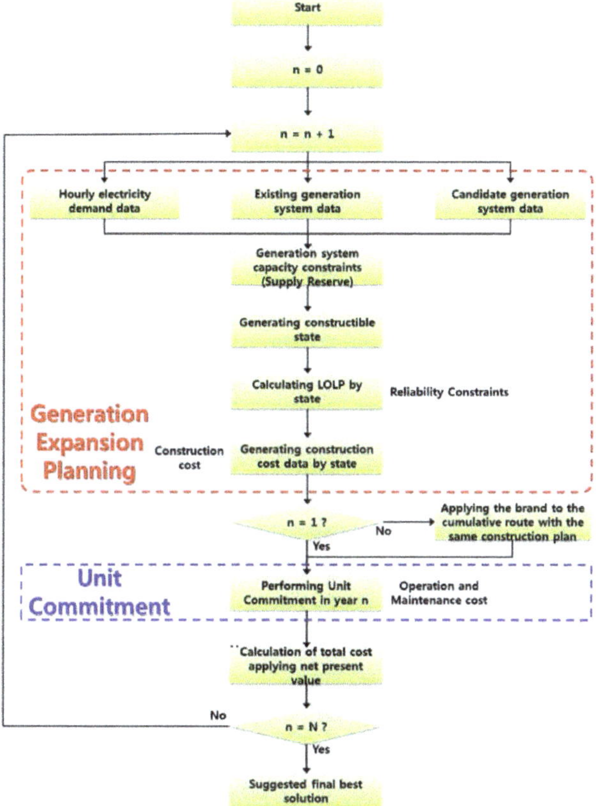

Figure 9. Flowchart of Integrated Model.

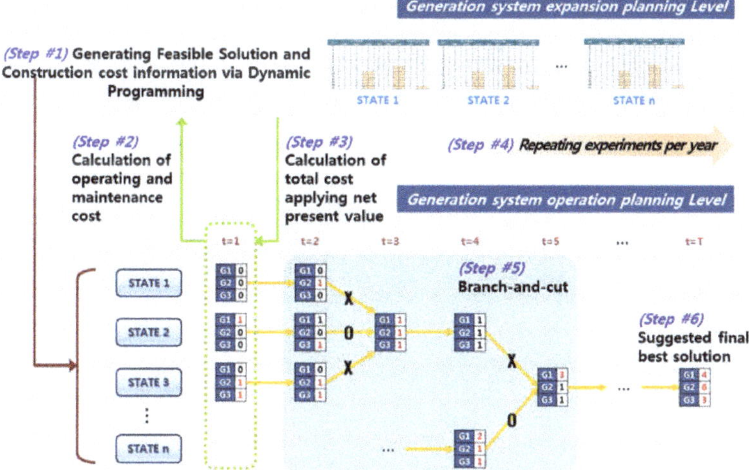

Figure 10. Simplified integrated model flow.

5.2. Results of Integrated Model Experiment and Comparative Analysis

5.2.1. Results of Integrated Model Experiment

Table 8 shows the results of the integrated model algorithm discussed in the previous section. First, the results showed that two nuclear, six coal, and five LNG power plants will be constructed by the final year of 2027 during the plan. In terms of costs, construction cost was $12.81 billion, the remaining value was $10.08 billion, and operation and maintenance costs were $96.95 billion, with the total cost being analyzed as $99.18 billion.

Table 8. Results and cost data derived from the integrated model.

| Year | Generation System Expansion Plan in the Current Study | | | Ratio of Reserve Rate to Maximum Demand | Cost Data ($100 Million) | | | |
	Nuclear (1500 MW)	Coal (1000 MW)	LNG (1000 MW)		Construction Costs	Remaining Value	Operation and Maintenance Costs	Total
2013				11.8			86.2	86.2
2014				17.8			80.7	80.7
2015				26.2			75.9	75.9
2016				37.6			73.0	73.0
2017				34.5			68.4	68.4
2018				33.2			68.4	68.4
2019				28.4			67.1	67.1
2020				25.7			65.3	65.3
2021				23.0			63.8	63.8
2022			2	21.6	9.2	5.3	59.7	63.6
2023		1	2	21.1	16.9	10.8	57.6	63.8
2024				22.6			54.4	54.4
2025		4	1	22.1	33.3	25.5	52.4	60.2
2026	1			22.1	18.3	15.6	49.1	51.7
2027	1	1		22.3	24.0	22.1	47.4	49.3
Total	6000	6000	1600		128.1	100.8	969.5	991.8

Considering that the operation and maintenance costs are more than twenty times that of the realistic construction cost (construction cost—remaining value), it would be most effective to propose a construction plan that can reduce operation and maintenance costs, in order to cut total costs.

5.2.2. Comparison with Generation System Expansion Plan (Stage II) Results

As shown in Table 9, a comparison of the experiment results from the integrated model and the previous expansion plan deduced from Stage II showed that new nuclear power plants were −2 and LNG was +3. Construction cost of the two plans were similar, and in terms of cost, a comparison with the previous expansion plan model (Stage II) shows that the operation and maintenance costs deduced through Stage IV (integrated model) are lower by $7.8 billion. This is analyzed to be an effect of the lower operation rate of nuclear power plants and the higher utility rate of LNG power plants. In fact, the cost difference of $7.8 billion between the two plans is large enough to construct an additional four nuclear power plants and five coal plants, which makes it important to remove the distortion effect of operation and maintenance costs in the previous expansion model. In addition, when establishing a long-term development plan, the effect of applying a power generation system operation plan (unit commitment plan) in Stage III to calculate operation and maintenance costs will be very large.

Table 9. Comparison of results and cost data derived from integrated model and Stage II model.

| Year | Experiment Results of Integrated Model | | | | | | | Experiment Results of Stage II | | | | | | |
| | Generation System Expansion Plan | | | Cost Data ($100 Million) | | | | Generation System Expansion Plan | | | Cost Data ($100 Million) | | | |
	Nuclear	Coal	LNG	Construction Cost	Remaining Value	Operation and Maintenance Cost	Total	Nuclear	Coal	LNG	Construction Cost	Remaining Value	Operation and Maintenance Cost	Total
2013						86.2	86.2						78.4	78.4
2014						80.7	80.7						77.5	77.5
2015						75.9	75.9						73.7	73.7
2016						73.0	73.0						72.6	72.6
2017						68.4	68.4						71.1	71.1
2018						68.4	68.4						71.5	71.5
2019						67.1	67.1						70.7	70.7
2020						65.3	65.3						70.1	70.1
2021						63.8	63.8						69.2	69.2
2022			2	9.2	5.3	59.7	63.6		1		8.6	5.0	67.9	71.5
2023		1	2	16.9	10.8	57.6	63.8		2	2	25.1	16.0	66.4	75.5
2024						54.4	54.4						64.6	64.6
2025		4	1	33.3	25.5	52.4	60.2	1	3		41.4	32.1	63.3	72.6
2026	1			18.3	15.6	49.1	51.7	1			18.3	15.6	60.6	63.3
2027	1	1		24.0	22.1	47.4	49.3	2			34.7	32.1	58.6	61.2
Total	6000	6000	1600	128.1	100.8	969.5	991.8	6000	6000	1600	128.1	100.8	1036.2	1063.4

5.2.3. Comparison with National Electricity Demand and Supply Plan

Lastly, Table 10 shows the comparison between the experiment results deduced through the integrated model (Stage IV) and the construction plan suggested in the national electricity demand and supply plan. Compared to the national electricity demand and supply plan, the results of the integrated model had −2 new nuclear plants, −3 coal plants, and +5 LNG plants. The proportion of LNG power plants increased in the experiment results of the integrated model because of the following reasons: First, the model is analyzed to select LNG combined thermal power plants as being more optimal when a high operation reserve rate is required when the rise and fall of demand is high, because they can turn on instantaneously in ramping up and down or stopped state. Next, the fuel cost of LNG is analyzed to be lower due to the introduction of shale gas, which increased the competitiveness of LNG combined thermal power plants.

Table 10. Comparison between results derived from integrated model and National 6th Plan.

| Year | Experiment Results of Integrated Model | | | | Results of National Electricity Demand and Supply Plan | | | |
	Nuclear	Coal	LNG	Capacity Reserve Rate	Nuclear	Coal	LNG	Capacity Reserve Rate
2013				11.8				11.8
2014				17.8				17.8
2015				26.2				26.2
2016				37.6				37.6
2017				34.5				34.5
2018				33.2				33.2
2019				28.4				28.4
2020				25.7				25.7
2021				23.0		1		24.0
2022			2	21.6		2		23.1
2023		1	2	21.1		2		21.9
2024				22.6		2		25.5
2025		4	1	22.1	1	1		21.6
2026	1			22.1	1	1		22.8
2027	1	1		22.3	2			23.1
Total	2	6	5		4	9	0	

6. Conclusions

The current study aims to establish the optimal Generation System Expansion Plan that can satisfy the increasing electricity demand while maintaining operational elements and the stability of the energy supply. Detailed goals following this objective are suggesting the optimal Generation System Expansion Plan methodology that integrates the planning level and operational level and verifying the validity of the model based on actual data in the national electricity demand and supply plan. Architecture meeting these objectives is divided into four stages. Stage I estimates the seasonal Holt-Winters time series function to generate future demand data to be used in establishing the Generation System Expansion Plan and executed statistical significant verification. In addition, power plant data to be used in the actual Generation System Expansion Plan were defined and verified data were collected.

Stage II involved the previous Generation System Expansion Plan model. In other words, based on the WASP-IV model, Korea's power supply and demand plan results were derived. The Korea Power Exchange has been using the WASP-IV model to derive the electricity supply plan from the first plan to the present 7th plan. Therefore, we used the results of the power development plan as a basic case to compare the results of the integrated model and power plants with similar characteristics were grouped to reduce the calculation complexity of unit commitment in Stage III.

Stage II served to generate a feasible solution considering the maximum demand and capacity reserve rate and produced construction costs excluding the remaining value of new power plants. In Stage III, a unit commitment model that can calculate power generation output and operation and maintenance cost was established, formalizing the problem of grouping power plants with similar characteristics to reduce the calculation complexity in a long-term expansion plan. Considering the limitations of the previous Generation System Expansion Plan, Stage IV proposed an integrated model in which the models proposed in Stage II and III interact. That is, this integrated model provides the optimal solution that minimizes the total cost among the results from multiple expansion plan candidates, by connecting the expansion plan candidates and construction cost information provided in Stage II and realistic operation and maintenance cost information. Through this model, the current study provided a quick method for finding the solution using a branch-and-cut technique on many possible combinations of feasible solutions. By establishing the same environment as the process of setting up a national electricity demand and supply plan, the current study proposed a result through the integrated model. In addition, the previous methodology provided in Stage II was used to provide comparison/analysis of the results of the national electricity demand and supply plan.

Based on the limitations of the long-term electricity demand and supply plan established by the government, the current study produced a power generation system operation plan model that can calculate realistic operation and maintenance costs. In addition, the study is meaningful in that it provided a methodology that can connect the Generation System Expansion plan and an operation plan. However, there is still a great deal of work to be done. Extensions are possible in the methodological aspects, technological aspects and the applications described in this research.

First, uncertainties must be considered when establishing plans for the future. Among the parameters used in the current study, ones with the highest uncertainty are demand data and fuel prices. The Generation System Expansion Plan produces solutions with a plan period of more than fifteen years. In such a case, it is impossible to precisely forecast long-term electricity demands and fuel prices based on the current point of time. As such, the research must expand into a stochastic optimization model or a robust optimization model that can solve such problems in this respect.

Next, the Korean government announced to the international community that it will cut national greenhouse gas emissions by 37% of business-as-usual (BAU) by 2030. Under such circumstances, greenhouse gas emissions in the Korean power generation industry accounted for 36.7% of national greenhouse gas emissions in 2011, which indicates that the industry must reduce its emissions to meet the government goal by 2030. As such, there is a need to transition to a model that can calculate the industry's yearly greenhouse gas emissions for the Korean government to meet its emissions goal. In addition, the industry must expand the model to consider fuel sources, such as new and renewable energy and CCS that can further reduce greenhouse gas emissions.

Finally, when the smart grid policy becomes commonplace, it will be necessary to consider a model of the power generation market that reflects the introduction of small scale distributed power sources and an integrated model that is linked to the power trading market.

Author Contributions: J.-y.K.: Developed the model and performed the experiments. K.S.K.: Created the overall idea and the basic outline of the paper.

Funding: This research received no external funding.

Acknowledgments: This work was supported by the Ministry of Education of the Republic of Korea and the National Research Foundation of Korea (NRF-2017S1A5B8060156). And this work was supported (in part) by the Yonsei University Research Fund (Post Doc. Researcher Supporting Program) of 2017 (project No.: 2017-12-0034).

Conflicts of Interest: The authors declare no conflict of interest.

Abbreviations

Indices

g	Generation Plant; $g \in \{CPP, OPP, LCPP, NPP, HPP, PHPP, REPP, RCPP, NEW_P\}$
l	Start-up type
CPP	Coal-fired Power Plant
OPP	Oil-fired Power Plant
$LCPP$	LNG-combined Cycle Power Plant
NPP	Nuclear Power Plant
HPP	Hydroelectric Power Plant
$PHPP$	Pumped-storage Hydroelectric Power Plant
$REPP$	Renewable Energy Power Plant
$RCPP$	Regional Cogeneration Power Plant
NEW_P	New Generation Plant
t, τ	Time Period (hour)
d, δ	Time Group (day)

Sets

G^1	Generation Plant for Frequency Regulating Reserve, $G^1 \in \{OPP, LCPP, HPP, PHPP\}$
G^2	Generation Plant for Replacement Reserve, $G^2 \in \{CPP, OPP, LCPP, HPP, PHPP, RCPP\}$
T^1	Peak Period, $T^1 \in \{\text{Winter, Summer}\}$
T^2	Normal Period, $T^2 \in \{\text{Spring, Fall}\}$

Optimization Variables

C^{total}	Total System Cost [\$]
$C^{oper}_{g,t}$	Variable costs [\$]
$C^{start}_{g,t}$	Start-up costs [\$]
C^{maint}_{g}	Maintenance Costs [\$]
$M_{g,t}$	Maintenance state (binary)
$M^{In}_{g,t}$	Starting maintenance indicator (binary)
$M^{Out}_{g,t}$	Ending maintenance indicator (binary)
$P_{g,t}$	Power output [MWh]
$U_{g,t}$	Commitment state (binary)
$I_{g,t}$	Startup indicator (binary)
$O_{g,t}$	Shutdown indicator (binary)
$R^{1,GF}_{g,t}$	Governor Free reserves [MWh]
$R^{1,AGC}_{g,t}$	Automatic generation control reserves [MWh]
$R^{2,up}_{g,t}$	Non-spinning or contingency reserves of ON state [MWh]
$R^{2,down}_{g,t}$	Non-spinning or contingency reserves of OFF state [MWh]

Parameters

$c^{fixO\&M}_{g}$	Fixed O&M cost [\$/MW]
s^{maint}_{g}	Time of required maintenance [hours]
c^{fuel}_{g}	Fuel cost [\$/Mcal]
$f_{g,t}(P_{g,t})$	Affine fuel use function [Mcal/MWh].
$c^{varO\&M}_{g}$	Variable operations and maintenance (O&M) costs [\$/MWh]
f^{outage}_{g}	Forced outage rate
$c^{fixstart}_{g}$	Fixed cost per startup [M\$]
n^{max}_{g}	Times the number units in the group
ed_t	Electricity demand (hourly) [MW]

p_g^{min}	Minimum power output per unit [MW]
p_g^{max}	Maximum power output per unit [MW]
Δp_g^{down}	Maximum down-ramp rate [MW/h]
Δp_g^{up}	Maximum up-ramp rate [MW/h]
$r^{1,GF}$	Regulation up reserves
$r^{1,AGC}$	Regulation down reserves
$r^{2,up}$	Load follow up reserve load fraction
$r^{2,down}$	Load follow down reserve load fraction
$s_g^{1,GF}$	Reserve capability of Governor Free [per unit]
$s_g^{1,AGC}$	Reserve capability of Automatic generation control [per unit]
$s_g^{2,up}$	Reserve capability of Non-spinning or contingency reserves of ON state [per unit]
$s_g^{2,down}$	Reserve capability of Non-spinning or contingency reserves of OFF state [per unit]
$s_g^{minup,down}$	Minimum up or down time [hours]
$s^{carbonlim}$	Limit on total carbon emissions
e_g	Carbon emission rate [tons/Mcal]

References

1. Palmintier, B. Incorporating Operational Flexibility into Electric Generation Planning: Impacts and Methods for System Design and Policy Analysis. Ph.D. Thesis, Massachusetts Institute of Technology, Cambridge, MA, USA, February 2013.
2. Bahrami, S.; Amini, M.H. A decentralized trading algorithm for an electricity market with generation uncertainty. *Appl. Energy* **2018**, *218*, 520–532. [CrossRef]
3. Bahrami, S.; Wong, V.W.; Huang, J. An Online Learning Algorithm for Demand Response in Smart Grid. *IEEE Trans. Smart Grid* **2017**. [CrossRef]
4. Bahrami, S.; Khazaeli, F.; Parniani, M. Industrial Load Scheduling in Smart Power Grids. In Proceedings of the 22nd International Conference on Electricity Distribution, Stockholm, Sweden, 10–13 June 2013. Available online: http://www.cired.net/publications/cired2013/pdfs/CIRED2013_0897_final.pdf (accessed on 10 July 2018).
5. Amini, M.H.; Frye, J.; Ilić, M.D.; Karabasoglu, O. Smart residential energy scheduling utilizing two stage mixed integer linear programming. In Proceedings of the IEEE North American Power Symposium (NAPS), Charlotte, NC, USA, 4–6 October 2015; pp. 1–6.
6. Hunt, S. *Making Competition Work in Electricity*; John Wiley & Sons: New York, NY, USA, 2002.
7. Mo, B.; Hegge, J.; Wangensteen, I. Stochastic generation expansion planning by means of stochastic dynamic programming. *IEEE Trans. Power Syst.* **1991**, *6*, 662–668. [CrossRef]
8. Ramos, A.; Perez-Arriaga, I.J.; Bogas, J. A nonlinear programming approach to optimal static generation expansion planning. *IEEE Trans. Power Syst.* **1989**, *4*, 1140–1146. [CrossRef]
9. Majumdar, S.; Chattopadhyay, D. A model for integrated analysis of generation capacity expansion and financial planning. *IEEE Trans. Power Syst.* **1999**, *14*, 466–471. [CrossRef]
10. Meza, J.L.C.; Yildirim, M.B.; Masud, A.S. A model for the multiperiod multiobjective power generation expansion problem. *IEEE Trans. Power Syst.* **2007**, *22*, 871–878. [CrossRef]
11. Park, Y.M.; Won, J.R.; Park, J.B.; Kim, D.G. Generation expansion planning based on an advanced evolutionary programming. *IEEE Trans. Power Syst.* **1999**, *14*, 299–305. [CrossRef]
12. Park, J.B.; Park, Y.M.; Won, J.R.; Lee, K.Y. An improved genetic algorithm for generation expansion planning. *IEEE Trans. Power Syst.* **2000**, *15*, 916–922. [CrossRef]
13. Firmo, H.T.; Legey, L.L. Generation expansion planning: An iterative genetic algorithm approach. *IEEE Trans. Power Syst.* **2002**, *17*, 901–906. [CrossRef]
14. Sirikum, J.; Techanitisawad, A.; Kachitvichyanukul, V. A new efficient GA-benders' decomposition method: For power generation expansion planning with emission controls. *IEEE Trans. Power Syst.* **2007**, *22*, 1092–1100. [CrossRef]
15. Kannan, S.; Slochanal, S.M.R.; Padhy, N.P. Application and comparison of metaheuristic techniques to generation expansion planning problem. *IEEE Trans. Power Syst.* **2005**, *20*, 466–475. [CrossRef]

16. Kannan, S.; Baskar, S.; McCalley, J.D.; Murugan, P. Application of NSGA-II algorithm to generation expansion planning. *IEEE Trans. Power Syst.* **2009**, *24*, 454–461. [CrossRef]

17. Meza, J.L.C.; Yildirim, M.B.; Masud, A.S. A multiobjective evolutionary programming algorithm and its applications to power generation expansion planning. *IEEE Trans. Syst. Man Cybern. Part A Syst. Hum.* **2009**, *39*, 1086–1096. [CrossRef]

18. Murugan, P.; Kannan, S.; Baskar, S. Application of NSGA-II algorithm to single-objective transmission constrained generation expansion planning. *IEEE Trans. Power Syst.* **2009**, *24*, 1790–1797. [CrossRef]

19. Malik, A.S.; Cory, B.J.; Wijayatunga, P.D.C. Applications of probabilistic peak-shaving technique in generation planning. *IEEE Trans. Power Syst.* **1999**, *14*, 1543–1548. [CrossRef]

20. Sepasian, M.S.; Seifi, H.; Foroud, A.A.; Hatami, A.R. A multiyear security constrained hybrid generation-transmission expansion planning algorithm including fuel supply costs. *IEEE Trans. Power Syst.* **2009**, *24*, 1609–1618. [CrossRef]

21. Tekiner, H.; Coit, D.W.; Felder, F.A. Multi-period multi-objective electricity generation expansion planning problem with Monte-Carlo simulation. *Electr. Power Syst. Res.* **2010**, *80*, 1394–1405. [CrossRef]

22. Ahmed, S.; Elsholkami, M.; Elkamel, A.; Du, J.; Ydstie, E.B.; Douglas, P.L. New technology integration approach for energy planning with carbon emission considerations. *Energy Convers. Manag.* **2015**, *95*, 170–180. [CrossRef]

23. Cheng, R.; Xu, Z.; Liu, P.; Wang, Z.; Li, Z.; Jones, I. A multi-region optimization planning model for China's power sector. *Appl. Energy* **2015**, *137*, 413–426. [CrossRef]

24. Flores, J.R.; Montagna, J.M.; Vecchietti, A. An optimization approach for long term investments planning in energy. *Appl. Energy* **2014**, *122*, 162–178. [CrossRef]

25. Botterud, A.; Ilic, M.D.; Wangensteen, I. Optimal investments in power generation under centralized and decentralized decision making. *IEEE Trans. Power Syst.* **2005**, *20*, 254–263. [CrossRef]

26. Lee, W.W.; Roh, D.S. *Study of the Effects of Electric Industry Restructuring on Power Generation Fuels*; Korea Energy Economics Institute: Ulsan, Korea, 2004.

27. Delarue, E.; Bekaert, D.; Belmans, R.; D'haeseleer, W. Development of a comprehensive electricity generation simulation model using a mixed integer programming approach. *Int. J. Electr. Comput. Syst. Eng.* **2007**, *1*, 92–97.

28. Delarue, E.D.; Luickx, P.J.; D'haeseleer, W.D. The actual effect of wind power on overall electricity generation costs and CO_2 emissions. *Energy Convers. Manag.* **2009**, *50*, 1450–1456. [CrossRef]

29. Andrianesis, P.; Biskas, P.; Liberopoulos, G. An overview of Greece's wholesale electricity market with emphasis on ancillary services. *Electr. Power Syst. Res.* **2011**, *81*, 1631–1642. [CrossRef]

30. Simoglou, C.K.; Biskas, P.N.; Bakirtzis, A.G. Optimal self-scheduling of a thermal producer in short-term electricity markets by MILP. *IEEE Trans. Power Syst.* **2010**, *25*, 1965–1977. [CrossRef]

31. Li, T.; Shahidehpour, M. Price-based unit commitment: A case of Lagrangian relaxation versus mixed integer programming. *IEEE Trans. Power Syst.* **2005**, *20*, 2015–2025. [CrossRef]

32. Vemuri, S.; Lemonidis, L. Fuel constrained unit commitment. *IEEE Trans. Power Syst.* **1992**, *7*, 410–415. [CrossRef]

33. Handschin, E.; Slomski, H. Unit commitment in thermal power systems with long-term energy constraints. *IEEE Trans. Power Syst.* **1990**, *5*, 1470–1477. [CrossRef]

34. Fu, Y.; Shahidehpour, M.; Li, Z. Long-term security-constrained unit commitment: Hybrid Dantzig-Wolfe decomposition and subgradient approach. *IEEE Trans. Power Syst.* **2005**, *20*, 2093–2106. [CrossRef]

35. Thorin, E.; Brand, H.; Weber, C. Long-term optimization of cogeneration systems in a competitive market environment. *Appl. Energy* **2005**, *81*, 152–169. [CrossRef]

36. Seki, T.; Yamashita, N.; Kawamoto, K. New local search methods for improving the Lagrangian-relaxation-based unit commitment solution. *IEEE Trans. Power Syst.* **2010**, *25*, 272–283. [CrossRef]

37. Wang, P.; Wang, Y.; Xia, Q. Fast bounding technique for branch-and-cut algorithm based monthly SCUC. In Proceedings of the 2012 IEEE Power and Energy Society General Meeting, San Diego, CA, USA, 22 July 2012.

38. Bai, Y.; Zhong, H.; Xia, Q.; Xin, Y.; Kang, C. Inducing-objective-function-based method for long-term SCUC with energy constraints. *Int. J. Electr. Power Energy Syst.* **2014**, *63*, 971–978. [CrossRef]

39. Chen, Z.; Wu, L.; Shahidehpour, M. Effective load carrying capability evaluation of renewable energy via stochastic long-term hourly based SCUC. *IEEE Trans. Sustain. Energy* **2015**, *6*, 188–197. [CrossRef]

40. Voorspools, K.R.; D D'haeseleer, W. Long-term unit commitment optimisation for large power systems: Unit decommitment versus advanced priority listing. *Appl. Energy* **2003**, *76*, 157–167. [CrossRef]
41. Holt, C.C. Forecasting seasonals and trends by exponentially weighted moving averages. *Int. J. Forecast.* **2004**, *20*, 5–10. [CrossRef]
42. Ministry of Knowledge Economy. *The 6th Basic Plan for Long-Term Electricity Supply and Demand (2013~2027)*; Ministry of Knowledge Economy: Sejong, Korea, 2013.
43. Gould, P.G.; Koehler, A.B.; Ord, J.K.; Snyder, R.D.; Hyndman, R.J.; Vahid-Araghi, F. Forecasting time series with multiple seasonal patterns. *Eur. J. Oper. Res.* **2008**, *191*, 207–222. [CrossRef]
44. Taylor, J.W.; Snyder, R.D. Forecasting intraday time series with multiple seasonal cycles using parsimonious seasonal exponential smoothing. *Omega* **2012**, *40*, 748–757. [CrossRef]
45. Dillon, T.S.; Edwin, K.W.; Kochs, H.D.; Taud, R.J. Integer programming approach to the problem of optimal unit commitment with probabilistic reserve determination. *IEEE Trans. Power Appar. Syst.* **1978**, *6*, 2154–2166. [CrossRef]
46. Pang, C.K.; Chen, H.C. Optimal short-term thermal unit commitment. *IEEE Trans. Power Appar. Syst.* **1976**, *95*, 1336–1346. [CrossRef]
47. Sheble, G.B.; Albuyeh, F. Evaluation of dynamic programming based methods and multiple area representation for thermal unit commitments. *IEEE Trans. Power Appar. Syst.* **1981**, *PAS-100*, 1212–1218.
48. Yang, H.T.; Huang, K.Y. Direct load control using fuzzy dynamic programming. *IEE Proc.-Gener. Transm. Distrib.* **1999**, *146*, 294–300. [CrossRef]
49. Lu, F.C.; Hsu, Y.Y. Fuzzy dynamic programming approach to reactive power/voltage control in a distribution substation. *IEEE Trans. Power Syst.* **1997**, *12*, 681–688.
50. Li, C.A.; Johnson, R.B.; Svoboda, A.J. A new unit commitment method. *IEEE Trans. Power Syst.* **1997**, *12*, 113–119.
51. Lai, S.Y.; Baldick, R. Unit commitment with ramp multipliers. *IEEE Trans. Power Syst.* **1999**, *14*, 58–64.
52. Abdul-Rahman, K.H.; Shahidehpour, S.M.; Aganagic, M.; Mokhtari, S. A practical resource scheduling with OPF constraints. *IEEE Trans. Power Syst.* **1996**, *11*, 254–259. [CrossRef]
53. Habibollahzadeh, H.; Frances, D.; Sui, U. A new generation scheduling program at Ontario Hydro. *IEEE Trans. Power Syst.* **1990**, *5*, 65–73. [CrossRef]
54. Wood, A.J.; Wollenberg, B.F.; Sheble, G.B. *Power Generation, Operation and Control*; Wiley: New York, NY, USA, 1996.
55. Fan, J.Y.; Zhang, L.; McDonald, J.D. Enhanced techniques on sequential unit commitment with interchange transactions. *IEEE Trans. Power Syst.* **1996**, *11*, 93–100. [CrossRef]
56. Roh, D.S. *Optimal Power Generation Mix with Economic and Social Costs of Nuclear Power*; Korea Energy Economics Institute: Ulsan, Korea, 2013.

Article

The Nexus Concept Integrating Energy and Resource Efficiency for Policy Assessments: A Comparative Approach from Three Cases

Floor Brouwer [1,*], Lydia Vamvakeridou-Lyroudia [2], Eva Alexandri [3], Ingrida Bremere [4], Matthew Griffey [5] and Vincent Linderhof [1]

[1] Wageningen Economic Research, Prinses Beatrixlaan 582-528, 2595 BM The Hague, The Netherlands; vincent.linderhof@wur.nl
[2] University of Exeter, Center for Water Systems, North Park Road, Exeter EX4 4QF, UK; L.S.Vamvakeridou-Lyroudia@exeter.ac.uk
[3] Cambridge Econometrics, Covent Garden, Cambridge CB1 2HT, UK; ea@camecon.com
[4] Baltic Environmental Forum, BEF Latvia, Antonijas 3-8, LV-1010 Riga, Latvia; ingrida.bremere@bef.lv
[5] South West Water Ltd., Peninsula House Rydon Lane, Exeter EX2 7HR, UK; mgriffey@southwestwater.co.uk
* Correspondence: floor.brouwer@wur.nl; Tel.: +31-(0)-70-3358-127

Received: 27 September 2018; Accepted: 14 December 2018; Published: 19 December 2018

Abstract: As the world increasingly runs up against physical constraints of energy, land, water, and food, there is a growing role for policy to reduce environmental pressures without adversely affecting increases in prosperity. There is therefore a need for policy makers to understand the potential trade-offs and/or synergies between the uses of these different resources, i.e., to encompass the water–energy–food–land nexus for policy and decision making, where it is no longer possible to ignore the limitations in land availability and its links to other natural resources. This paper proposes a modelling approach to help to assess various policies from a nexus perspective. The global macro-econometric model (E3ME) explores a low-carbon transition through different sets of energy and climate policies applied at different spatial scales. The limitations of the E3ME model in assessing nexus interactions are discussed. The paper also argues and offers an explanation for why no single traditional or classic model has the potential to cover all parts of the nexus in a satisfactory way, including feedback loops and interactions between nexus components. Other approaches and methodologies suitable for complexity science modelling (e.g., system dynamics modelling) are proposed, providing a possible means to capture the holistic approach of the nexus in policy-making by including causal and feedback loops to the model components. Based on three case studies in Europe, the paper clarifies the different steps (from policy design towards conceptual model) in modelling the nexus linkages and interactions at the national and regional levels. One case study (The Netherlands) considers national low-carbon transitions at national level. Two other case studies (Latvia and southwest UK) focus on how renewable energy may impact the nexus. A framework is proposed for the generic application of quantitative modelling approaches to assess nexus linkages. The value of the nexus concept for the efficient use of resources is demonstrated, and recommendations for policies supporting the nexus are presented.

Keywords: nexus concept; energy modelling; resource efficiency; renewable energy; low-carbon economy

---

1. Introduction

Impact assessments for policy support in the areas of energy, food, the sustainable management of natural resources (e.g., water), the use of biomass, and climate change are partly based on projections delivered by models. Integrated assessment modelling approaches usually have a limited focus on these topics, and current approaches cannot adequately take into account these different contexts [1]. Moreover [1] states that "there is little clarity on how models should be evaluated and compared, both with individual disciplines and as components of larger integrated assessment modelling". Energy and climate policies are widely supported by impact assessments [2–5]. Policies have had to widen the scope of their main objectives to take account new challenges, such as climate change, which, in turn, should also be informed by model outputs [6]. With sectoral policies (e.g., energy, water, and agriculture) becoming more and more interrelated, policy coherence is becoming paramount, achieving synergies among energy policies to the benefit of other policies (e.g., agriculture, climate, water, land). The availability and use of (renewable and non-renewable) energy is increasingly being linked with other natural resources (water, land, and food), and the coherent system of interlinkages between natural resources is also called the "nexus" [7].

The nexus concept aims to develop a holistic and comprehensive understanding of how the use of energy interacts with the provision and consumption of food and water, all within the context of a changing climate. The introduction of the nexus concept in policy support has three features [8]:

- Interlinkages between natural resources (i.e., water, energy, food, land and climate) are taken into account, trade-offs among them are made explicit, and potential synergies are exploited.
- Natural resources are managed sustainably and in an integrated manner. The optimization of food production, for example, might cause trade-offs with other natural resources (e.g., energy, water and land). The nexus concept allows us to seek for synergies and overcome trade-offs.
- Governance processes, including policy coherence, are an essential part of the nexus concept. Policy coherence is an attribute of policy that reduces conflicts and exploits synergies within and across policy areas at different spatial scales [9]. One may argue in favor of interlinkages between resources, but the lack of policy coherence has the risk of trade-offs from inadequate decision-making. Governance processes could be targeted on different objectives, including resource efficiency and circular economy. Resource efficiency (e.g., producing more of a given service, while using fewer natural resources), for example, is a major policy area in EU, but circular economy could also be linked to the nexus.

Transdisciplinary research approaches, where practitioners work with the scientific community, are needed to implement the nexus concept. This includes using state-of-the-art scientific approaches and integrating the involvement of stakeholders from policy, business, and civil-society organisations. The current paper integrates different modelling approaches (using the E3ME model and system dynamics modelling with causal loops) with knowledge from local experts (representing business and policy) to incorporate energy and resource efficiency targets with the nexus concept.

The objective of this paper is two-fold. Firstly, a modelling tool is proposed to help to assess various policies from a nexus perspective. The global E3ME model provides detailed outputs for energy aspects of the nexus. The knowledge arising from such a model is presented with interaction mechanisms through complexity science methods (i.e., system dynamics modelling). The second objective is to apply such a modelling tool to specific cases with a view to integrate energy and resource efficiency by identifying sector drivers, relevant key policies, and how sectors and policies interact [10].

The paper introduces the nexus concept that integrates energy and resource efficiency for policy assessment. The analysis includes three subsequent steps:

1. Key energy and economic indicators for three case studies (Latvia, The Netherlands, and southwest UK) are proposed using an existing macroeconomic model, providing detailed outputs for specific aspects of the nexus with a focus on energy and climate. The E3 (energy–environment–economy) macroeconomic model is well placed to provide a detailed analysis of the macroeconomic impacts of energy policy. The outcomes of a baseline scenario are compared with a two-degree C scenario.
2. In addition to the detailed outputs for specific aspects of the nexus, the paper makes explicit which synergies could be created between the energy sector and other parts of the

water–food–land–climate nexus. In addition, trade-offs are noticed between the nexus sectors and policies that could potentially enhance coherence among the nexus sectors.

The macroeconomic modelling approach does not allow full integration of the nexus concept for policy assessment. System dynamics modelling (SDM) is therefore proposed as a methodology to analyze, study, and manage complex systems, especially when formal analytical methods do not exist or are hard to apply.

2. Materials and Methods

2.1. Terminology Used in this Paper

Efficient use of the resource energy enables the economic output to increase while reducing energy use. Modelling tools are available to assess the resource efficiency, notably the economic impact of investments in the energy sector as well as climate and energy policies. They provide detailed outputs for specific aspects of the nexus of energy and climate. The energy model presented in this paper, E3ME, is a global energy, environment, and economy model. It does not fully capture the nexus sectors, largely ignoring their interactions with water, food, and land. However, resource efficiency in energy also depends on the availability and use of other resources (e.g., water, food, land and climate). Alternative approaches are proposed to integrate the outputs from models like E3ME with other numerical approaches. Such so-called complexity science approaches are adopted for interactive development with stakeholder participation, and presented in the next section.

Three case studies are implemented to test the nexus concept integrating energy and resource efficiency for policy assessments. By working with stakeholders (from policy, business, and civil society organisations), these case studies (e.g., Latvia, The Netherlands, and southwest UK) adopt transdisciplinary approaches that are driven by their needs.

2.2. Selection of Case Studies

Three case studies were selected, which are all part of an ongoing (2016–2020) EU-funded project "Sustainable Integrated Management for the nexus of water–land–food–energy–climate for a resource efficient Europe" (SIM4NEXUS). SIM4NEXUS uses advanced integration methodologies based on SDM to bridge the knowledge gap related to the complex interactions between the nexus. Three case studies with similar objectives (low-carbon economy and/or increasing renewable energy use) were selected. They were implemented to showcase this methodology as a test bed for achieving resource efficiency through successful policy initiatives.

The three case studies draw on transdisciplinary research methods, with knowledge partners working with end-users (policy makers, business, small and medium-sized enterprises (SMEs), and civil society organisations) and adopting participatory approaches. The case studies respect the data security rules according to the General Data Protection Regulation (GDPR) 2016/679 of the European Parliament and of the Council from 27 April 2016 on the protection of natural persons with regard to the processing of personal data and the free movement of such data.

The information was collected through 15 semi-structured interviews with representatives from the private sector, policy, and civil society (case of The Netherlands), and each case study was organized as a workshop (with approximately 15 participants) to discuss the critical issues at stake and to identify synergies between the nexus sectors.

2.3. Complexity Science Approaches

Complexity science approaches are used to combine existing energy knowledge models (e.g., E3ME) with interaction mechanisms. The application of such approaches to policy assessments provides a means of exploring the effects of various types (legal, political, business, and financial) of spatial and temporal drivers and constraints on the behavior of society. System dynamics is used as a holistic approach to nexus decision and policy making. SDM links the main feedback mechanisms (loops and iterations), breaking down problems into sub-systems and submodels. In a way, this is similar to the conceptual thinking of non-programmers, as reflected by conceptual models. Each SDM model consists of (i) stocks/compartments (levels-state variables), (ii) connectors (arrows), (iii) flows or influences (rates), (iv) converters (auxiliaries/parameters), and (v) decision processes (priorities, allocation and relations) [10,11].

2.4. E3ME Model

E3ME is a macroeconomic model that is applied for a variety of regions and is continually updated and maintained [12]. It is used to assess the linkages between the economy and energy systems, and their impacts on environmental emissions.

The E3ME model integrates energy systems with the economy and so is very suitable for assessing policies that affect energy demand and supply. The model has been used to provide policy makers with information regarding different energy policy options, for example, in several related Impact Assessments for the European Commission, including for the 2030 climate and energy framework and the Energy Efficiency Directive [3]. The model is able to provide a wide range of results at a disaggregated level, including changes to energy demand, electricity supply, and emissions as well as a range of macroeconomic and sectoral economic indicators.

The economic structure of E3ME is based on the system of national accounts, with further linkages to energy demand and environmental emissions. E3ME is based on a post-Keynesian, demand-driven framework, which sets it aside from the more standard computable general equilibrium (CGE) approach [13,14]. Assumptions common to CGE models, such as perfect knowledge and rational behavior, are replaced with equations based on real-world relationships, as determined by the historical data. The financial sector is a key part of the system [15].

E3ME's historical database covers the period 1970–2016, and the model projects forward annually to 2050. The econometric specifications of E3ME give the model a strong empirical grounding. E3ME uses a system of error corrections, allowing short-term dynamic (or transition) outcomes, moving towards a long-term trend. The dynamic specifications are important when considering short and medium-term analysis (e.g., up until 2020) and rebound effects, which are included as standard in the model's results (see [16]).

The labor market is also covered in detail, including both voluntary and involuntary unemployment. In total, there are 33 sets of econometrically estimated equations, also including the components of the gross domestic product (GDP) (consumption, investment, international trade), prices, energy demand, and material demand. Each equation set is disaggregated by country and by sector.

The main dimensions of E3ME are:

- Fifty-nine countries—all major world economies, the EU28, and candidate countries plus groupings of other countries' economies;
- Seventy industry sectors for EU countries and 44 industry sectors for non-EU countries, based on standard international classifications;
- Forty-three consumer expenditure categories for EU countries and 28 for non-EU countries;
- Twenty-three different fuel users of 12 different fuel types; and
- Fourteen types of airborne emission (where data are available), including the six greenhouse gases monitored under the Kyoto Protocol.

The most recent applications of E3ME [17,18] assess the impacts of a global set of policies that are designed to limit temperature change to 2 °C. The modelling approach is more generally described in [19]. The model manual provides an overview of the model's structure [12,20].

The E3ME model is used in the paper to provide detailed information about the impacts of a transition towards two-degree on the energy part of the nexus. The model does not explicitly cover the other nexus components; however, some model results can be used to infer potential impacts in other components of the nexus. For example, the demand for bioenergy and food consumption are represented in E3ME; however, changes to these model outcomes allows us to infer potential changes to land-use and water demand.

3. Results

3.1. The Baseline and Two-Degree Scenario

A baseline scenario was introduced to represent the current trends of the systems being modelled. It was assumed not to include future policies, but only the ones implemented up until the base year of the analysis. Energy consumption for the coming two decades, for example, was assumed to follow the same annual growth rate of the recent past (e.g., 5 years). The E3ME baseline for the EU was calibrated to the PRIMES Reference scenario 2016 [21]. The PRIMES Reference scenario focuses on the EU energy system, transport, and greenhouse gas (GHG) emission developments, including specific sections on emission trends not related to energy and on the various interactions among policies in these sectors. The Reference Scenario is used by the European Commission (DG Energy) as a benchmark of current policy and market trends. For non-EU regions, the E3ME model uses the International Energy Agency's World Energy Outlook (IEA WEO) Current Policy Scenario (CPS) for 2016 [22]. The CPS takes into account only those policies for which implementing measures had been formally adopted as of mid-2016.

In the two-degree scenario, all baseline policies are included. In addition to the baseline policies, the following additional energy and climate mitigation polices have also been added:

- Moderate carbon prices levels (see Table 1), usually collected as a tax, which are set at a global level but applied on a national basis.
- Exogenous improvements to energy efficiency in final use sectors (such as households and buildings, but also other industry sectors, as outlined by information taken from the International Energy Agency World Energy Outlook 450ppm scenario (IEA 450ppm), except for road transport which is covered separately below. The rates of improvement are derived from the IEA 450PPM scenario.
- The required investment is funded through public programs, using the revenues from the carbon pricing. This means that all revenues raised from the carbon tax or emissions trading scheme are used by the government to finance energy efficiency investments. Within each country, if there is any shortfall in (carbon) revenue, other taxes (split evenly between VAT, income taxes, and labor costs) are increased to meet the cost of investment.
- Power sector: A combination of feed-in-tariffs and direct subsidies are implemented to promote the uptake of renewables, in particular, wind and solar power (see below). Some assumptions are made about increased availability of storage or demand management to support the increase in intermittent power technologies.
- Decarbonization of the road transport sector using policies to encourage the uptake of electric vehicles, including registration taxes. Introduction of biofuel mandates is also required in some countries.
- A biofuel mandate is applied to aviation, possibly to the extent that about 18% of aviation fuel will be derived from biofuels by 2050.

Table 1. CO_2 prices in the baseline and the two-degree scenario, US$/t$CO_2$.

	2010	2020	2030	2040	2050
Baseline	10.7	14.5	25.8	41.4	66.6
Two-degree scenario	10.7	18.9	85.8	173.0	359.9

The following sections summarize the macroeconomic and energy impacts of the policies implemented above.

3.1.1. Impacts on GDP

The GDP impacts of moving towards the two-degree target were positive in all case studies as compared with baseline levels (Table 2). Table 2 captures the overall GDP impact of the transition to a low carbon system. All case studies are expected to benefit from the additional investment in energy efficiency and, in most cases, power generation (increased support for renewables) as well as a lower dependency on fossil fuel imports (as is the case in Latvia, for example) and lower fossil fuel costs on what is still being imported.

Table 2. Gross domestic product (GDP) impact by case study, % difference from baseline.

	2010	2020	2030	2040	2050
EU28	0	0.3	0.5	0.7	1.6
Latvia	0	1.3	2.5	3.3	6.8
The Netherlands	0	1.0	1.6	1.9	2.7
UK [1]	0	0.3	0.4	0.4	0.6

Note: [1] Results presented for the UK as whole.

Here, it is important to note that while the energy efficiency investment is paid for in the year it is made, the power generation investment is paid for over the lifetime of the plant. This means that some of the investment will still be paid for beyond 2050.

3.1.2. Impact on Energy Consumption

The policies specifically concerned with electricity generation lead to a substantial decrease in the use of coal for power generation and increased used of renewables, in particular, solar and wind. Some electricity generation based on gas is expected to remain, mainly as back-up to ensure grid stability. The energy efficiency policies, particularly those targeting buildings (both residential and offices) are expected to lead to a decrease in demand for electricity and, in particular, gas. In road transport, the increased support for electric vehicles is expected to lead to higher demand for electricity in this sector by 2050. As such, total primary energy consumption would be expected to decrease with the two-degree scenario compared to the baseline (Table 3). Table 3 highlights the changes in the energy systems resulting from the decarbonization policies: a transition to a low carbon system, lower energy use, and what is used is mainly from renewable sources.

As expected, fossil fuel primary energy consumption is expected to decrease considerably, as it is the primary target of policies implemented across the countries (e.g., higher carbon prices, higher support for renewable technologies and policies encouraging the uptake of more efficient or alternative-fuel vehicles). Nonfossil fuel primary energy consumption is expected to increase in most case studies, but this is dependent on the level of energy efficiency uptake in each country. For example, in Latvia, the nonfossil fuel primary energy consumption is projected to increase by 2030 followed by a sharp decrease, reaching a 7.5% reduction by 2050 (as compared to the baseline). Bioenergy consumption is expected to increase in the scenario compared to baseline in the EU28, The Netherlands, and in the UK. The rate of increase is expected to slow down closer to 2050, as most bioenergy used in road transport will be replaced by electric vehicles. A somewhat different pattern is observed in Latvia where the bioenergy consumption is expected to increase by 2030 and then reverse to a decrease by 8.9% below the baseline consumption by 2050. However, this increase may lead to further pressures on land-use.

Electricity demand in the two-degree scenario is affected by three main trends: (1) higher energy efficiency which contributes to a decrease in demand for electricity; (2) household switch from gas to electricity appliances as they are more efficient, which increases the demand for electricity, and (3) increased uptake of electric vehicles, which again is expected to lead to increased demand. Depending on the scale of the three effects mentioned above, electricity demand is expected to decrease or increase in each case. Generally, a substantial decrease compared to baseline is predicted to occur by 2030, with the rate of decrease slowing down afterwards as the electric vehicle uptake increases.

Table 3. Total primary energy consumption by case study, % difference from baseline.

	2010	2020	2030	2040	2050
EU28					
Fossil fuels	0	−3.3	−13.8	−19.2	−28.1
Nonfossil fuels	0	0.7	1.9	3.1	4.2
Bioenergy [1]	0	3.1	6.1	8.7	6.3
Latvia					
Fossil fuels	0	−4.6	−15.3	−18.6	−25.1
Nonfossil fuels	0	2.4	4.7	0.0	−7.5
Bioenergy [1]	0	3.0	5.8	3.6	−8.9
The Netherlands					
Fossil fuels	0	−6.0	−15.8	−22.6	−29.2
Nonfossil fuels	0	1.8	−2.6	3.9	9.8
Bioenergy [1]	0	5.8	3.7	14.9	18.8
UK [2]					
Fossil fuels	0	−2.1	−11.7	−18.9	−27.4
Nonfossil fuels	0	4.2	9.0	12.4	22.3
Bioenergy [1]	0	13.0	17.8	26.1	25.1

Note: [1] Part of nonfossil fuels. [2] Results presented for the UK as whole.

Table 4 summarizes final electricity demand results for the case studies.

Table 4. Electricity consumption by case study, % difference from baseline.

	2010	2020	2030	2040	2050
EU28	0	−1.3	−4.8	−3.7	−0.3
Latvia	0	−0.5	−4.5	−11.4	−9.3
The Netherlands	0	−5.3	−16.1	−16.4	−12.6
UK [1]	0	0.2	–0.7	1.5	5.2

Note: [1] Results presented to the UK as whole.

As expected, emissions are predicted to substantially decrease in the two-degree scenario compared to baseline in all case studies (Table 5), while Table 6 summarizes baseline CO_2 emission trends.

Table 5. CO_2 emissions by case study, % difference from baseline.

	2010	2020	2030	2040	2050
EU28	0	−4.0	−19.8	−27.5	−44.0
Latvia	0	−5.4	−19.3	−35.0	−59.0
The Netherlands	0	−7.8	−23.2	−33.8	−42.9
UK [1]	0	–2.3	–16.7	−27.8	–41.9

Note: [1] Results presented to the UK as whole.

Table 6. Baseline energy-related CO_2 emission levels, mTCO$_2$.

	2010	2020	2030	2040	2050
EU28	3906.6	3293.2	2673.2	2200.0	1835.0
Latvia	7.5	6.0	5.3	4.7	4.2
The Netherlands	176.4	156.8	127.5	109.4	101.0
UK[1]	517.5	420.3	334.4	276.1	243.0

Note: [1] Results presented to the UK as whole.

By policy design, the two-degree scenario is more energy-efficient compared to the baseline. There are number of energy efficiency policies that ensure energy savings, while the policies specifically implemented in the road transport sector are design to lead to encourage the uptake of more efficient vehicles. In E3ME, it is difficult to determine the impacts that these may have on land-use and water demand, as these components are not explicit in the model. It is likely that the higher demand for biofuels may cause pressure with respect to food production. Another convert that may impact land-use and food production is the increase in consumer demand predicted by the model. The increased economic activity because of the additional investment in renewables and energy efficiency, is expected to lead to increased consumer expenditure (see Table 7).

Table 7. Changes to consumer expenditure on food, drink and tobacco, % difference from baseline.

	2010	2020	2030	2040	2050
EU28	0	0.2	1.4	1.3	2.6
Latvia	0	0.5	0.9	1.6	1.8
The Netherlands	0	0.2	0.5	0.4	0.6
UK[1]	0	0.3	0.4	0.5	1.4

Note: [1] Results presented to the UK as whole.

In particular, EU28 expenditure on food, drink and tobacco is expected to increase by 2.6% in 2050 compared to baseline, indicating that food production in EU and imports are also expected to increase. This increase in food demand combined with higher bioenergy requirements are expected to put more pressure on land.

3.2. The Energy Policies Covered in the Case Studies

Countries and regions implement different strategies involving the integration of energy and resource efficiency, and these largely vary depending on policy targets and related measures. The main energy policies are summarized for the three case studies in Latvia, The Netherlands, and southwest UK. The opinions presented in this section largely draw from interviews with policy makers and civil society organizations, complemented by a review of grey literature including national assessments.

3.2.1. Latvia

Latvia has a high potential for renewable energy but remains largely dependent on imported fossil fuels and electricity. Thus, energy security is of a key concern and ensuring the energy supply, competitiveness, energy efficiency, and the use of renewable energy are the main energy challenges for the coming years. The country aims to reach at least an 80% reduction in GHG emissions by 2050 in comparison to 1990. The intermediate target to be reached by 2040 is increasing the carbon sequestration to fully cover the total amount of anthropogenic GHG emissions of the country and to achieve carbon neutrality [23]. About 30% of electricity is imported, and hydro-power is by far the predominant renewable energy source in electricity production in Latvia. Renewable energy regarding the production of heat in centralized systems is fully dominated (share of 97%) by biomass wood fuel.

3.2.2. The Netherlands

The Netherlands has the potential for the use of renewable energy sources like wind, solar, and biomass. Hydropower generation is limited and does not have a high potential for use. Biomass is seen as the major primary renewable energy source as it can substitute the use of large-scale fossil fuels, like coal and oil. However, the use of biomass as a renewable energy source does not necessarily contribute to the reduction of greenhouse gas emissions. Moreover, the supply of biomass is insufficient for large-scale energy production in the Netherlands, and it has to be imported from other EU member states, like Sweden and Latvia. Another important aspect is that energy efficiency will increase gradually over time.

In 2015, 5.6% of total energy was generated from renewable resources [24]. Nearly two-thirds of this renewable energy was generated from biomass. Biomass produced in the Netherlands is composed of waste (35%), wood from various sources (39%), and biogas from manure and sewage waste (13%). There is potential to increase the production of biomass in the Netherlands from 80 PJ in 2015 [24] to a maximum of 200 PJ [25].

With existing measures on the energy transition towards renewable energy, the share of renewable energy in total energy use is expected to be 16.7% in 2023 [24]. Between 2020 and 2023, the production of renewable energy is predicted to increase by 30% due to major investments in large-scale wind power at sea and solar energy. The energy generated with biomass is predicted to increase slightly between 2020 and 2023 because of the restrictions to physical capacity of cofiring and financial capacity in the renewable energy subsidies [24].

In October 2017, the Dutch government agreed to reduce GHG emissions by 2050 by 80–95% compared to the 1990 level [26]. Their ambition, presented in the Climate Law, is that the Netherlands should achieve the Paris targets by 2030. Greenhouse gas emissions, therefore, need to be reduced by 49% by 2030 compared to the level of emissions in 1990. Biomass is one of the major sources of renewable energy, but it should be combined with particular forms of carbon storage. A major public debate regarding the options for large-scale storage of carbon remains to be launched. There is resistance in society regarding storage on land, and storage at sea is considered. The government has already urged to shut down large coal power plants in 2030 at the latest. Most electricity producers owning a coal power facility are now considering switching to biomass. As there is not sufficient biomass produced in the Netherlands that can be used in large electricity producing facilities, biomass has to be imported on a large-scale.

Five mitigation and adaptation options were explored to achieve a low-carbon economy in the Netherlands by 2050 [27]): (i) the electrification of energy, (ii) the production of carbon neutral energy, (iii) energy saving, (iv) the use of renewable energy sources, such as wind power, solar power, and energy from biomass, and (v) carbon capture and storage (CCS). The report argues that a mix of all these measures is necessary to reach policy targets. However, it is undefined as to which policy mix is required to transform the economy into a low-carbon economy. All of these technological measures will have different consequences for GHG emissions and the production and use of energy and food as well as for water and land. These technological options will have socioeconomic consequences. Moreover, policies and socioeconomic interventions can contribute to the reduction of GHG emissions. These societal consequences and their cost-effectiveness have not yet been evaluated. For instance, energy saving can be realised by replacing energy-intensive technologies with less energy-intensive technologies or, in some cases, by energy neutral technologies.

3.2.3. Southwest UK

The primary energy challenge is one of economics, with primary energy consumption estimated to decrease in the two-degree scenario compared to the baseline. The economic regulators of the UK utility sectors have been instructed by the government to minimize the unit cost of all utilities to domestic customers, while at the same time requiring an increase in service level, resilience, and environmental performance [28]. This has been aggravated by the general tendency toward time inconsistency in strategic planning at the government and utility levels [29].

The southwest is the chosen location for the next major nuclear energy installation, which will potentially act as a bottleneck that limits the capacity of the transmission/distribution network to accept more renewable energy generation. However, this situation may be mitigated by reinforcement of the network [30].

3.3. Synergies and Trade-Offs in the Case Studies between Energy and the Nexus of Water–Food–Land–Climate

This section presents an overview of interlinkages between energy and the nexus sectors water, food, land, and climate. Interviews with stakeholders were held in the three case studies (Latvia, The Netherlands, and southwest UK), including policy, business, research and civil society organizations.

3.3.1. Latvia

There are synergies between energy supply and the use of wood-based fuels for heat production and hydro-power for electricity production. Already in 2001, the share of renewable energy sources (RES) based energy accounted for 34% of the total energy supply. Latvia will reach a 40% share of energy from renewable sources in gross final consumption in 2020. The main challenges in the RES sector relate to the efficient utilization of historically developed sources, i.e., wood-based biomass and hydro-power and to ensure energy diversification through the application of other renewable energy sources (e.g., wind, solar, other types of biomass).

There are potential conflicts regarding the use of renewable energy (mainly bioethanol and biodiesel) in the transport sector. It is still underdeveloped in Latvia and constitutes less than 3% of the total use (2016) as compared with the 10% target to be reached by 2020. The transport sector is obliged to sell petrol and diesel blended with biofuels, and there are tax regulation mechanisms [31]. While the sustainability criteria for biofuels have been set, there is no obligation to sell blends only with biofuels corresponding to the sustainability criteria. There are also no exemptions for applications for reduced excise duty tax for all biofuels. Amendments in legislation may overcome such trade-offs between energy resources, land management, and climate policies.

The use of solid biomass (e.g., wood fuels) for energy production potentially creates synergy with climate mitigation, but, at the same time, puts pressure on forests in regard to biodiversity, sequestration of carbon, and competition with the production of wood-based products with high added-value. On top of this, wood is also exported both as wood timber and as a fuel for combustion plants. Continuation of averting the share of renewable energy sources from domestic use may intensify the pressure on forests to ensure the production capacity of raw materials. Moreover, the growth of energy plants may compete with arable land to be used for food production as well as affect water quality in water bodies from the application of fertilizers and pesticides.

Replacing fossil fuels with biofuels helps to reduce GHG emissions from the transport sector. However, the production processes need to change to keep this energy source feasible. The supply of energy crops (e.g., rapeseed) is increasing, and this is foreseen to remain in the coming years. This however may result in indirect land-use change (biofuels compete for land for food production). Unfolding a potential of the second generation biofuels could be the way forward, but the following conditions have to be assessed: availability of resources for production (e.g., biodegradable waste), technological readiness level, affordability of investments.

Installed technologies for energy production from RES help to reduce GHG emissions, but the use of solar panels and wind turbines for power generation, etc., involves direct impacts on land, such as the removal of vegetation and soil and alterations to topography. At the same time, meteorological conditions directly govern the actual outputs of thermal solar panels, photovoltaics, and wind turbines. Currently, wind and solar energy do not play important roles in the energy balance in Latvia, although recent developments show a good prospect for penetration of the respective technologies at a broader scale.

3.3.2. The Netherlands

Based on the literature review and stakeholder consultation, the main nexus challenges related to the use of biomass for energy generation are as follows [32]:

- Biomass should be produced and collected sustainably, which means without harm to food availability and biodiversity [25], although such biomass is scarce in the Netherlands (and also is imported);
- Large-scale application of biomass for energy production will affect the availability and quality of land, water, food, and energy and will affect the climate;

- There is debate as to whether the use of biomass for energy generation contributes to a net reduction in GHG emissions or not. This is only feasible if biomass use for energy is combined with carbon capture and storage. The sustainability criteria for biomass are also under debate [31].
- In addition, biomass has a negative image because it is often associated with the use of coal for energy production (cofiring) and with large-scale deforestation. It is also associated with land grabbing and competition with local food production;
- In addition, there are knowledge gaps between politicians and the public about the diversity of biomass and the best application of these different types.

The increase in the use of solid biomass for energy production creates a synergy with climate change, although this is mainly effective if biomass for energy production is combined with carbon capture and storage (see [33]). However, the supply of solid biomass that can replace the use of fossil fuels for energy production is significantly larger than the amount of solid biomass that can be produced in the Netherlands. An increase in land-use for the generation of solid biomass would require a large share of mainly agricultural land, which then would replace food production (meat and food crops).

More recently, [34] showed that renewable energy production, such as solar or wind power, also requires land in order to produce carbon neutral electricity at a large-scale. Given that land is already scarce in the Netherlands, the land requirements of solar and wind power might restrict the application of these types of renewable energy.

Biomass has multiple applications. It can be used for energy production, but the chemical sector also tries to green their resources for production, so there is competition for biomass as well (see [33]).

3.3.3. Southwest UK

A major trade-off to be explored is centered on the aim of energy decarburization, and the prioritization of either nuclear or renewable energy, which have both been identified as low-carbon solutions [35]. The southwest region of the UK has England's largest natural supply of wind and solar energy, with the greatest installed capacity [36]. The southwest peninsula also has the most accessible offshore renewable resources in England, including wave, tidal, and wind power, which are largely unexploited. The conflict arises as the southwest has additionally been chosen as the location for the next major nuclear energy installation. Nuclear energy, while excellent at providing very consistent baseload output, has a very limited ability to respond rapidly to fluctuations in demand. This is incongruent with the government's objective of creating a flexible energy network and the intermittent nature of renewables, which fluctuate with the available resource. Further, the baseload output of the nuclear energy station will potentially act as a bottleneck that limits the capacity of the transmission/distribution network to accept more renewable energy generation. It is believed that the grid capacity challenge can be mitigated by reinforcement of the network but at significant capital cost [30]. For nuclear and renewables to coexist in the southwest, there is a greatly heightened need for mechanisms to attenuate the temporal disparities between supply and demand and increase the network capacity [37]. To compound the complexity of the problem, at a national level, both new nuclear and renewables are subsidized via the same funding mechanism "Contracts for Difference", and they access the same budgetary resources [38]. Therefore, both economic and technical dimensions play a role in the trade-off between nuclear and renewable energy.

Synergies with water management are created through raw water resources, providing the opportunity for hydro-power generation. This option is suitable for the region, although new plants have high capital costs, and the economically viable resources are largely being fully exploited.

Synergies could also be created by changing land-use and water management practices. Upstream catchment management and paid ecosystem services, for example, would improve the surface water quality and reduce the energy demands from drinking water treatment. A pioneer demonstration program of rewetting moorlands and improving farming practices has potential benefits for surface water quality and biodiversity [39]. However, there might be challenges to establish and maintain such paid ecosystem agreements [40]. Similarly, synergies could be established through sustainable urban drainage systems (SUDS), reducing surface flood risks, sewer flood risk, and sewer storm flow [41]. This would reduce wastewater pumping and treatment and, consequently, energy demand. Southwest water has an engagement program with local authorities and housing developers in regard to the implementation of SUDS. The main barriers to full exploitation are due to the high capital cost for retrofit, and complex issues surrounding the responsibility of ownership and maintenance. Also, there is a significant challenge since economic benefits are not usually seen by those financing SUDs and/or payback periods can be long and difficult to calculate [42].

There are potential synergies from water to land and energy, since the anaerobic digestion of sewage sludge generates methane gas that is suitable for energy use, and composted sludge cake from anaerobic digestion of sludge is rich in phosphates and nitrogen [43]. When disposed to agricultural land, composted sludge cake can provide valuable fertilizer, offsetting the need for fertilizers from other sources and reducing energy consumption. Sludge passed to anaerobic digestion remains at a relatively low proportion within the southwest, and the majority of sludge is 'limed', which is of lower agricultural value. The main barriers to further exploitation are the logistic challenge of sludge transport to centralized anaerobic digested treatment and the capital costs to build a treatment plant.

Synergies between water and energy could be created by improving the resilience or security of the energy supply. Energy supply in the southwest UK is critical to the water services in this region. It would improve the resilience and security of water, but high capital costs are a barrier to further exploitation.

Large land resources are suitable for the onshore production of wind energy. Such synergies from land to energy can be deployed on land also used for cattle and crop cultivation. So far, most economically viable resources are being exploited, and further development remains limited due to planning restrictions and the grid connection capacity [44].

3.4. System Dynamics Modelling as a Methodology of Integration

3.4.1. A Methodology of Integration

System dynamics modelling (SDM) was introduced as a methodology to take into consideration the interactions generated in the nexus of water, land, energy, food and climate. It merges top-down and bottom-up approaches. The top-down learning draws from the E3ME model with expert validated decisions and how they are implemented in such a modelling framework. The bottom-up learning focuses on new relations between systems drawn from expert knowledge and participatory approaches with local stakeholders. Causal loops were introduced to integrate knowledge and create causal feedback mechanisms, offering a holistic approach to the nexus. An additional advantage of such causal loops is that they are fully transparent and hence, can be easily understood by nonexperts and therefore, be suitable for interactive development with stakeholders.

The SDM was used in the Latvia case to focus on low-carbon development by respecting thresholds for greenhouse gas emissions and increased use of RES. Interlinkages are presented between water, land (e.g., forestry), food (e.g., agriculture), and energy production from RES and fossil sources. The SDM in the Netherlands focused on competing claims regarding the use of biomass for the bio-economy, including food, fiber, feedstock for the chemical and manufacturing industry, and energy. Imports of biomass were included as well. The sustainability aspects of biomass use were also taken into account. The SDM was used in the UK to enable detailed scenario-based analysis and eventually also support both business planning and stakeholder engagement. Resource efficiency and de-carbonization were the two main objectives, and the two key metrics to track performance were the total CO_2 emissions and the ratio between the total resources supplied by each nexus sector (including energy) and those directly consumed by societal demand.

3.4.2. Energy Sector Submodel in Southwest UK

The energy sector submodel seeks to examine the balance between supply and demand within the regions of electrical and thermal energy. In this study, all forms of renewable energy generation within the southwest region were included as were all forms of fossil fuel and grid electricity import.

The energy sector submodel is the most notable example of a supply-led philosophy, replacing the more common demand-led philosophy seen elsewhere within the nexus SDM (see Figure 1).

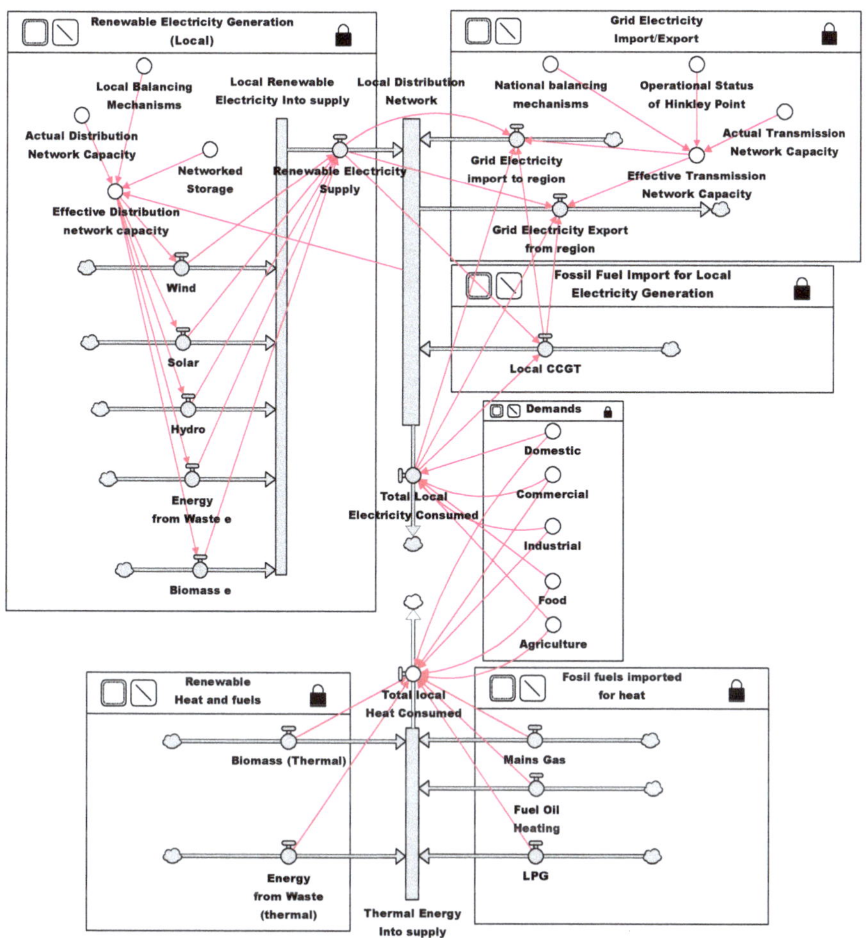

Figure 1. The energy sector submodel in the southwest UK case study.

The supply-led approach was used to account for the nature of renewable energy generation, which is largely driven by resource availability. For example, photovoltaic solar energy can only be generated during daylight hours, and output is governed by light intensity. In its current state, the distribution network operator (DNO) conducts a very limited generation curtail of renewable energy suppliers to balance supply and maintains the network capacity. This, however, is projected to increase dramatically in the coming years as DNOs switch to a distribution system operator model, whereby they become responsible for balancing arrangements. The SDM, therefore, provides an opportunity to examine strategies for enhancing the utilization of renewable energy generation by including load and generation curtailment and dynamic network capacity controls.

A renewable generation local module is part of the energy sector submodel (see Figure 2). The model nominally assumes that all renewable energy generated by the available resources is supplied into the local distribution network without curtailment or constraint by offtake demand. This situation is only possible while renewable energy generation is nominally lower than demand and while sufficient capacity within the distribution network exists.

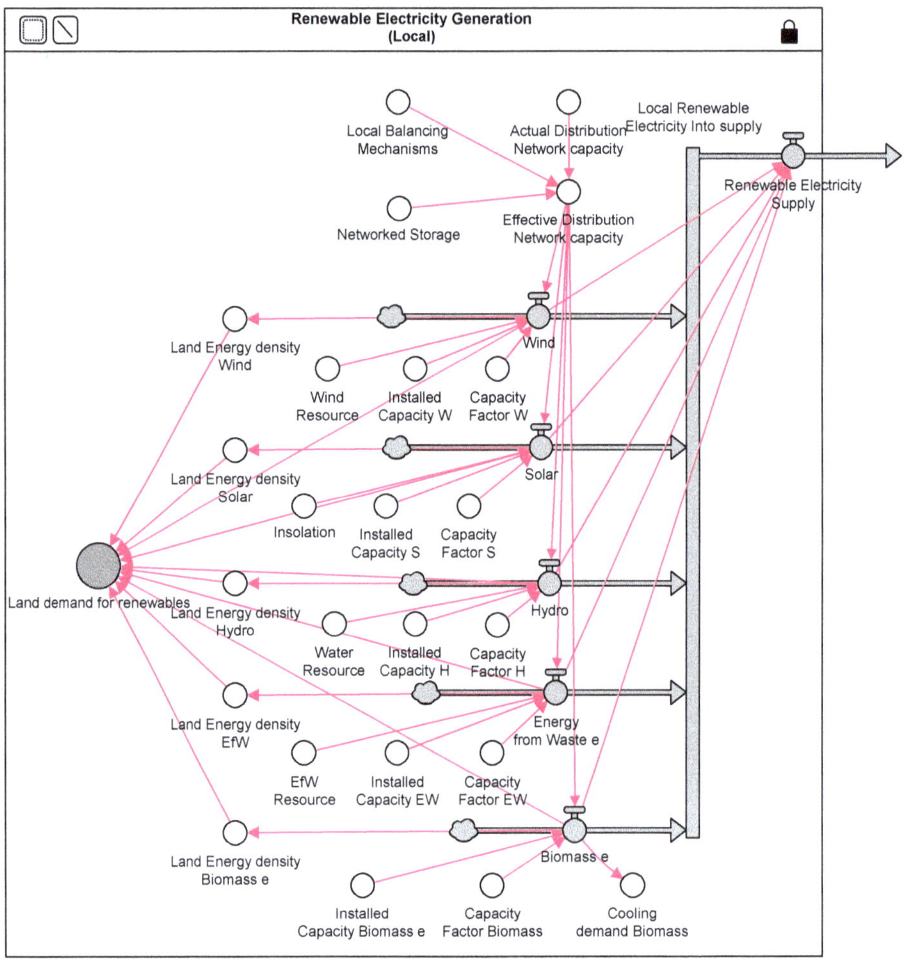

Figure 2. A renewable generation local module in southwest UK.

It is worth noting that major constraints within the distribution and transmission networks do exist and are the major limiting factors in the development of generating new capacity within the region. The model therefore includes network capacity coefficients as management variables. This takes the form of an *Actual Network Capacity* and an *Effective Network Capacity*, which is driven by coefficients for *Network Storage* and the presence of *Local Balancing Mechanisms* (dynamic supply/demand activities). The objective of these variables is to seek to identify the benefits (in terms of increased utilization of renewables) of increasing capacity within the distribution network or implementing generation curtailment. All forms of renewable energy generation are connected to the *Local Renewable Electricity into Supply* stock and are described in terms of *Available Resources, Installed Capacity,* and *Capacity Factor*.

This approach enables all three variables to change over time. In the case of *Available Resources,* this may change as a result of climate change or land-use restrictions. Installed capacity describes the total megawatt generating capacity of all aggregated assets of that type; policies to deploy renewables will impact this variable directly. The capacity factor describes the relationship between actual generation and potential generation of the asset; therefore, with improved efficiency or different management philosophy, the capacity factor may improve or decline over time. At present, it is not clear if this approach is too complex; it may become easier to use a purely data driven method that relies entirely on the thematic models to provide volume of energy generated by technology type.

4. Discussion and Conclusions

This paper introduced the nexus concept in relation to achieving resource efficiency in Europe through three specific cases in Latvia, The Netherlands and southwest UK. In these three specific cases, which were used as examples, a macroeconomic model E3ME was applied, allowing for the estimation of energy demands for a baseline scenario and the two-degree scenario. The outcomes were compared with the use of system dynamics modelling, a complexity science methodology where feedback loops and interactions between various nexus components are implemented in the conceptual model.

Traditional modelling approached (e.g., E3ME) do not take into account the direct and indirect linkages between energy and related resources (e.g., water, land, food and climate). Also, the existing modelling capacity of E3ME) does not implement impacts of land-use and water demand, as follows:

- There might be numerous technical opportunities for the water sector to synergies with the objective of the energy sector (generating energy, reducing or controlling demand) but their feasibility is challenged by the high capital costs, especially in the case of the southwest of the UK, as detailed in Section 3.3.3. Similar to this, nonengineering, management-based solutions that address land-use practices offer multi-partied benefits to land, water, and energy, but they are contractually complex, again as detailed in Section 3.3.3.
- The energy and chemical sectors compete for sustainable biomass. The penetration of alternative fuels in the transport sector has implications to land-use and agriculture. Cross-sectoral interactions in the economy and with future land-use patterns are barely covered in the current modelling capacity, which makes traditional models like E3ME less likely to implement these aspects in regions where these aspects are important (e.g., for Latvia, as detailed in Section 3.3.1).
- Land-use implications could create synergies with energy demand. The lower demand for energy and increased use of renewables has implications for land-use in the sense that if some energy is produced domestically (e.g., coal mines), these may be closed, and the land may become available. Moreover, reduced demand for energy may also translate in fewer power plants being required, so again, this would have implications for land-use. Similarly, land is required for the supply of renewable energy, including biomass, wind, and solar power.
- The diversity of biomass and its particular applications: Generally (with the exception of hydro and bioenergy), renewables are less water-intensive than conventional power plants. The scenarios presented in this paper do encourage the uptake of solar and wind, which have very small water requirements so are likely to have less pressure on water availability. However, the increased demand for food and bioenergy would likely lead to increased use of water/land resources, so the interesting question is here is: which impact is larger? This particular linkage is significant for The Netherlands and Latvia.

In principle, energy is a critical resource, and the existing modelling capacity enables the projection of energy demand and related emissions of greenhouse gases. Energy saving and investments to finance improvements in energy efficiency are key components to substantially decrease energy-related emissions. Water and land have a limited focus in the current energy modelling tools, such as E3ME. It is also hard to estimate the deviations in the two-degree scenario with regard to land-use and water demand, as detailed in Section 3.1.2. However, the availability of such resources is critical to resource efficiency, strengthening a transition to a low-carbon economy with an increase in non-renewable energy sources (e.g., for The Netherlands and the southwest of the UK).

Water is an important resource. For power generation, for example, water use requirements will reduce over time, because of increased renewables and less water being needed for cooling. However, increased bioenergy and food demand will probably require more water. Land can be an important source for renewable energy supply, as is currently debated in the Netherlands. An increase in renewable energy generation requires land for either wind and solar power, or for biomass production.

Integrating the knowledge from existing modelling (e.g., E3ME) with System Dynamics Modelling approaches will allow the nexus concept to be implemented in policy assessments integrating energy and related natural resources. This approach does merge learning from well-established modelling tools (e.g., E3ME) with expert knowledge from participatory approaches, leading to integrated conceptual approaches, as detailed in Figures 1 and 2 for the southwest UK.

In general, the efficient use of energy resources needs to be supported by policies and decision-making. In this context, the nexus concept provides integrated knowledge for cross-sectoral decision-making and planning, with an explicit focus on biophysical, socioeconomic, and policy interactions across sectors (including trade-offs and synergies), leading to more efficient strategies for a resource efficient Europe. Such aspects are given high priority by the stakeholders in the three case studies (e.g., Latvia, The Netherlands and Southwest UK) with an interest in climate and renewable energy and can be taken into account through SDM.

Author Contributions: F.B. and L.V.-L. worked on the concept and methodology for this paper. F.B. wrote the paper with contributions from L.V.-L., E.A., I.B., V.L., and M.G., E.A. wrote the part on the E3ME model. I.B. analyzed the case study on Latvia, V.L. analyzed the case study on the Netherlands, and M.G. analyzed the case study on southwest UK. I.B. drafted the text for Latvia, V.L. drafted the text on the Netherlands, and M.G. drafted the text on southwest UK.

Funding: The work described in this paper was conducted within the project SIM4NEXUS. This project received funding from the European Union's Horizon 2020 research and innovation programme under Grant Agreement No 689150 SIM4NEXUS. This paper and the content included in it do not represent the opinion of the European Union, and the European Union is not responsible for any use that might be made of its content.

Acknowledgments: The authors thank three anonymous reviewers whose comments contributed to significantly improve the quality of this paper.

Conflicts of Interest: The authors declare no conflicts of interest. The funders had no role in the design of the study; in the collection, analyses, or interpretation of data; in the writing of the manuscript, and in the decision to publish the results.

References

1. Kling, C.L.; Arritt, R.W.; Calhoun, G.; Keiser, D.A. Integrated assessment models of the food, energy, and water Nexus: A review and an outline of research needs. *Annu. Rev. Resour. Econ.* **2017**, *9*, 143–163. [CrossRef]
2. Climate Strategies. The EU's 2030 Climate and Energy Framework and Energy Security. London, UK. Available online: https://bit.ly/2qj3Sjp (accessed on 30 October 2018).
3. European Commission. The Macroeconomic and other Benefits of Energy Efficiency (EED Directive). Study for the European Commission, Directorate-General for Energy, Contract No. ENER/C3/2013-484/03/FV2015-523 under the Multiple Framework Service Contract ENER/C3/2013-484. Available online: https://bit.ly/2mf0jux (accessed on 31 October 2018).
4. Cambridge Econometrics. Employment Effects of Selected Scenarios from the Energy Roadmap 2050. Report was Carried under a Specific Contract within DG ENER's Framework Contract ENER A2 360-2010 Regarding Impact Assessment and Evaluations (Ex-Ante, Intermediate and Ex-Post). Available online: https://bit.ly/2DeYkyy (accessed on 31 October 2018).
5. European Commission. *Commission Staff Working Document Impact Assessment, Accompanying the document "Proposal for a Directive of the European Parliament and of the Council amending Directive 2012/27/EU on Energy Efficiency"*; European Commission: Brussels, Belgium, 2016.
6. National Grid PLC. Future Energy Scenarios. National Grid PLC: Warwick, UK. Available online: https://ngrid.com/2CPquzf (accessed on 30 October 2018).

7. UNECE. Deployment of Renewable Energy: The Water-Energy-Food-Ecosystem Nexus Approach to Support the Sustainable Development Goals. Geneva, United Nation Economic Commission for Europe. Available online: https://bit.ly/2DaMkhR (accessed on 30 October 2018).
8. Brouwer, F.; Avgerinopoulos, G.; Fazekas, D.; Laspidou, C.; Mercure, J.-F.; Pollitt, H.; Ramos, E.P.; Howells, M. Energy modelling and the Nexus concept. *Energy Strategy Rev.* **2018**, *19*, 1–6. [CrossRef]
9. Munaretto, S.; Witmer, M. *Water-Land-Energy-Food-Climate Nexus: Policies and Policy Coherence at European and International Scale*; Netherlands Environmental Assessment Agency (PBL): The Hague, The Netherlands, 2017.
10. Sušnik, J.; Chew, C.; Domingo, X.; Mereu, S.; Trabucco, A.; Evans, B.; Vamvakeridou-Lyroudia, L.; Savić, D.A.; Laspidou, C.; Brouwer, F. Multi-Stakeholder Development of a Serious Game to Explore the Water-Energy-Food-Land-Climate Nexus: The SIM4NEXUS Approach. *Water* **2018**, *10*, 139. [CrossRef]
11. Trabucco, A.; Sušnik, J.; Vamvakeridou-Lyroudia, L.; Evans, B.; Masia, S.; Blanco, M.; Roson, R.; Sartori, M.; Alexandri, E.; Brouwer, F.; et al. Water-Food-Energy Nexus under Climate Change in Sardinia. *Proceedings* **2018**, *2*, 609. [CrossRef]
12. Cambridge Econometrics. E3ME: Out Global Macro-econometric Model. Available online: https://bit.ly/2LkkWPF (accessed on 17 December 2018).
13. Lavoie, M. *Post-Keynesian Economics: New Foundations*; Edward Elgar: Cheltenham, UK; Northampton, MA, USA, 2014; ISBN 978-1-84720-483-7.
14. King, J.E. *Advanced Introduction to Post Keynesian Economics*; Edward Elgar: Cheltenham, UK; Northampton, MA, USA, 2015; ISBN 978-1-78254-842-3.
15. Pollitt, H.; Mercure, J.-F. The role of money and the financial sector in energy-economy models used for assessing climate and energy policy. *Clim. Policy* **2017**, *18*, 184–197. [CrossRef]
16. Barker, T.; Dagoumas, A.; Rubin, J. The macroeconomic rebound effect and the world economy. *Energy Effic.* **2009**, *2*, 411–427. [CrossRef]
17. Mercure, J.-F.; Pollitt, H.; Viñuales, J.E.; Edwards, N.R.; Holden, P.B.; Chewpreecha, U.; Salas, P.; Sognnaes, I.; Lam, A.; Knobloch, F. Macroeconomic impact of stranded fossil fuel assets. *Nat. Clim. Chang.* **2018**, *8*, 588–593. [CrossRef]
18. Holden, P.B.; Edwards, N.R.; Ridgwell, A.; Wilkinson, R.D.; Fraedrich, K.; Lunkeit, F.; Pollitt, H.; Mercure, J.-F.; Salas, P.; Lam, A.; et al. Climate—Carbon cycle uncertainties and the Paris Agreement. *Nat. Clim. Chang.* **2018**, *8*, 609–613. [CrossRef]
19. Mercure, J.-F.; Pollitt, H.; Bassi, A.M.; Viñuales, J.E.; Edwards, N.R. Modelling complex systems of heterogeneous agents to better design sustainability transitions policy. *Glob. Environ. Chang.* **2016**, *37*, 102–115. [CrossRef]
20. Mercure, J.-F.; Pollitt, H.; Edwards, N.R.; Holden, P.B.; Chewpreecha, U.; Salas, P.; Lam, A.; Knobloch, F.; Vinuales, J.E. Environmental impact assessment for climate change policy with the simulation-based integrated assessment model E3ME-FTT-GENIE. *Energy Strategy Rev.* **2018**, *20*, 195–208. [CrossRef]
21. European Commission. EU Reference Scenario 2016—Energy, Transport and GHG Emissions Trends to 2050. Brussels, European Commission, Directorate-General for Energy, Directorate-General for Climate Action and Directorate-General for Mobility and Transport. Available online: https://ec.europa.eu/energy/sites/ener/files/documents/20160713%20draft_publication_REF2016_v13.pdf (accessed on 10 August 2018).
22. IEA. World Energy Outlook 2013. OECD/IEA: Paris, France. Available online: https://www.iea.org/publications/freepublications/publication/WEO2013.pdf (accessed on 10 August 2018).
23. Prūse, I. Developments Related to Low Carbon Development Strategy 2050 and Climate Change Adaptation Strategy 2030 (in Latvian). Available online: https://bit.ly/2PJEwW6 (accessed on 10 August 2018).
24. Schoots, K.; Hekkenberg, M.; Hammingh, P. Nationale Energieverkenning 2017. ECN, Amsterdam/Petten, ECN-O-17-018. 2017. Available online: https://bit.ly/2EydEHx (accessed on 17 December 2018).
25. Planbureau voor de Leefomgeving. Biomass: Wishes and Limitations. Planbureau voor de Leefomgeving: The Hague, The Netherlands. Available online: https://themasites.pbl.nl/biomass/ (accessed on 17 December 2018).
26. VVD, CDA, D66, ChristenUnie. Vertrouwen in de Toekomst; Regeerakkoord 2017–2021 VVD, CDA, D66 en ChristenUnie. Den Haag. Available online: https://bit.ly/2i1wmgo (accessed on 10 September 2018).
27. PBL. *Opties voor Energie-en Klimaatbeleid*; Netherlands Environmental Assessment Agency (PBL): The Hague, The Netherland, 2016. Available online: https://bit.ly/2ao3d7n (accessed on 10 September 2018).

28. OFGEM. Our Strategy for Regulating the Future Energy System. 2017. Available online: https://www.ofgem.gov.uk/system/files/docs/2017/08/our_strategy_for_regulating_the_future_energy_system.pdf (accessed on 24 August 2018).

29. Fankhauser, S. A Practitioner's Guide to a Low-Carbon Economy: Lessons from the UK. Centre for Climate Change Economics and Policy (CCCEP). Available online: http://www.lse.ac.uk/GranthamInstitute/wp-content/uploads/2014/03/PP_low-carbon-economy-UK.pdf (accessed on 24 August 2018).

30. Western Power Distribution. South West 132kV Network Capacity Restriction UPDATE March 2016. Available online: https://www.westernpower.co.uk/docs/connections/Generation/Generation-capacity-map/Distributed-Generation-EHV-Constraint-Maps/WPD-Southwest-network-capacity-restriction.aspx (accessed on 24 August 2018).

31. RES LEGAL Europe. Available online: http://www.res-legal.eu/search-by-country/latvia (accessed on 10 August 2018).

32. Linderhof, V.; Polman, N.; Selnes, T.; Witmer, M.; Munaretto, S.; Susnik, J. The Main Nexus Challenges in 'The Netherlands'. Deliverable 5.2 for the Netherlands. Available online: https://bit.ly/2O53uyW (accessed on 1 September 2018).

33. Strengers, B.; Eerens, H.; Smeets, W.; van den Born, G.J.; Ros, J. *Negatieve Emissies: Technisch Potentieel, Realistisch Potentieel en Kosten voor Nederland*; Planbureau voor de Leefomgeving: The Hague, The Netherlands, 2018. Available online: https://bit.ly/2Me6Wpf (accessed on 1 September 2018).

34. Van Zalk, J.; Behrens, P. The spatial extent of renewable and non-renewable power generation: A review and meta-analysis of power densities and their application in the U.S. *Energy Policy* **2018**, *123*, 83–91. [CrossRef]

35. Department of Energy & Climate Change (DECC). Overarching National Policy Statement for Energy (EN-1). Available online: https://assets.publishing.service.gov.uk/government/uploads/system/uploads/attachment_data/file/47854/1938-overarching-nps-for-energy-en1.pdf (accessed on 24 August 2018).

36. Regen. Renewable Energy. A Local Progress Report for England 2016. Available online: https://www.regen.co.uk/wp-content/uploads/2016_Progress_Report_2016-1.pdf (accessed on 24 August 2018).

37. Western Power Distribution. DSO-Transition-Strategy. 2017. Available online: https://www.westernpower.co.uk/docs/About-us/Our-business/Our-network/Strategic-network-investment/DSO-Strategy/DSO-Transition-Strategy.aspx (accessed on 18 December 2018).

38. Department of Energy & Climate Change (DECC). Annual Energy Statement. 2014. Available online: https://assets.publishing.service.gov.uk/government/uploads/system/uploads/attachment_data/file/371388/43586_Cm_8945_print_ready.pdf (accessed on 24 August 2018).

39. South West Water. Annual Report and Financial Statements. 2014. Available online: https://www.southwestwater.co.uk/globalassets/document-repository/annual-reports/sww_annual_report___financial_statements_2014.pdf (accessed on 24 August 2018).

40. Water LIFE. Opportunities and Barriers to Using Payments for Ecosystem Services and Supply Chain Measures. Participatory Research in WaterLIFE Demonstration Catchments. Available online: https://www.catchmentbasedapproach.org/media/attachments/2017/07/03/09_complete_opps_barriers_pes.pdf (accessed on 18 December 2018).

41. Zhou, Q. A review of sustainable urban drainage systems considering the climate change and urbanization impacts. *Water* **2014**, *6*, 976–992. [CrossRef]

42. Ossa-Moreno, J.; Smith, K.; Mijic, A. Economic analysis of wider benefits to facilitate SuDS uptake in London, UK. *Sustain. Cities Soc.* **2017**, *28*, 411–419. [CrossRef]

43. Wiechmann, B.; Dienemann, C. Sewage Sludge Management in Germany. Umweltbundesamt, 2013. Available online: https://www.umweltbundesamt.de/sites/default/files/medien/378/publikationen/sewage_sludge_management_in_germany.pdf (accessed on 24 August 2018).

44. Distribution Future Energy Scenarios A generation and demand study Technology Growth Scenarios to 2032 Wester Power Distribution & Regen SW 2018. Available online: https://bit.ly/2Gnekkx (accessed on 17 December 2018).

Article

Hybrid Neural Fuzzy Design-Based Rotational Speed Control of a Tidal Stream Generator Plant

Khaoula Ghefiri [1,2,*], Izaskun Garrido [2], Soufiene Bouallègue [1], Joseph Haggège [1] and Aitor J. Garrido [2]

[1] Laboratory of Research in Automatic Control—LA.R.A, National Engineering School of Tunis (ENIT), University of Tunis El Manar, BP 37, Le Belvédère, 1002 Tunis, Tunisia; soufiene.bouallegue@issig.rnu.tn (S.B.); joseph.haggege@enit.rnu.tn (J.H.)

[2] Automatic Control Group—ACG, Department of Automatic Control and Systems Engineering, Engineering School of Bilbao, University of the Basque Country, 48012 Bilbao, Spain; izaskun.garrido@ehu.es (I.G.); aitor.garrido@ehu.es (A.J.G.)

* Correspondence: kghefiri001@ikasle.ehu.eus; Tel.: +34-94-601-4443

Received: 20 August 2018; Accepted: 12 October 2018; Published: 17 October 2018

Abstract: Artificial Intelligence techniques have shown outstanding results for solving many tasks in a wide variety of research areas. Its excellent capabilities for the purpose of robust pattern recognition which make them suitable for many complex renewable energy systems. In this context, the Simulation of Tidal Turbine in a Digital Environment seeks to make the tidal turbines competitive by driving up the extracted power associated with an adequate control. An increment in power extraction can only be archived by improved understanding of the behaviors of key components of the turbine power-train (blades, pitch-control, bearings, seals, gearboxes, generators and power-electronics). Whilst many of these components are used in wind turbines, the loading regime for a tidal turbine is quite different. This article presents a novel hybrid Neural Fuzzy design to control turbine power-trains with the objective of accurately deriving and improving the generated power. In addition, the proposed control scheme constitutes a basis for optimizing the turbine control approaches to maximize the output power production. Two study cases based on two realistic tidal sites are presented to test these control strategies. The simulation results prove the effectiveness of the investigated schemes, which present an improved power extraction capability and an effective reference tracking against disturbance.

Keywords: fuzzy logic control; artificial neural networks control; tidal stream generator; swell effect disturbance; doubly fed induction generator; maximum power point tracking

---

1. Introduction

Renewable energy technologies are being increasingly exploited worldwide. Countries around the world are resorting to integrating renewable energy resources into their energy policy to reduce fossil fuel usage and carbon emissions [1–4]. Electricity demand is growing rapidly as countries develop, with increased use of electricity to meet a range of needs. According to the projections of the International Energy Agency (IEA), the global energy need has increased by about 40% since 1990, and a 53% increase is expected by 2030 [5]. This fast-rising energy demand will require some US $45 trillion in new infrastructure investment by 2030 [6]. The renewable energy technologies can improve energy security and decrease dependence on fossil fuels. The International Energy Outlook (IEO2016) confirms that these systems are related to a growing renewable energy converters. Renewable energy consumption increases by an average of 2.6% per year between 2012 and 2040 [7]. The technical potential for renewable energy is far greater than current human energy use, and studies suggest it could supply 95% of global energy demand by 2050, and double its current share by 2030, at a relatively low net cost [8,9]. Marine renewable energy has become the focus of national research and development because of its abundant, renewable, and non-polluting characteristics [10–12]. Tidal energy represents an important energy source as the tidal energy potential is estimated to be around 450 TWh/year, with about 24 TWh/year on the European coasts [13].

Among marine renewable energy converters, Tidal Stream Generator (TSG) promises to be an environmentally friendly way to generate renewable electric energy with no emission of greenhouse gases during normal operation [14,15]. The horizontal axis tidal stream turbine has apparent similarities with the wind turbine. Nevertheless, they have different operational behaviors. In normal conditions, the fluid is over eight hundred times $(1025/1.225 = 837)$ denser than the air [16]. This is due to the huge kinetic energy density of the water. Therefore, at equal power, the tidal turbine will be more compact than a wind turbine. This will lead to a significant difference in the rotor size [17]. Consequently, the advantages of these opposing views will appear in construction, transportation, and charge of installation. In addition, the differences are mainly in the load design, size, and the inertia of the rotor. These characteristics are figured in different operating conditions. In effect, studies demonstrate that variations of the rotational speed for a TST are higher than for a windmill system, despite the perturbation in the wind speed is much higher than that of marine current [16]. Concerning the tidal stream converters, the swell effect is supposed to be the most disturbing phenomenon for the tidal current speed input [18]. This fluctuation will affect the harnessed output power.

In this area of research, control strategies have a valuable role to enhance the dynamic behavior of the TSG plant. In this context, several control approaches have been used. The Maximum Power Point Tracking (MPPT) strategy is employed to search the maximum harvested power from tides and tracking the Optimal Regimes Characteristic (ORC) operation [19]. Other research focused on the control of the active and reactive powers through the Doubly Fed Induction Generator (DFIG) by the use of the Rotor Side Converter (RSC) and the Grid Side Converter (GSC). The RSC control is used to keep the generator speed at its reference signal and the GSC control is dedicated to ensuring that the DC-link voltage remains constant [20,21]. In the literature, the PI controller has been used to control the operation of the marine current turbine through the back-to-back power converter aiming to maximize the captured energy [22]. One can envisage two designs: the torque and rotational speed control loops. However, the torque control loop is sensitive to the parametric variations and the turbulent tidal resource [23]. In addition, advanced control approaches can be employed to provide better performance especially to ensure the robustness under the modeling uncertainties [24,25]. In this framework, the sliding mode control approach is a suitable method for nonlinear systems [26]. It has been used in the field of marine energy conversion [27]. Its robustness to the disturbances and parametric uncertainties renders unnecessary a precise knowledge of the system. However, the main drawback of this method is the chattering phenomenon which is the high-frequency oscillations. This can negatively affect the generator because of discontinuous control. Many approaches were proposed to deal with this drawback as presented in [28]. Furthermore, the artificial intelligence techniques are capable of handling nonlinear problems in various signal processing applications, from pattern recognition and extended to renewable energy converters [29]. An artificial neural network is considered a technique which is well accepted for nonlinear statistical adjustment applications [30]. As discussed in [31], the Artificial Neural Networks (ANN) technique is used to more accurately determine the wind speed distribution law of a site, enabling the better assessment of wind energy potential and wind generator performances. The approach enables wind speed prediction with less errors. An application of using neural networks in wave energy systems is detailed in [32], where the assessment of the wave energy potential in near shore coastal areas is investigated by means of ANNs. The ANN model developed forecast wave energy potential with great accuracy. The Fuzzy Gain Scheduling (FGS) technique has been used as well for renewable energy converters. The MPPT controller for photovoltaic systems using an FGS strategy has been studied in [33]. This approach creates an adaptive MPPT controller and achieves better overall system performance. Furthermore, the proposed technique detailed in [34] is applied to design FGS-PID (Proportional Integral Derivative) controllers of superconducting magnetic energy storage for power system stabilization. The study confirms that the controller provides high robustness under various operating conditions and large disturbances. From the benefits of both advanced approaches, which are the ANN and FGS, this study focuses on the power output improvement of the TSG system by implementing a hybrid neural fuzzy design.

The main application of the proposed control is affected by the change in velocities that are not predictable by astronomical tide—in particular, the swell effect phenomenon. For that reason, the robustness of the investigated control strategies has been compared by acquiring data from realistic tidal site in order to show how much energy will be saved. In particular, the novel hybrid Neural Fuzzy design is investigated to reach the power output improvement extraction by varying the rotational speed. The Artificial Neural Networks (ANN) based-MPPT approach has the advantage to approximate and interpolate multi-variate data that require huge databases. Furthermore, the fuzzy gain scheduling based control eliminates the fixed gains during operation. Therefore, the proposed fuzzy block will provide the adaptive change of controller gains which adequately vary according the variable tidal speed. In the operation in variable speed, the FGS-PI-based control is applied to the RSC. This enables the TSG to track the MPPT strategy. The MPPT approach uses a multilayer feed-forward ANN that enhances a fuzzy rotational speed controller. The aim of this command is to control the TSG plant, which, at each tidal velocity, must follow the optimal rotational speed where the maximum generated power is satisfied. The analysis of the investigated control approaches has been tested in the case of an irregular tidal resource and the occurrence of a disturbance during normal operation.

The rest of this article is structured as follows. In Section 2, the TSG system is described and modeled. Section 3 is devoted to the control design of the MPPT-based ANN and the FGS-based rotational speed control. In Section 4, the control robustness and disturbance rejection are investigated. Two study cases are presented and discussed using numerical input and real measured input. Finally, concluding discussions are drawn in Section 5.

2. Model Statement

The hydrodynamic turbine design is complex due to the changes in the non-constant current tidal and the direction, and the effect of the fluid depth. Therefore, the modeling assumptions are related to tidal turbine hydrodynamics, considering a constant thrust loading at the disk. In addition, the current speed at the disk is the average of the upstream and downstream currents [15,35]. The topology of the TSG system is shown in Figure 1. It consists of a DFIG based on a Tidal Stream Turbine (TST). The configuration of the DFIG allows variable speed operation on a specific operation range. The connection to the grid is done through the stator of the DFIG via a transformer, while the DFIG is connected to the grid through the power electronic converters.

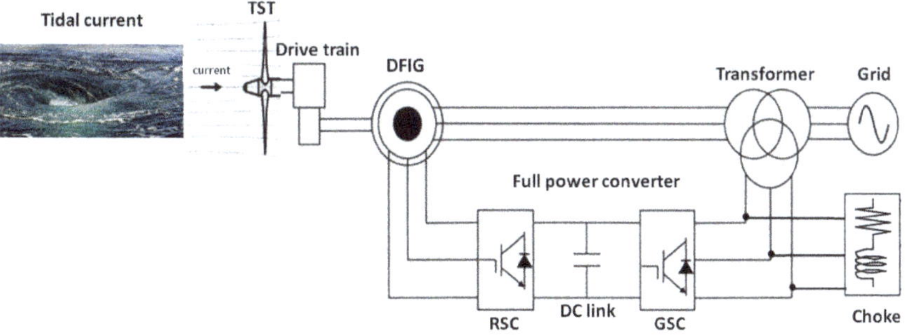

Figure 1. Tidal stream generator global scheme.

2.1. Tidal Turbine Model

The power harnessed from the tidal current speed V, can be expressed as follows [36]:

$$P_t = \frac{1}{2}C_p(\lambda, \beta)\rho\pi R^2 V^3,\tag{1}$$

where P_t denotes the harnessed power from the tidal speed in (W), and R is the radius of the rotor blades expressed in (m).

However, the power captured from the tides cannot be used totally by the turbine because of Betz limit. The power coefficient C_p is defined as a function of the blade pitch angle β expressed in (deg) and the tip-speed ratio λ, given as the following equation [37,38]:

$$\lambda = \frac{\omega_t R}{V},\tag{2}$$

where ω_t is the rotor speed of the tidal turbine defined in (rad/s).

The generated torque of the tidal turbine, in (Nm), is expressed as follows:

$$T_{tst} = \frac{P_t}{\omega_t}.\tag{3}$$

2.2. Shaft Model

The torque produced by the tidal turbine is transmitted to the generator using the drive train, which ensures the connection between the rotor and the generator via a flexible shaft. The model of the drive train shaft used is the two-mass model which is developed using the stiffness coefficient K_{sh} in (Nm/rad) and the damping coefficient D_{sh} in (Nms/rad). The expressions of the developed model are expressed as [39]:

$$T_{tst} - T_t = 2H_t\frac{d\omega_t}{dt},\tag{4}$$

$$T_t = D_{sh}(\omega_t - \omega_g) + K_{sh}\int(\omega_t - \omega_g)dt,\tag{5}$$

$$T_t - T_{em} = 2H_g\frac{d\omega_g}{dt},\tag{6}$$

where T_t is the produced torque by the rotor shaft given in (Nm), T_{em} is the electromagnetic torque of the generator in (Nm), ω_g is the generator speed expressed in (rad/s), and H_t and H_g are the turbine and the generator inertia constants defined in s.

2.3. DFIG Model

The functioning in variable speed mode of the tidal turbine based on a DFIG has proven robustness due to the ability to achieve a higher power quality, reduced cost, and improved system efficiency. In addition, the DFIG with a four-quadrant operation enables a decoupled control of the active and reactive powers of the generator [40–42]. The dynamical model of the DFIG is defined using the Park's transformation in $d - q$ as explained in [43]. The equations of the stator voltages and flux, expressed in (V) and in (Wb) are given as:

$$\begin{cases} U_{sd} = R_s I_{sd} + \frac{d\varphi_{sd}}{dt} - \omega_s\varphi_{sq}, \\ U_{sq} = R_s I_{sq} + \frac{d\varphi_{sq}}{dt} - \omega_s\varphi_{sd}, \end{cases}\tag{7}$$

$$\begin{cases} \varphi_{sd} = L_s I_{sd} + L_m I_{rd}, \\ \varphi_{sq} = L_s I_{sq} + L_m I_{rq}. \end{cases}\tag{8}$$

The voltages and flux of the rotor are given as follows:

$$\begin{cases} U_{rd} = R_r I_{rd} + \frac{d\varphi_{rd}}{dt} - \omega_r \varphi_{rq}, \\ U_{rq} = R_r I_{rq} + \frac{d\varphi_{rq}}{dt} - \omega_r \varphi_{rd}, \end{cases} \tag{9}$$

$$\begin{cases} \varphi_{rd} = L_r I_{rd} + L_m I_{sd}, \\ \varphi_{rq} = L_r I_{rq} + L_m I_{sq}. \end{cases} \tag{10}$$

The generator electromagnetic torque is defined in $d-q$ by Equation (11):

$$T_{em} = \frac{3}{2} p L_m \left(I_{sq} I_{rd} - I_{sd} I_{rq} \right), \tag{11}$$

where I_{sd}, I_{sq} are the currents of the stator, I_{rd}, I_{rq} are the currents of the rotor given in $d-q$ in (A), R_s and R_r are the resistances of the stator and rotor given in (Ω), ω_s and ω_r are the pulsations of the stator and rotor expressed in (rad/s), L_s and L_r are the inductances of the stator and rotor given in (H), L_m is the magnetizing inductance defined in (H) and p is the number of the pole pair.

2.4. Back-to-Back Converter Model

The configuration of the DFIG-based TST with the back-to-back power converters allows for decoupling the control for both GSC and RSC components [44,45]. In this subsection, a model of the power converters is presented. The power converters that consist of the RSC and GSC coupled by means of the DC-link as shown in Figure 1. The GSC is controlled thanks to the vector control scheme in order to adjust the voltage of the DC-link—in addition to controlling the flow of the reactive power to ensure the DC voltage adjustments [46]. The control of RSC is proposed to regulate the system in order to achieve the maximum output power production. For that reason, the rotational speed control is investigated using the vector control strategy [43].

The active and reactive powers of the plant, which are defined in (W) and (VAR), are expressed as:

$$P_g = \frac{3}{2} \left(U_{dg} I_{dg} - U_{qg} I_{qg} \right), \tag{12}$$

$$Q_g = \frac{3}{2} \left(U_{qg} I_{dg} - U_{dg} I_{qg} \right), \tag{13}$$

where U_{dg}, U_{qg} in (V) and I_{dg}, I_{qg} in (A) are the voltages and currents of the grid expressed in the $d-q$.

The voltages of the d-axis and the grid are aligned, i.e., $U_{dg} = U_g$ and $U_{qg} = 0$, to reach the voltage oriented control. Thus, the expressions of the active and reactive powers can be given as follows:

$$P_g = \frac{3}{2} U_g I_{dg}, \tag{14}$$

$$Q_g = -\frac{3}{2} U_g I_{qg}. \tag{15}$$

The expression that links the power saved in the DC-link and the power transmitted to the grid is given as follows:

$$P_g = \frac{3}{2} U_g I_{dg} = U_{dc} I_{dc}, \tag{16}$$

where U_{dc} is the voltage of the DC-link given in (V) and I_{dc} is the current flowing in the DC-link in (A).

3. Control Statement

Tidal current speed may vary in two ways; on a large time scale by means of the gravitational effort of the sun and the moon. In this case, the current speed profile is composed of spring and neap tides and the period occurs each 12 h and 25 min. On a small time scale, these currents are affected by climate disturbance as the case of swell effect phenomenon which can occur during a period of few seconds. Both time scales should be taken into consideration when implementing the control scheme to drive the TSG in the variable speed mode. The investigated control block for the tidal stream converter system is depicted in Figure 2.

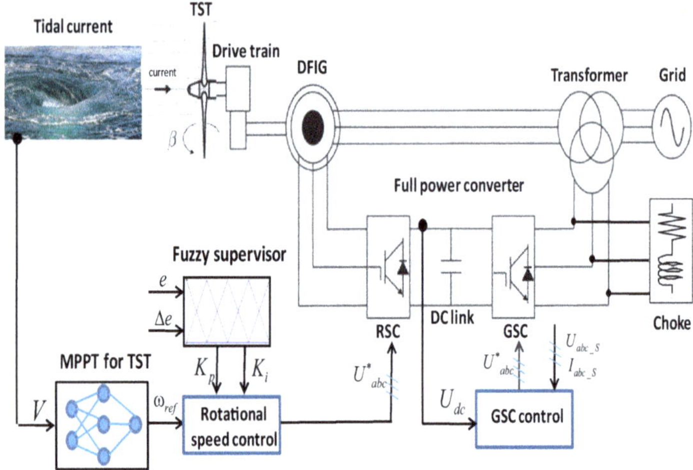

Figure 2. Control strategies scheme description for TSG.

In order to perform the output power's improvement under different input conditions, a novel hybrid rotational speed control is investigated. The MPPT-based ANN approach is designed to provide the suitable rotor speed according to the change of the tidal input. The reference rotational speed is adjusted using a multilayer feed-forward neural network. Then, the proposed fuzzy gain scheduled PI controller is used to regulate the rotational speed. Such a proposed fuzzy supervisor adequately modifies the controller gains providing the control a novel adaptative mechanism to the input parameter changes.

3.1. ANN-Based Maximum Power Point Tracking Approach

The MPPT approach has been strongly used to optimize the efficiency of renewable energy plants [47,48]. However, this strategy must be adapted to the tidal energy environment according to the control requirements in the TSG system. The MPPT algorithm used in this study is based on an ANN-based approach. The aim of this control strategy is to control the TSG system on the way to follow the optimal rotational speed, which corresponds to the maximum generated power. Figure 3 shows the variation of the generated power as a function of the rotational speed and the tidal current speed.

The system under study consists of a DFIG-based tidal turbine for which the maximum harnessed power is $P_n = 1.5$ MW. In order to harness the maximum power from the tides, the power coefficient and thus the tip-speed ratio should be maintained at their maximum values as $C_{p\,max} = 0.44$ and $\lambda_{opt} = 6.34$, respectively [49]. The structure of the proposed feed-forward neural network is depicted in Figure 4.

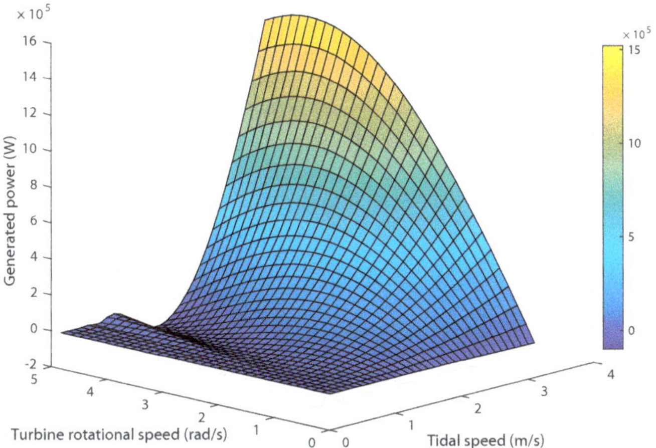

Figure 3. Harnessed output power as a function of the rotor speed for given velocities.

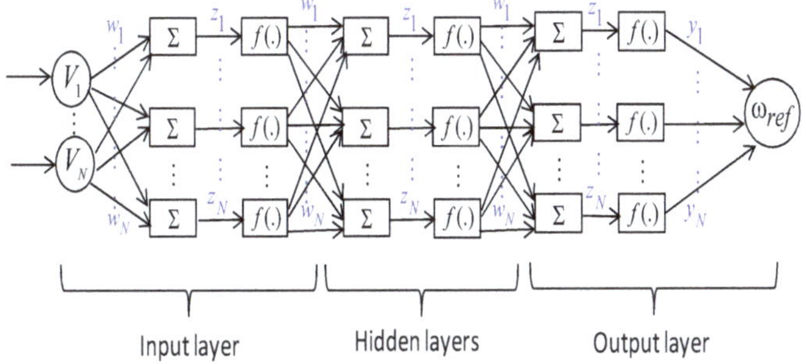

Figure 4. Layout of feed-forward neural network.

The model of a common neuron is given by the following expression:

$$y_i = \sum_{j=1}^{N} w_{ij}\, x_j + T_i^h,$$
$$z_i = f_i^{(l)}\, (y_i), \tag{17}$$

where z_i is the output of the ith neuron in the lth hidden layer, w_{ij} are the synaptic weights linking the jth neurons to ith neurons, $f_i^l(.)$ is the activation function of the ith neuron of layer l, i is the number of the input neurons, x_j are the input neurons, and T_i^h are the threshold terms of the hidden layer.

The resulting MPPT block adequately adapts the rotational speed reference at each tidal velocity input, so as to maximize the power extracted from the system. Once the network is created and configured, the weights are randomly set at first. The used training method is a Levenberg–Marquardt algorithm [50,51]. The used learning algorithm is conceived to adjust the weights to minimize the error for each calculated output and the solution given by the ANN for the adequate input.

The used method is trial-and-error based on a forward strategy process. This begins with an undersized number of neurons in the hidden layer until the training and testing results are ameliorated. This rule uses a statistical analysis to prove the best performance criteria reached. Several tests have been investigated to choose the suitable number of neurons in the hidden layer. The criterion was set according to the smallest Mean Squared Error (MSE) determined. Figure 5 depicts the variation of the result of the MPPT block versus the tidal current speed for a various value of neurons in the hidden layer.

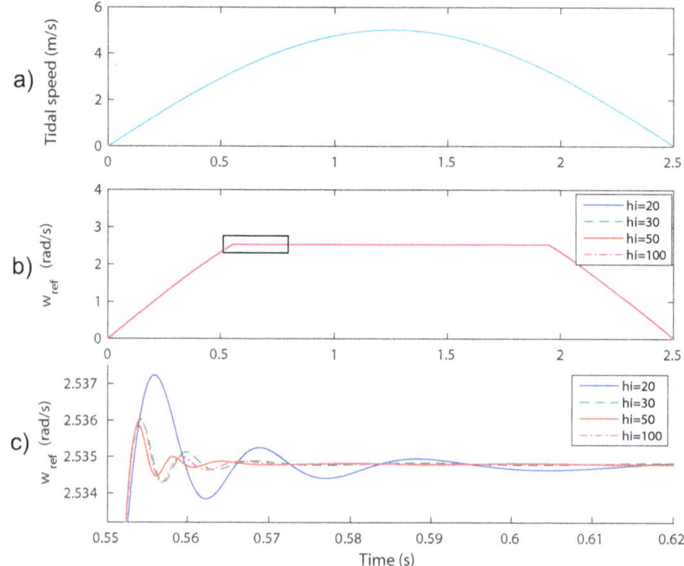

Figure 5. Control performances: (**a**) tidal speed input; (**b**) response of the reference rotational speed for different number of neurons; (**c**) zoom into the reference rotational speed variation for 0.07 s.

The range of variation of V is from 0 to 5 m/s, which represents a high tidal velocity that can be recorded at the high energetic sites [52]. As it may be seen that, for all tidal velocities less than 3.2 m/s, the response is regulated to follow the reference. When the tidal velocity is superior to the threshold limit of the rotor speed, it is kept at its maximum value, that is, 2.53 rad/s. The number of neurons in the hidden layer will be chosen to be $h_i = 50$ as a trade-off between a low number and a transient with small oscillations.

It may be seen in Figure 6 the iteration for which the validation performance is minimal. The best validation performance is 1.407×10^{-7} at epoch 1000. This character displays that the training data admit a good fit.

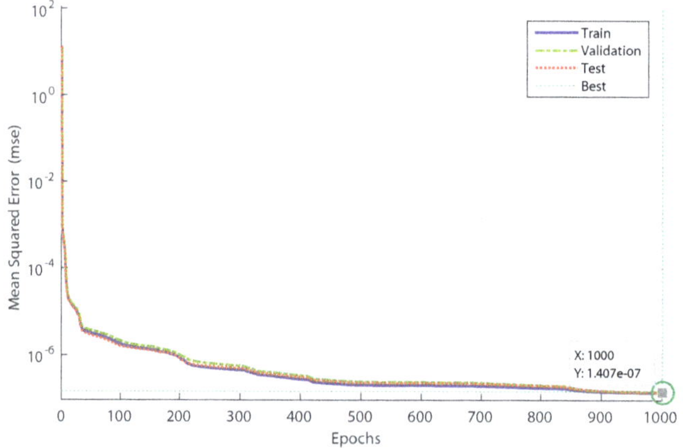

Figure 6. Training performance of the ANN block.

3.2. FGS-PI Based-Rotational Speed Control

The developed block control which is conceived to the RSC is shown in Figure 7. The adequate rotational speed ω_{ref} is acquired from the developed ANN based MPPT approach. The control loop of the rotational speed is conceived by means of a fuzzy gain scheduling approach since it is considered as a robust control strategy against uncertainties [53]. Thus, the principle of FGS-PI based control is to refine the parameters of the tuned PI controller to satisfy the required performance [33,54,55].

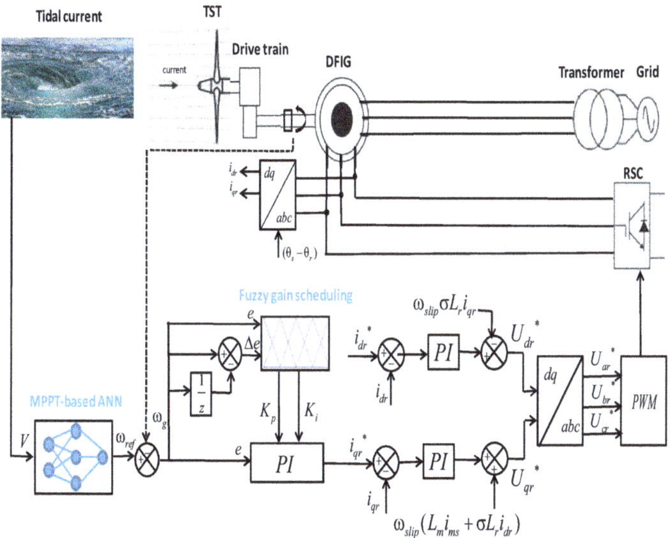

Figure 7. Control scheme of the RSC component.

The discrete-time expression of the PI control law is expressed by the following equation:

$$u(k) = K_p \Delta e(k) + K_i \, T_s \, e(k) + u(k-1), \tag{18}$$

where $e(k)$ is the error from the adequate rotational speed acquired from the ANN-based MPPT strategy and the actual rotor speed, $\Delta e(k) = e(k) - e(k-1)$ is the variation of the error, K_p and K_i are the PI controller parameters and T_s is the sampling time.

In effect, the fuzzy logic controller acquires the inputs and outputs which should be actual numbers to satisfy the actuators conditions. Thus, the fuzzification and defuzzification process for the input and outputs variables is essential. The aim of the fuzzification is to convert the crisp values to fuzzy linguistic terms in order to apply the fuzzy inferences using the rules. For that reason, the inputs and outputs are normalized in the universe of discourse [56,57].

The fuzzy supervisor includes $e(k)$ and $\Delta e(k)$ the two inputs and two outputs K_p and K_i as the proportional and integral gains, respectively. These gains are normalized using the linear transformation as follows [53]:

$$\begin{cases} K'_p = (K_p - K_{p\,min}) / (K_{p\,max} - K_{p\,min}), \\ K'_i = (K_i - K_{i\,min}) / (K_{i\,max} - K_{i\,min}), \end{cases} \tag{19}$$

where $\left[K_{p\,min}, K_{p\,max}\right]$ and $\left[K_{i\,min}, K_{i\,max}\right]$ are the tolerable range of the parameters of the controller.

The gain scheduling of the PI block is obtained adopting the fuzzy rules described by the following equation:

$$if \ e(k) \ is \ A_i \ and \ \Delta e(k) \ is \ B_i,$$
$$then \ K'_p \ is \ C_i \ and \ K'_i \ is \ D_i, \tag{20}$$

where A_i, B_i, C_i and D_i are the fuzzy sets on the corresponding supporting sets, which $i = 1, 2, ..., m$.

The types of membership functions considered in this study are triangular and trapezoidal, which are uniformly distributed and symmetrical in the universe of discourse. The used linguistic levels are Negative Big (NB), Negative (N), Zero (Z), Positive (P) and Positive Big (PB). The corresponding membership functions related to the inputs e and Δe of A_i and B_i fuzzy sets and to the outputs K'_p and K'_i of C_i and D_i fuzzy sets are illustrated in Figures 8 and 9.

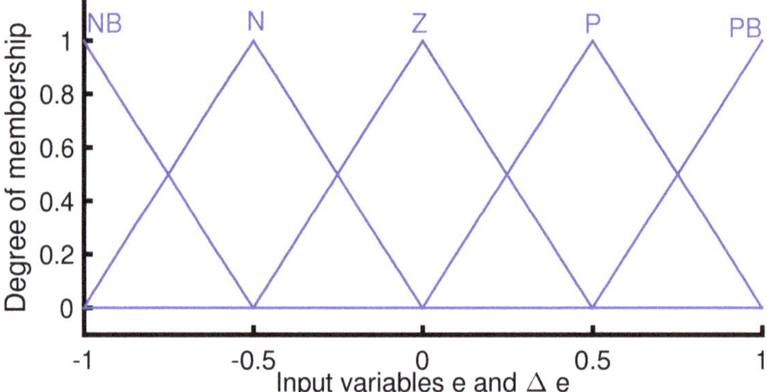

Figure 8. Membership functions for inputs e and Δe.

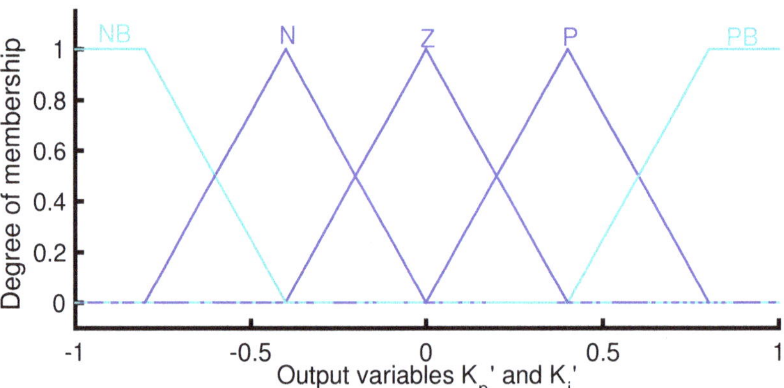

Figure 9. Membership functions for outputs K'_p and K'_i.

The grade of the membership functions μ_{A_i} and μ_{B_i} are defined as:

$$
\mu_{NB}(X) = \begin{cases} (-2X-1) & if\ X \in (-1,\ -^1/_2) \\ 0 & if\ X \geq\ -^1/_2, \end{cases}
$$

$$
\mu_{N}(X) = \begin{cases} 2(X+1) & if\ X \in (-1,\ -^1/_2] \\ -2X & if\ X \in (-^1/_2, 0) \\ 0 & if\ X \geq\ 0, \end{cases}
$$

$$
\mu_{Z}(X) = \begin{cases} 0 & if\ X \leq -^1/_2 \\ (2X+1) & if\ X \in (-^1/_2,\ 0] \\ (1-2X) & if\ X \in (0,\ ^1/_2) \\ 0 & if\ X \geq\ ^1/_2, \end{cases}
\tag{21}
$$

$$
\mu_{P}(X) = \begin{cases} 0 & if\ X \leq 0 \\ 2X & if\ X \in (0,\ ^1/_2] \\ 2(1-X) & if\ X \in (^1/_2,\ 1), \end{cases}
$$

$$
\mu_{PB}(X) = \begin{cases} 0 & if\ X \leq ^1/_2 \\ (2X-1) & if\ X \in (^1/_2,\ 1], \end{cases}
$$

where X represents e or Δe.

In addition, the grade of the membership functions μ_{C_i} and μ_{D_i} are defined as follows:

$$
\mu_{NB}(Z) = \begin{cases} 1 & if\ Z \in \left[-1,\ -^4/_5\right] \\ (-1-{}^{5Z}/_2) & if\ Z \in \left[-^4/_5,\ -^2/_5\right] \\ 0 & if\ Z >\ -^2/_5, \end{cases}
$$

$$
\mu_{N}(Z) = \begin{cases} 0 & if\ Z \leq -^4/_5 \\ ({}^{5Z}/_2+2) & if\ Z \in (-^4/_5,\ -^2/_5] \\ -{}^{5Z}/_2 & if\ Z \in (-^2/_5, 0) \\ 0 & if\ Z \geq\ 0, \end{cases}
$$

$$
\mu_{Z}(Z) = \begin{cases} 0 & if\ Z \leq -^2/_5 \\ ({}^{5Z}/_2+1) & if\ Z \in (-^2/_5,\ 0] \\ (1-{}^{5Z}/_2) & if\ Z \in (0,\ ^2/_5) \\ 0 & if\ Z \geq\ ^2/_5, \end{cases}
\tag{22}
$$

$$
\mu_{P}(Z) = \begin{cases} 0 & if\ Z \leq 0 \\ {}^{5Z}/_2 & if\ Z \in (0,\ ^2/_5] \\ (2-{}^{5Z}/_2) & if\ Z \in (^2/_5,\ ^4/_5) \\ 0 & if\ Z \geq\ ^4/_5, \end{cases}
$$

$$
\mu_{PB}(Z) = \begin{cases} 0 & if\ Z \leq ^2/_5 \\ ({}^{5Z}/_2-1) & if\ Z \in \left(^2/_5,\ ^4/_5\right] \\ 1 & if\ Z \in \left(^4/_5,\ 1\right], \end{cases}
$$

where Z represents K'_p or K'_i.

The set of fuzzy rules considered is given in Tables 1 and 2. The proposed rules are gathered in order to adjust the behaviour of the PI controller in accordance with the error $e(k)$ and the error change $\Delta e(k)$.

Table 1. Fuzzy rules for K_p parameter.

$e(k)/\Delta e(k)$	NB	N	Z	P	PB
NB	NB	NB	NB	N	Z
N	NB	N	N	N	Z
Z	NB	N	Z	P	PB
P	Z	P	P	P	PB
PB	Z	P	PB	PB	PB

Table 2. Fuzzy rules for K_i parameter.

$e(k)/\Delta e(k)$	NB	N	Z	P	PB
NB	PB	PB	PB	N	NB
N	PB	P	P	Z	NB
Z	P	P	Z	N	NB
P	Z	P	N	N	NB
PB	Z	N	NB	NB	NB

The truth value of the ith rule is obtained by the product of the truth value of the components of the antecedent clauses as:

$$\mu_i = \mu_{A_i}(e(k)) . \mu_{B_i}(\Delta e(k)). \tag{23}$$

By using the membership functions, we obtain:

$$\sum_{i=1}^{m} \mu_i = 1, \tag{24}$$

thus the defuzzification scheme is defined as:

$$\begin{cases} K'_p = \sum_{i=1}^{m} \mu_i \mu_{C_i}, \\ K'_i = \sum_{i=1}^{m} \mu_i \mu_{D_i}. \end{cases} \tag{25}$$

The decision-making output is acquired using a Max-Min fuzzy inference where the crisp outputs are obtained using the method of defuzzification and the center of gravity given as follows:

$$\begin{cases} K_p = K_{p\,min} + (K_{p\,max} - K_{p\,min})K'_p, \\ K_i = K_{i\,min} + (K_{i\,max} - K_{i\,min})K'_i. \end{cases} \tag{26}$$

By implementing the fuzzy block using the set of fuzzy rules, the fuzzy surfaces for the outputs K_p and K_i parameters are illustrated in Figures 10 and 11.

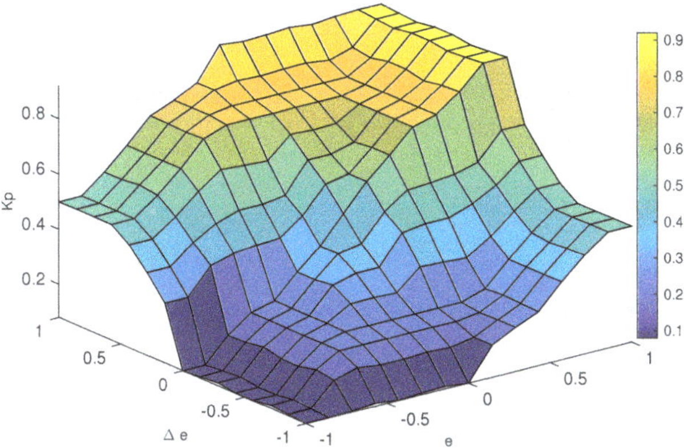

Figure 10. Fuzzy surface for K_p gain.

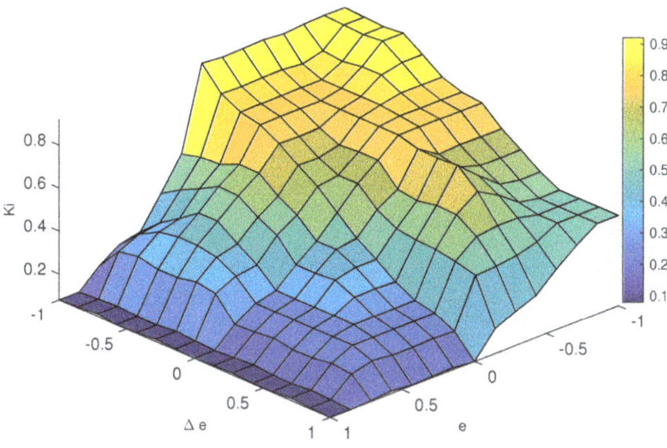

Figure 11. Fuzzy surface for K_i gain.

The inner current loops find the rotor voltage reference in $d - q$ frame. The equations that rely on the rotor voltages and currents defined in (V) and in (A) are expressed by Equation (27) as detailed in [43]:

$$\begin{cases} U_{dr} = R_r i_{dr} + \sigma L_r \frac{di_{dr}}{dt}, \\ U_{qr} = R_r i_{qr} + \sigma L_r \frac{di_{qr}}{dt}, \end{cases} \tag{27}$$

where σ is the leakage factor.

In addition, the terms of decoupling are joined to the expressions of U_{dr}^* and U_{qr}^* in order to enhance the transient response of the plant [58]. Thus, the reference voltages of the rotor are expressed by:

$$\begin{cases} U_{dr}^* = -\omega_{slip}\sigma L_r i_{qr} + (K_{Pi} e_d + K_{Ii} \int e_d \, dt), \\ U_{qr}^* = \omega_{slip}(L_m i_m + \sigma L_r i_{dr}) + (K_{Pi} e_d + K_{Ii} \int e_d \, dt), \end{cases} \tag{28}$$

where ω_{slip} is the angular frequency of slip given in (rad/s) and i_m is the stator magnetizing current supposed as constant. K_{Pi} and K_{Ii} are the parameters of the controllers.

The method of tuning of the PI controllers is the well-known Ziegler–Nichols method [59]. After that, another refinement of the tuned controller parameters is conceived by means of the robust response time algorithm [60]. The voltage references of the rotor are converted to the *abc* frame which will be affected to the RSC through the Pulse Width Modulation (PWM) block.

3.3. GSC Control

The control of the GSC is developed using the voltage oriented control strategy as shown in Figure 12. This approach admits two PI current controllers and one outer PI voltage controller [36]. The developed block diagram regulates the DC-link voltage U_{dc} and the reactive power Q_g. The use of the Phase Locked Loop (PLL) block is to save the input signal phase which denotes θ_g. The currents and voltages expressed in *dq* frame are achieved using the Park's transformation.

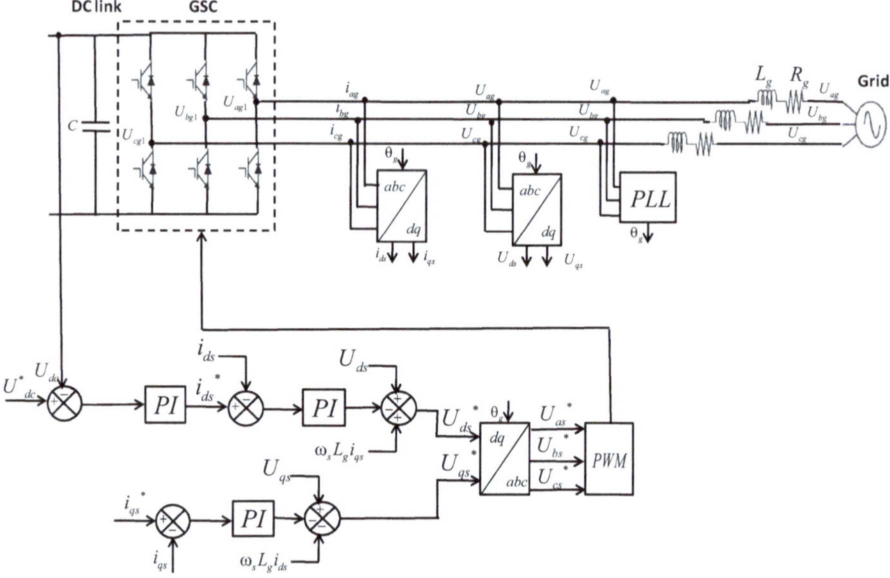

Figure 12. Block control scheme for the GSC component.

The voltages of the grid are expressed using a *dq* frame as:

$$\begin{cases} U_{gd} = i_{ds}R_g + L_g\frac{di_{ds}}{dt} - \omega_s L_g i_{qs} + U_{gd1}, \\ U_{gq} = i_{qs}R_g + L_g\frac{di_{qs}}{dt} - \omega_s L_g i_{ds} + U_{gq1}, \end{cases} \tag{29}$$

where R_g and L_g are the coupling resistance and inductance of the grid, and U_{gd1} and U_{gq1} are the terminal voltages of the converter in the *dq* frame.

The active and reactive powers are regulated through the *dq* axis currents. The current loops are similar and provide the voltage references of the grid U_{ds}^* and U_{qs}^* as defined by Equation (30). Thus, as to improve the transient response of the plant, the terms of compensator and feedforward voltages are joined to the command signals:

$$\begin{cases} U_{gd}^* = U_{gd} + \Omega_g L_g i_q - (K_{Pi}e_d + K_{Ii}\int e_d\,dt), \\ U_{gq}^* = U_{gq} - \Omega_g L_g i_d - (K_{Pi}e_q + K_{Ii}\int e_q\,dt). \end{cases} \tag{30}$$

The voltage control loop is intended to regulate the voltage of the DC-link so as to keep it constant around its reference. The current control loops adjust the currents i_{ds} and i_{qs} in *dq* frame. The current i_{qs} aims to control the reactive power and the current reference in the q-axis is assumed zero. Likewise, the RSC control and the PI controller gains are determined using the empirical Ziegler–Nichols method [61].

After that, the voltage references are converted to the *abc* stationary frame, which will acquire the PWM signals for the GSC.

4. Validation Tests and Discussion

In this part, based on realistic tidal sites, two study cases are given to test the robustness of the developed control schemes. The demonstrative studies are set to enhance the harnessed power under irregular tidal current speed input. In addition, the fuzzy gain scheduling supervisor was analyzed to favor disturbance rejection. The simulation results have been executed using the TSG system parameters given in Table 3.

Table 3. TSG system parameters.

Turbine	Drive-train	DFIG	Converter
$\rho = 1027 \ \text{kg/m}^3$	$H_t = 3 \ \text{s}$	$P_n = 1.5 \ \text{MW}$	$V_{dc} = 1150 \ \text{V}$
$R = 8 \ \text{m}$	$H_g = 0.5 \ \text{s}$	$U_{rms} = 690 \ \text{V}$	$C = 0.01 \ \text{F}$
$C_{p\,\max} = 0.44$	$K_{sh} = 2 \times 10^6 \ \text{Nm/rad}$	$freq = 50 \ \text{Hz}$	
$\lambda_{opt} = 6.96$	$D_{sh} = 3.5 \times 10^5 \ \text{Nms/rad}$	$R_s = 2.63 \ \text{m}\Omega$	
$V_n = 3.2 \ \text{m/s}$		$R_r = 2.63 \ \text{m}\Omega$	Choke
		$L_s = 0.168 \ \text{mH}$	$R_g = 0.595 \ \text{m}\Omega$
		$L_r = 0.133 \ \text{mH}$	$L_g = 0.157 \ \text{mH}$
		$L_m = 5.474 \ \text{mH}$	
		$p = 2$	

4.1. Control Robustness against Irregular Tidal Speed with Numerical Input

So as to examine the robustness of the developed control approaches, a first study case based on the characteristics of a sea state in the winter of the western coast of Europe is investigated. The data of the velocities are based on a mathematical model of the swell disturbance [62]. The swell model is calculated using the first-order Stokes model [52,63]. The average height of the highest one-third waves is 3 m and the average period of these one-third waves is 13.2 s. The sea depth is 30 m. The average tidal current $V_{avr} = 2 \ \text{m/s}$ is chosen in a small time scale. Figure 13 illustrates the tidal velocity profile with a lower speed of 0.6 m/s and an upper speed of 3.1 m/s.

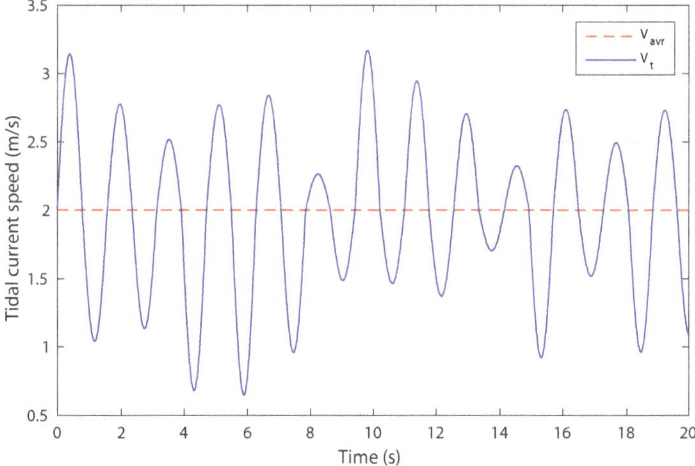

Figure 13. Study case 1: Tidal current speed input.

Figure 14 shows the power coefficient response. In this experiment, it is obvious that the system has good behavior, thus the power coefficient is maintained around its optimum value $C_p = 0.4373$.

The 5% settling time is reached at 0.01 s; it is obvious that the controlled system is able to track fast the desired value in the steady-state regime.

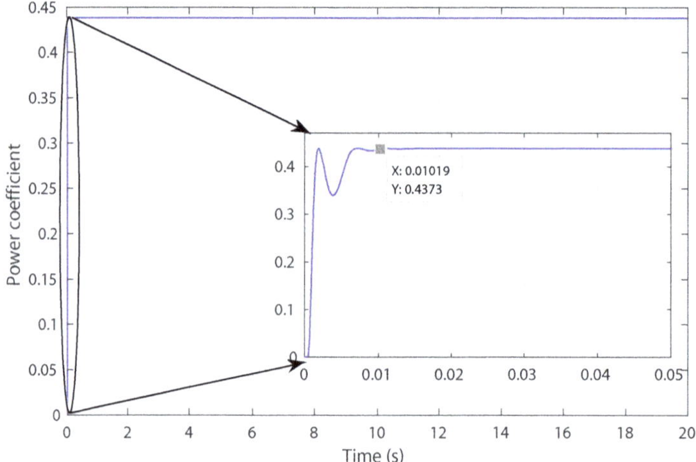

Figure 14. Study case 1: Power coefficient response.

Figure 15 depicts the response of the rotor speed and the adequate signal acquired from the implemented ANN-based MPPT approach. The controller displays a good tracking performance of the adequate rotor speed. This indicates that the FGS-PI controller has a decreased steady-state error because the integral action is adequately changing in accordance to the changes of the input.

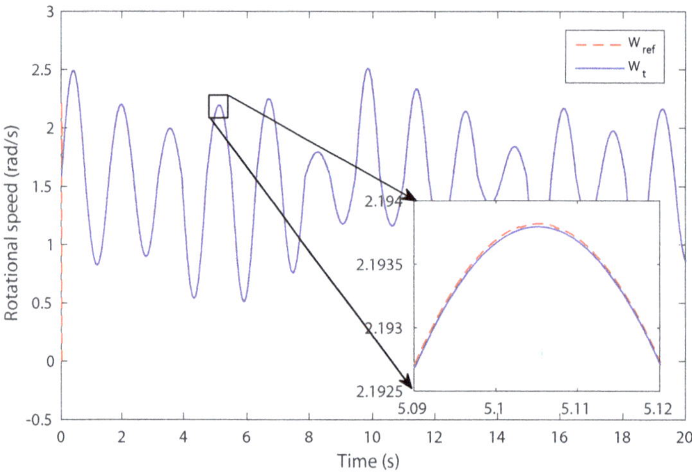

Figure 15. Study case 1: Rotor speed curve and its reference.

The output torque and power variations are shown in Figures 16 and 17. An uncontrolled study case is set to compare the output power. The resulting torque and power change according to the variation of the tidal velocity. The average values of the extracted power are 425 kW and 549 kW corresponding to the uncontrolled case and the hybrid neural fuzzy control, respectively. It can be seen that using the developed FGS control provides a good speed tracking performance, which leads to a power generation improvement with a 29.18%. Therefore, the TSG system is capable of augmenting the recuperated output power in case of the swell effect.

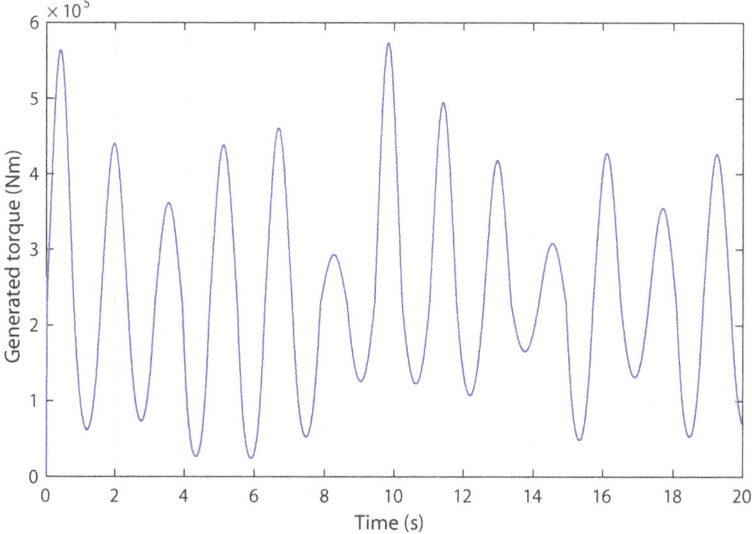

Figure 16. Study case 1: Response of the hydrodynamic torque.

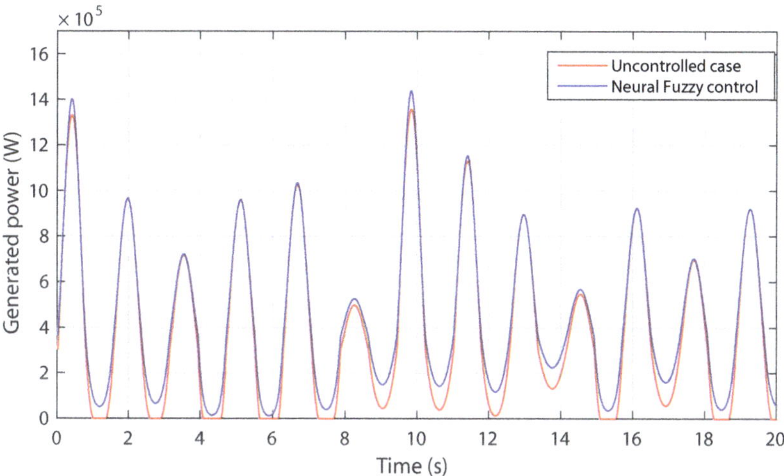

Figure 17. Study case 1: Generated turbine's power curve.

4.2. Control Robustness against Irregular Tidal Speed with Real Measured Input

A second realistic tidal site is considered to test the developed control strategies. The station under study is named Middle Ground Shoal which is located in Cook Inlet, USA. According to the National Oceanic and Atmospheric Administration (NOAA), this tidal site can reach important tidal velocities up to 2.3 m/s as mentioned in the Cook Inlet 2012 Current Survey as described in [64]. The used data from this tidal site is recorded from 25 June 2012 at 00:00:00 to 26 June 2012 at 23:59:00.

These data are extracted using an Acoustic Doppler Current Profiler [65] with an approximated water depth of 31.15 m within an interval of 6 min. Figure 18 illustrates the real instantaneous measured tidal currents over two days. Each semidiurnal tide related to spring and neap tides is corresponding to an approximately period of 7 h. It is obviously clear that the oscillation in the current speed profile is important with respect to the average tidal speed which is 1.038 m/s.

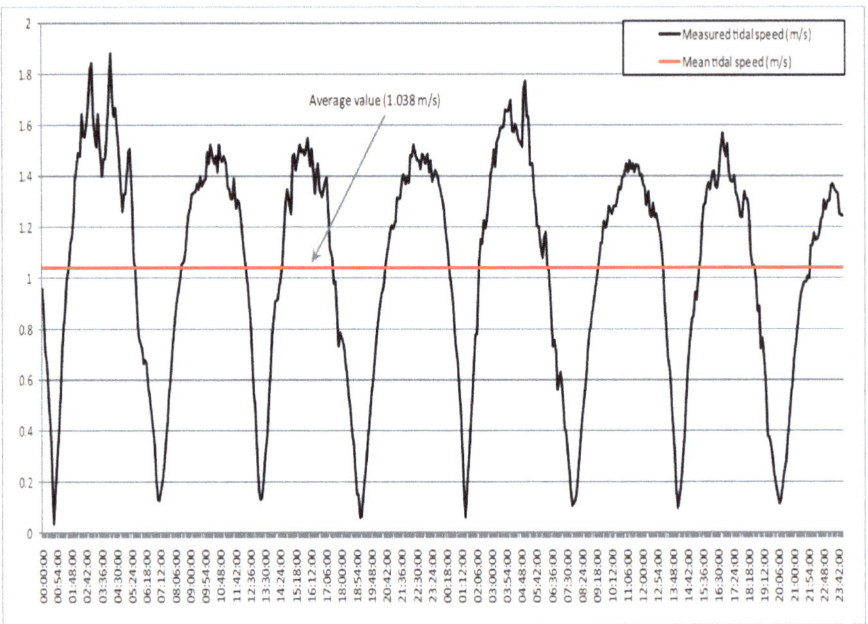

Figure 18. Study case 2: Measured tidal current speed in Cook Inlet from 25 June 2012 at 00:00:00 to 26 June 2012 at 23:59:00.

In the comparative study between the uncontrolled case and the proposed control approaches, the realistic tidal current velocity over the period 00 h:00 min to 07 h:00 min of 25 June 2012 is chosen as the input to the implemented model. The response of the power coefficient is depicted in Figure 19. The power coefficient is maintained constant and in the steady state regime reaches a value of 0.4382 at 0.01 s.

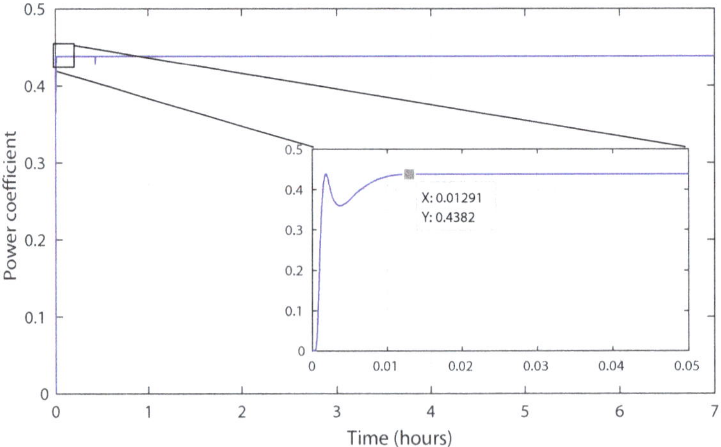

Figure 19. Study case 2: Power coefficient response.

Figure 20 shows the rotational speed curve changing according to the tidal current speed. This result proves that the controller successfully manages to follow the optimal reference provided by the MPPT strategy.

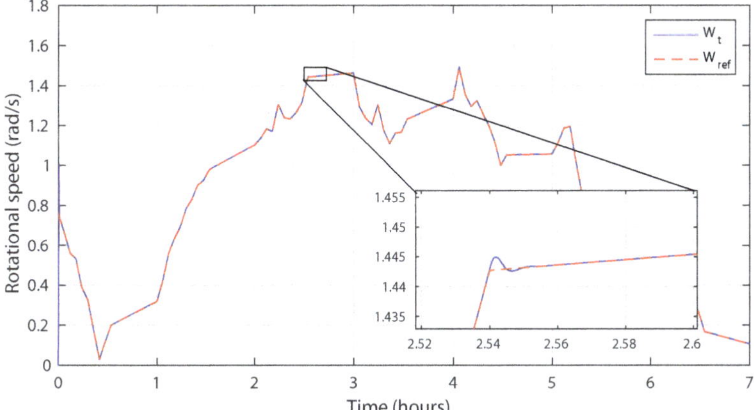

Figure 20. Study case 2: Rotational speed response and its reference.

The response of the hydrodynamic torque is depicted in Figure 21. The torque increases according to the tidal current speed input variation. Figure 22 shows the generated power in the uncontrolled case and the controlled case using the neural fuzzy techniques. The output generated power is improved in terms of average values with 22.4%, which leads to maximizing the energy harnessed by the tidal velocity even in the case of high disturbance.

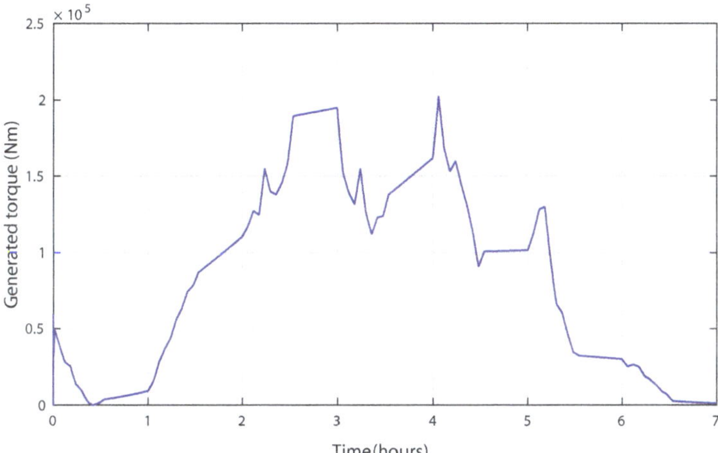

Figure 21. Study case 2: Response of the hydrodynamic torque.

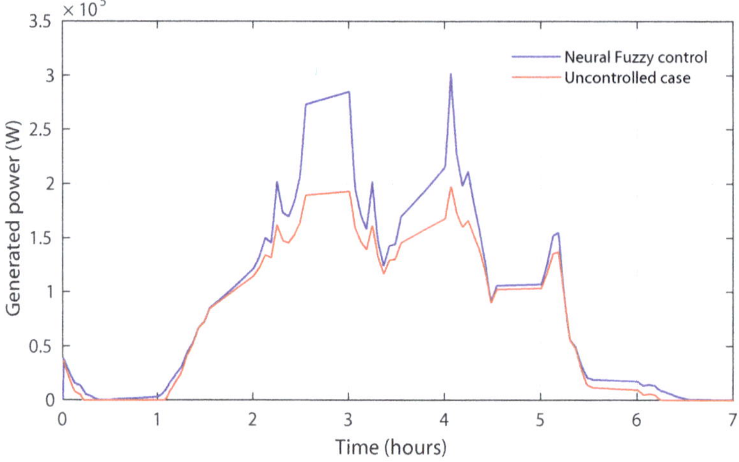

Figure 22. Study case 2: Generated turbine power curve.

4.3. Disturbance Rejection

To test the robustness of the proposed control, a disturbance rejection experiment was carried out. The disturbance is injected into the measured signal, which is the rotational speed. The tidal speed input versus time considered starts from 1.5 m/s and steps to 3.2 m/s at $t = 5$ s as shown in Figure 23.

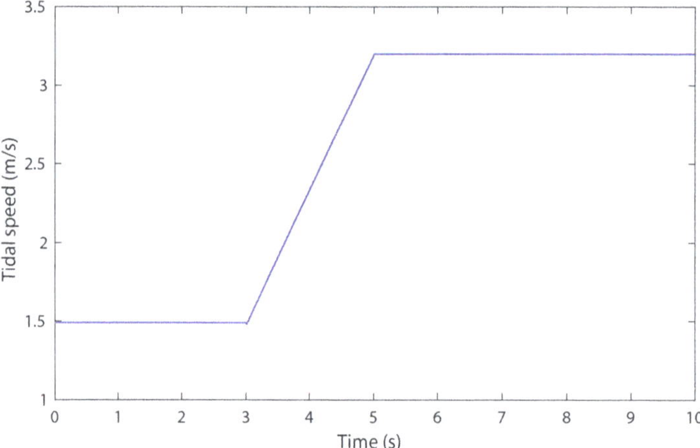

Figure 23. Step tidal speed input.

The measured rotational speed increases according to the tidal current speed from 1.18 rad/s to 2.53 rad/s as illustrated in Figure 24. The curve shows an effective tracking performance of the rotational speed.

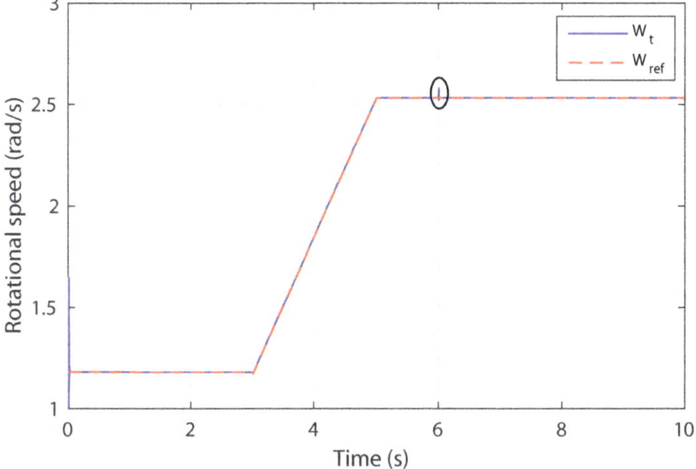

Figure 24. Response of the rotational speed.

The occurrence of the disturbance is injected at $t = 6s$ with a 10% of the average rotational speed as depicted in Figure 25a. In fact, when zooming in on the curve, it can be clearly seen that the controller is able to reject the disturbance in $0.015\,s$ and allows the system to be stable in the steady-state operation as depicted in Figure 25b. The resulting power coefficient adequately adapts to the disturbance condition and it is noted that the response is closer to the optimum value with 0.4379.

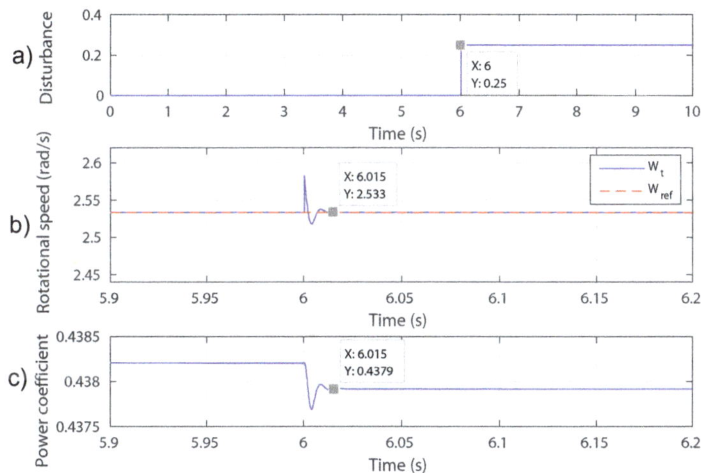

Figure 25. Disturbance rejection at $t = 6\,s$. (**a**) disturbance occurrence; (**b**) zoom-in rotational speed versus time; (**c**) zoom-in power coefficient response.

The generated output power is depicted in Figure 26. The TSG system is is capable of enhancing the output power of 1.48 MW according to a tidal speed of 3.2 m/s. The implemented control enhances the generated power with a decreased error of approximately 2% compared to the tolerable supported power by the system. It can be noted that the developed control approaches lead to a power improvement against disturbances.

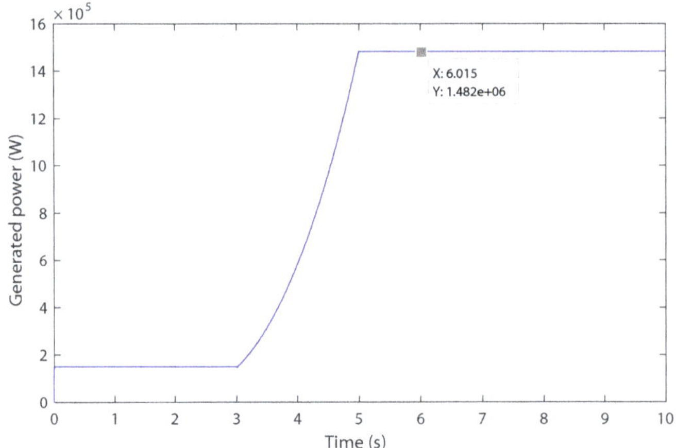

Figure 26. Generated power curve.

5. Conclusions

In this article, a tidal stream generator system has been designed and controlled. A hybrid neural fuzzy design has been developed to deal with the power disturbances due to the swell effect.

The hybrid design consists of an ANN-based MPPT approach which adequately generates the reference rotational speed in order to drive coupled with a fuzzy gain schedule that drives the system in the maximum power. The ANN design adaptively changes its weights to provide the suitable trajectory for each marine velocity. The block design is the fuzzy gain scheduling which controls the rotational speed control loop. The fuzzy controller adaptively changes its gains using the designed fuzzy supervisors.

To test the effectiveness of the novel hybrid FGS PI-controller, two realistic tidal sites were investigated. The first scenario is proposed with a variable spring and neap marine velocity provoking swell effect disturbances on the Western coast of Europe. The results found prove that approaches successfully deal with these perturbations that enable the TSG plant to harness the maximum output power. A second scenario based on the realistic data from the Cook Inlet, USA was considered. Comparing with the uncontrolled case, the hybrid neural fuzzy controller shows the power generation improvement offered by the developed control schemes.

Another case of study was considered to assess the robustness of the implemented control strategies under disturbances with an excellent reference tracking. The proposed hybrid FGS-PI control that has been enhanced with an ANN provides very good output power performance improvement from the tidal stream generator system.

This study consists of enhancing the operation of the tidal stream generator system for several study cases by varying the input profile. In effect, the plant is controlled in a way to maximize the harnessed power. The work proves that regulating the tidal turbine in variable speed functioning lead to a high energy yield by operating with a maximum power coefficient.

Author Contributions: Conceptualization, K.G., I.G., S.B., J.H. and A.J.G.; Formal Analysis, K.G.; Investigation, K.G.; Methodology, K.G.; Software, K.G.; Supervision, I.G. , S.B., J.H. and A.J.G.; Validation, I.G. and A.J.G.; Writing, Review and Editing, K.G.

Funding: This research received no external funding.

Acknowledgments: This work was supported by the MINECO through the Research Project DPI2015-70075-R (MINECO/FEDER, UE). The authors would like to thank the collaboration of the Basque Energy Agency (EVE) through Agreement UPV/EHUEVE23/6/2011, the Spanish National Fusion Laboratory (EURATOM-CIEMAT) through Agreement UPV/EHUCIEMAT08/190 and EUSKAMPUS—Campus of International Excellence.

Conflicts of Interest: The authors declare no conflict of interest.

Abbreviations

The following abbreviations are used in this manuscript:

ANN   Artificial Neural Network
DFIG   Doubly Fed Induction Generator
GSC   Grid Side Converter
IEA   International Energy Agency
FGS   Fuzzy Gain Scheduling
IEO   International Energy Outlook
MPPT   Maximum Power Point Tracking
MSE   Mean Square Error
NOAA   National Oceanic and Atmospheric Administration
ORC   Optimal Regime Characteristic
PI   Proportional Integral
PLL   Phase Locked Loop
PTO   Power Take Off
PWM   Pulse Width Modulation
RSC   Rotor Side Converter
TSG   Tidal Stream Generator
TST   Tidal Stream Turbine
NB   Negative Big
N   Negative
Z   Zero
P   Positive
PB   Positive Big

Notations

$P_{t,g,n}$ Turbine, generator and nominal powers (W).
C_p, C_{pmax} Power coefficient and its maximum.
λ, λ_{opt} Optimal speed ratio and its optimal value.
β, ρ, R Blade pitch angle (deg), fluid density (kg/m^3) and blade radius (m).
V, V_n Tidal current speed and its nominal value (m/s).
$\omega_{t,g}, \omega_{s,r}$ Rotational speed of turbine and generator, pulsations of the stator and rotor (rad/s).
ω_{ref} Reference rotational speed (rad/s).
T_{tst}, T_t, T_{em} Turbine, rotor shaft and electromagnetic torques (Nm).
$H_{t,g}, T_s$ Turbine and generator inertia constants, sampling time (s).
$D_{sh}, K_{sh}, p, \sigma$ Stiffness coefficient (Nm/rad), damping coefficient (Nms/rad), leakage factor.
ω_{slip}, p Angular frequency of slip (rad/s), pole pair numbers.
$U_{sd,sq}, U_{rd,rq}$ Stator and rotor voltages in $d - q$ frame (V).
$I_{sd,sq}, I_{rd,rq}, i_m$ Stator and rotor currents in $d - q$ frame, stator magnetizing current (A).
$\varphi_{sd,sq}, \varphi_{rd,rq}$ Stator and rotor flux in $d - q$ frame (Wb).
$L_{s,r}, L_m, L_g$ Stator and rotor inductances, magnetizing inductance, grid coupling inductance (H).
$R_{s,r}, R_g$ Stator and rotor resistances, grid coupling resistance (Ω).
$U_{gd,gq}, U_{gd1,gq1}$ Grid voltages and terminal voltages of the converter in $d - q$ frame (V).
$I_{gd,gq}, I_{dc}$ Grid currents in $d - q$ and DC-link current (A).
U_{dc}, c DC-link voltage (V), DC-link capacitor (F).
z_i, x_i, T_i Output neurons, input neurons, threshold terms of the hidden layer.
$\omega_{i,j}, h_i$ synaptic weights, number of neurons in the hidden layer.
u_k Fuzzy control law.
$e(k), \Delta e(k)$ The error and the error change.
K_p, K_i Fuzzy PI gains.
K_p', K_i' Normalized fuzzy PI gains.
$\mu_{Ai,Bi,Ci,Di}$ Grades of the membership functions.

References

1. Segura, E.; Morales, R.; Somolinos, J.A.; Lopez, A. Techno-economic challenges of tidal energy conversion systems: Current status and trends. *Renew. Sustain. Energy Rev.* **2017**, *77*, 536–550. [CrossRef]
2. World Energy Council. *World Energy Resource Marine Energy 2016*; Technical Report; World Energy Council: London, UK, 2016; pp. 1–76.
3. Zhang, Y.L.; Lin, Z.; Liu, Q.L. Marine renewable energy in China: Current status and perspectives. *Water Sci. Eng.* **2014**, *7*, 288-305.
4. Grabbe, M.; Lalander, E.; Lundin, S.; Leijon, M. A review of the tidal current energy resource in Norway. *Renew. Sustain. Energy Rev.* **2009**, *13*, 1898–1909. [CrossRef]
5. Kadiri, M.; Ahmadian, R.; Bockelmann-Evans, B.; Rauen, W.; Falconer, R. A review of the potential water quality impacts of tidal renewable energy systems. *Renew. Sustain. Energy Rev.* **2012**, *16*, 329–341. [CrossRef]
6. Stern, N.; Calderon, F. *Better Growth, Better Climate: The New Climate Economy Report*; The Global Commission on the Economy and Climate: New York, NY, USA, 2014. Available online: http://newclimateeconomy. report/ (accessed on 12 October 2018).
7. EIA, U. *International Energy Outlook 2016 with Projections to 2040*; Energy Department, Energy Information Administration, Office of Energy Analysis: Washington, DC, USA, 2016.
8. International Energy Agency OECD/IEA. *World Energy Outlook 2013, Chapter 6: Renewable Energy Outlook*; International Energy Agency OECD/IEA: Paris, France, 2013.
9. IRENA. *REmap 2030: A Renewable Energy Roadmap*; IRENA: Dhabi, United Arab Emirates, 2014. Available online: www.irena.org/remap (accessed on 12 October 2018).
10. Uihlein, A.; Magagna, D. Wave and tidal current energy—A review of the current state of research beyond technology. *Renew. Sustain. Energy Rev.* **2016**, *58*, 1070–1081. [CrossRef]
11. Borthwick, A.G. Marine renewable energy seascape. *Engineering* **2016**, *2*, 69–78. [CrossRef]
12. Garrido, A.J.; Garrido, I.; Otaola, E.; Lekube, J.; MZoughi, F.; Ghefiri, K.; Mundackamattam, D.G.; Oleagordia, I. Capture chamber modelling and validation in OWC on-shore devices. In Proceedings of the Region 10 Conference (TENCON), Singapore, 22–25 November 2016; pp. 1682–1685.
13. El Tawil, T.; Charpentier, J.F.; Benbouzid, M. Tidal energy site characterization for marine turbine optimal installation: Case of the Ouessant Island in France. *Int. J. Mar. Energy* **2017**, *18*, 57–64. [CrossRef]
14. Bryden, I.G.; Couch, S.J. ME1-marine energy extraction: Tidal resource analysis. *Renew. Energy* **2006**, *31*, 133–139. [CrossRef]
15. Myers, L.; Bahaj, A.S. Simulated electrical power potential harnessed by marine current turbine arrays in the Alderney Race. *Renew. Energy* **2005**, *30*, 1713–1731. [CrossRef]
16. Winter, A.I. Differences in fundamental design drivers for wind and tidal turbines. In Proceedings of the 2011 IEEE-Spain OCEANS, Santander, Spain, 6–9 June 2011; pp. 1–10.
17. Whitby, B.; Ugalde-Loo, C.E. Performance of pitch and stall regulated tidal stream turbines. *IEEE Trans. Sustain. Energy* **2014**, *5*, 64–72. [CrossRef]
18. Hammons, T.J. Tidal power. *Proc. IEEE* **1993**, *81*, 419–433. [CrossRef]
19. Choi, J.S.; Jeong, R.G.; Shin, J.H.; Kim, C.K.; Kim, Y.S. New Control Method of Maximum Power Point Tracking for Tidal Energy Generation System. In Proceedings of the International Conference on Electrical Machines and Systems, Seoul, Korea, 8–11 October 2007; pp. 165-168.
20. Rahman, M.L.; Oka, S.; Shirai, Y. Hybrid power generation system using offshore-wind turbine and tidal turbine for power fluctuation compensation (HOT-PC). *IEEE Trans. Sustain. Energy* **2010**, *1*, 92–98. [CrossRef]
21. Xiang, D.; Ran, L.; Tavner, P.J.; Yang, S. Control of a doubly fed induction generator in a wind turbine during grid fault ride-through. *IEEE Trans. Energy Convers.* **2006**, *21*, 652–662. [CrossRef]
22. Sousounis, M.C.; Shek, J.K.H.; Mueller, M.A. Modelling and control of tidal energy conversion systems with long distance converters. In Proceedings of the 7th IET International Conference on Power Electronics, Machines and Drives (PEMD 2014), Manchester, UK, 8–10 April 2014; pp. 1–6.
23. Munteanu, I.; Bratcu, A.L.; Cutululis, N.A.; Ceanga, E. *Optimal Control of Wind Energy Systems: Towards a Global Approach*; Springer: Berlin, Germany, 2008.
24. Marzband, M.; Azarinejadian, F.; Savaghebi, M.; Pouresmaeil, E.; Guerrero, J.M.; Lightbody, G. Smart transactive energy framework in grid-connected multiple home microgrids under independent and coalition operations. *Renew. Energy* **2018**, *126*, 95–106. [CrossRef]

25. Tavakoli, M.; Shokridehaki, F.; Marzband, M.; Godina, R.; Pouresmaeil, E. A Two Stage Hierarchical Control Approach for the Optimal Energy Management in Commercial Building Microgrids Based on Local Wind Power and PEVs. *Sustain. Cities Soc.* **2018**, *41*, 332–340. [CrossRef]
26. Utkin, V.I. *Sliding Modes in Control and Optimization*; Springer: Berlin, Germany, 1992.
27. Elghali, S.E.B.; Benbouzid, M.E.H.; Charpentier, J.F.; Ahmed-Ali, T.; Munteanu, I. Experimental Validation of a Marine Current Turbine Simulator: Application to a Permanent Magnet Synchronous Generator-Based System Second-Order Sliding Mode Control. *IEEE Trans. Ind. Electron.* **2011**, *58*, 118–126.
28. Feng, Y.; Han, F.; Yu, X. Chattering free full-order sliding-mode control. *Automatica* **2014**, *50*, 1310–1314. [CrossRef]
29. Kalogirou, S.A. Artificial neural networks in renewable energy systems applications: A review. *Renew. Sustain. Energy Rev.* **2001**, *5*, 373–401. [CrossRef]
30. Morgan, N.; Bourlard, H.A. Neural networks for statistical recognition of continuous speech. *Proc. IEEE* **1995**, *83*, 742–772. [CrossRef]
31. Bilgili, M.; Sahin, B.; Yasar, A. Application of artificial neural networks for the wind speed prediction of target station using reference stations data. *Renew. Energy* **2007**, *32*, 2350–2360. [CrossRef]
32. Castro, A.; Carballo, R.; Iglesias, G.; Rabunal, J.R. Performance of artificial neural networks in nearshore wave power prediction. *Appl. Soft Comput.* **2014**, *23*, 194–201. [CrossRef]
33. Dounis, A.I.; Kofinas, P.; Alafodimos, C.; Tseles, D. Adaptive fuzzy gain scheduling PID controller for maximum power point tracking of photovoltaic system. *Renew. Energy* **2013**, *60*, 202–214. [CrossRef]
34. Chaiyatham, T.; Ngamroo, I. Optimal fuzzy gain scheduling of PID controller of superconducting magnetic energy storage for power system stabilization. *Int. J. Innov. Comput. Inf. Control* **2013**, *9*, 651–666.
35. Bahaj, A.S.; Molland, A.F.; Chaplin, J.R.; Batten, W.M.J. Power and thrust measurements of marine current turbines under various hydrodynamic flow conditions in a cavitation tunnel and a towing tank. *Renew. Energy* **2007**, *32*, 407–426. [CrossRef]
36. Ghefiri, K.; Bouallègue, S.; Garrido, I.; Garrido, A.J.; Haggège, J. Complementary Power Control for Doubly Fed Induction Generator-Based Tidal Stream Turbine Generation Plants. *Energies* **2017**, *10*, 862. [CrossRef]
37. Ghefiri, K.; Bouallègue, S.; Haggège, J. Modeling and SIL simulation of a Tidal Stream device for marine energy conversion. In Proceedings of the 2015 6th International Renewable Energy Congress (IREC), Sousse, Tunisia, 24–26 March 2015; pp. 1–6.
38. Muljadi, E.; Gevorgian, V.; Wright, A.; Donegan, J.; Marnagh, C.; McEntee, J. *Turbine Control of a Tidal and River Power Generator: Preprint (No. NREL/CP-5D00-66867)*; National Renewable Energy Lab.(NREL): Golden, CO, USA, 2016.
39. Fernandez, L.M.; Jurado, F.; Saenz, J.R. Aggregated dynamic model for wind farms with doubly fed induction generator wind turbines. *Renew. Energy* **2008**, *33*, 129–140. [CrossRef]
40. Benelghali, S.; Benbouzid, M.E.H.; Charpentier, J.F. Generator systems for marine current turbine applications: A comparative study. *IEEE J. Ocean. Eng.* **2012**, *37*, 554–563. [CrossRef]
41. Amundarain, M.; Alberdi, M.; Garrido, A.J.; Garrido, I. Modeling and simulation of wave energy generation plants: Output power control. *IEEE Trans. Ind. Electron.* **2011**, *58*, 105–117. [CrossRef]
42. Fan, L.; Kavasseri, R.; Miao, Z.L.; Zhu, C. Modeling of DFIG-based wind farms for SSR analysis. *IEEE Trans. Power Deliv.* **2010**, *25*, 2073–2082. [CrossRef]
43. Pena, R.; Clare, J.C.; Asher, G.M. Doubly fed induction generator using back-to-back PWM converters and its application to variable-speed wind-energy generation. *IEE Proc.-Electr. Power Appl.* **1996**, *143*, 231–241. [CrossRef]
44. Zhou, D.; Blaabjerg, F.; Lau, M.; Tonnes, M. Optimized reactive power flow of DFIG power converters for better reliability performance considering grid codes. *IEEE Trans. Ind. Electron.* **2015**, *62*, 1552–1562. [CrossRef]
45. Muller, S.; Deicke, M.; De Doncker, R.W. Doubly fed induction generator systems for wind turbines. *IEEE Ind. Appl. Mag.* **2002**, *8*, 26–33. [CrossRef]
46. Alberdi, M.; Amundarain, M.; Garrido, A.J.; Garrido, I.; Casquero, O.; De la Sen, M. Complementary control of oscillating water column-based wave energy conversion plants to improve the instantaneous power output. *IEEE Trans. Energy Convers.* **2011**, *26*, 1021–1032. [CrossRef]
47. Rizzo, S.A.; Scelba, G. ANN based MPPT method for rapidly variable shading conditions. *Appl. Energy* **2015**, *145*, 124–132. [CrossRef]

48. Makarynskyy, O.; Makarynska, D.; Rusu, E.; Gavrilov, A. Filling gaps in wave records with artificial neural networks. *Marit. Transp. Exploit. Ocean Coast. Resour.* **2005**, *2*, 1085–1091.

49. Ghefiri, K.; Bouallègue, S.; Garrido, I.; Garrido, A.J.; Haggège, J. Modeling and MPPT control of a Tidal Stream Generator. In Proceedings of the 2017 4th International Conference on Control, Decision and Information Technologies (CoDIT'17), Barcelona, Spain, 5–7 April 2017; pp. 1003–1008.

50. Hagan, M.T.; Menhaj, M.B. Training feedforward networks with the Marquardt algorithm. *IEEE Trans. Neural Netw.* **1994**, *5*, 989–993. [CrossRef] [PubMed]

51. Wilamowski, B.M.; Yu, H. Improved computation for Levenberg-Marquardt training. *IEEE Trans. Neural Netw.* **2010**, *21*, 930–937. [CrossRef] [PubMed]

52. Lewis, M.J.; Neill, S.P.; Hashemi, M.R.; Reza, M. Realistic wave conditions and their influence on quantifying the tidal stream energy resource. *Appl. Energy* **2014**, *136*, 495–508. [CrossRef]

53. Zhao, Z.Y.; Tomizuka, M.; Isaka, S. Fuzzy gain scheduling of PID controllers. *IEEE Trans. Syst. Man Cybern.* **1993**, *23*, 1392–1398. [CrossRef]

54. Tursini, M.; Parasiliti, F.; Zhang, D. Real-time gain tuning of PI controllers for high-performance PMSM drives. *IEEE Trans. Ind. Appl.* **2002**, *38*, 1018–1026. [CrossRef]

55. Bouallègue, S.; Haggège, J.; Ayadi, M.; Benrejeb, M. PID-type fuzzy logic controller tuning based on particle swarm optimization. *Eng. Appl. Artif. Intell.* **2012**, *25*, 484–493. [CrossRef]

56. Chen, Y.Y.; Perng, C.F. Input scaling factors in fuzzy control systems. In Proceedings of the 1994 3rd International Fuzzy Systems Conference, Orlando, FL, USA, 26–29 June 1994; pp. 1666–1670.

57. Bedoud, K.; Ali-rachedi, M.; Bahi, T.; Lakel, R. Adaptive fuzzy gain scheduling of PI controller for control of the wind energy conversion systems. *Energy Procedia* **2015**, *74*, 211–225. [CrossRef]

58. Qu, L.; Qiao, W. Constant power control of DFIG wind turbines with supercapacitor energy storage. *IEEE Trans. Ind. Appl.* **2011**, *47*, 359–367. [CrossRef]

59. Astrom, K.J.; Hagglund, T. *Advanced Pid Control*; ISA-The Instrumentation, Systems, and Automation Society: Research Triangle Park, NC, USA, 2006.

60. Vilanova, R.; Visioli, A. *PID Control in the Third Millennium*; Springer: London, UK, 2012.

61. Blaabjerg, F.; Teodorescu, R.; Liserre, M.; Timbus, A.V. Overview of control and grid synchronization for distributed power generation systems. *IEEE Trans. Ind. Electron.* **2006**, *53*, 1398–1409. [CrossRef]

62. Zhou, Z.; Benbouzid, M.; Charpentier, J.F.; Scuiller, F.; Tang, T. A review of energy storage technologies for marine current energy systems. *Renew. Sustain. Energy Rev.* **2013**, *18*, 390–400. [CrossRef]

63. Alves, J.H.G. Numerical modeling of ocean swell contributions to the global wind-wave climate. *Ocean Model.* **2006**, *11*, 98–122. [CrossRef]

64. National Oceanic and Atmospheric Administration (NOAA). Available online: https://tidesandcurrents. noaa.gov/ (accessed on 30 March 2018).

65. Kostaschuk, R.; Best, J.; Villard, P.; Peakall, J.; Franklin, M. Measuring flow velocity and sediment transport with an acoustic Doppler current profiler. *Geomorphology* **2005**, *68*, 25–37. [CrossRef]

Article

Single-Phase Active Power Harmonics Filter by Op-Amp Circuits and Power Electronics Devices

Emad Samadaei [1], Mina Iranian [2], Mohammad Rezanejad [3], Radu Godina [4,*] and Edris Pouresmaeil [5,*]

[1] Department of Electronics Design (EKS), Mid Sweden University, Holmgatan 10, 85170 Sundsvall, Sweden; emad.samadaei@miun.se
[2] Department of Engineering, Atlas Danesh Co, Ghaemshahr 47658-37449, Iran; Iranian_mina@yahoo.com
[3] Engineering Faculty, University of Mazandaran, Babolsar 47416-13534, Iran; m.rezanejad@umz.ac.ir
[4] C-MAST, University of Beira Interior, 6201-001 Covilhã, Portugal
[5] Department of Electrical Engineering and Automation, Aalto University, 02150 Espoo, Finland
*   Correspondence: rd@ubi.pt (R.G.); edris.pouresmaeil@aalto.fi (E.P.)

Received: 2 November 2018; Accepted: 23 November 2018; Published: 26 November 2018

Abstract: This paper introduces a new structure for single-phase Active Power Harmonics Filter (APHF) with the simple and low-cost controller to eliminate harmonics and its side effects on low voltage grid. The proposed APHF includes an accurate harmonic detector circuit, amplifier circuit to trap tiny harmonics, switching driver circuit for precise synchronization, and inverter to create injection current waveform, which is extracted from reference signal. The control circuits are based on electrostatic devices consist of Op-Amp circuits. Fast dynamic, simplicity, low cost, and small size are the main features of Op-Amp circuits that are used in the proposed topology. The aim is removing the all grid harmonic orders in which the proposed APF injects an appropriate current into the grid in parallel way. The proposed control system is smart enough to compensate all range of current harmonics. A prototype is implemented in the power electronics laboratory and it is installed as parallel on a distorted grid by the non-linear load (15 $A_{Peak-Peak}$) to verify the compensating of harmonics. The harmonics are compensated from THD% = 24.48 to THD% = 2.86 and the non-sinusoidal waveform is renovated to sinusoidal waveform by the proposed APHF. The experimental results show a good accurate and high-quality performance.

Keywords: active power harmonics filter; electrostatic devices; hysteresis switching; op-amp; power electronics

---

1. Introduction

The growth of applying for semiconductor devices and nonlinear loads in industrial, residential, and commercial areas has led to the destruction of power grid voltage and current waveforms in which they cause harmonic distortion in the electrical system [1,2]. Harmonics in the electricity network make harmful damages, such as power losses, the overload in transmission lines, the reduction of the power quality, lower efficiency in equipment, and disturbance in the performance of devices [3–6].

Therefore, the detection of harmonics and finding a strategy are essential to eliminate and reduce them down to standard allowed. During many years, passive filters have been the conventional solution to minimize harmonics pollution [7–11]. There are many power factor correctors (PFC) converters with the ability to reduce harmonics as well. Family of single-phase and hybrid PFC buck-boost converters are introduced in [12,13]. In [14], three-level unidirectional single-phase PFC rectifier topologies are presented. Some other topologies discussed the range of output to develop PFC converters [15].

The PFC converter usually has low power factor (PF) and poor harmonic performance due to the inherent dead angle of the input current, especially at low input/output ranges [13]. Active power harmonic filters (APHFs) have been proposed as a power electronics solution, since passive filters have quite a few disadvantages [16–19]. The smart control ability of the active harmonic filters is a very prominent advantage [20]. As a best harmonic detective device, it can be installed at various scales along with harmonic loads to prevent the spread of harmonics into the grid, so that the network remains in its sinusoidal waveform [21–24]. The vast applications of this device are effective to increase network power quality [25–30]. The active filters produce the same amount (but opposite) of harmonic by monitoring the harmonics of electrical load current that forbid the current harmonics to flow through the power line [31,32].

The active power harmonic filter as compensator is divided into two parts: (1) power circuit; and, (2) control circuit (Harmonic Detector Algorithm for control of switching). The defect in each section and unsuitable connection not only lead to compensative performance but also increase the harmonic components that reduce the power quality [33]. The accurate algorithm of the harmonic detector and switching pattern can lead to a reduction in the cost of the power part structure. Hence, there are many articles in the control, harmonic detection, and switching pattern. Most articles discuss the control circuits based on programming, especially the transfer function and DQ-axis (direct-quadrature) transformations [34–38]. Search algorithms have been used as well [39,40]. The authors in [41] investigate the prediction on the harmonic load for control of the power quality. Although microprocessors will reduce the complexity of control circuits, but the decrease in quality of sampling signals due to the analog/digital converting, some inefficient coding algorithm, slower response to non-complex calculations, and expensive cost than electrostatic circuits should be considered [42–44]. On the other hand, fast response in electrostatic circuits for non-complex calculations and low-cost than microprocessors can increase the performance of the control circuit. In addition, the removing of microprocessors makes a simpler circuit in which the decreasing in the total cost of the implementation would be expected [45–48]. Another important challenge in APHF is switching and synchronizing with the current grid. More damages in the network are expected without the accurate synchronization of the reference signal. Fast switching and accurate synchronization are some features of the hysteresis switching technique that are used in the high-speed electrostatic circuit [49,50].

In this paper, the control strategy of active harmonics filter is presented by electrostatic devices and Op-Amp circuits that cause the removal of microprocessors and programming devices. The removing of microprocessors declines the complication of programming and analog/digital converting and its quality distortion in sampled signals for APHF. The voltage sensor (sampling) is also removed in the proposed control strategy due to hysteresis switching with a precise synchronization. Thus, the cost of implementation can be reduced, although the performance quality is increased by the fast dynamic response and accurate harmonics elimination. Proposed topology can be used by residential, commercial, and small industries electricity consumers. The operation of the proposed controller follows as: the load current is being sensed by a current sensor that is infected by harmonics. Then, harmonics are extracted by proposed Op-Amp circuits with a fast dynamic response. The extracted signals are boosted in amplifier circuit since the tiny high-order harmonics can be considered in switching pattern. It significantly increases the quality of the APHF compensation. Section 2 illustrates these issues. Hysteresis and synchronization are described in Section 3. Experimental results are shown in Section 4 to verify the high performance of different parts of the proposed controller circuit and the compensation of APHF.

2. The Strategy of Harmonics Detection

Some properties in control circuit should be considered, such as the smart extraction of harmonics, the pitch adjustment in magnitude and phase for the reference signal, the remaining quality in fundamental harmonic order (1st) after APFH operation, and simple structure. The perfect performance of each these properties lead to proper overall results in APHF.

Figure 1 presents a simple and high quality schematic for a high-pass filter with the exact pitch adjustment to extract grid harmonics. The sampled signal (from the grid by the current sensor) has been sent to the proposed circuit through V_{in} and the circuit acts as a high-pass filter that is based on the values of the capacitor and resistors. Consequently, it separates all of the harmonics that are higher than the cutoff frequency and the circuit reveals them in output (V_{out}). The output signal will be used as a reference signal for switching part.

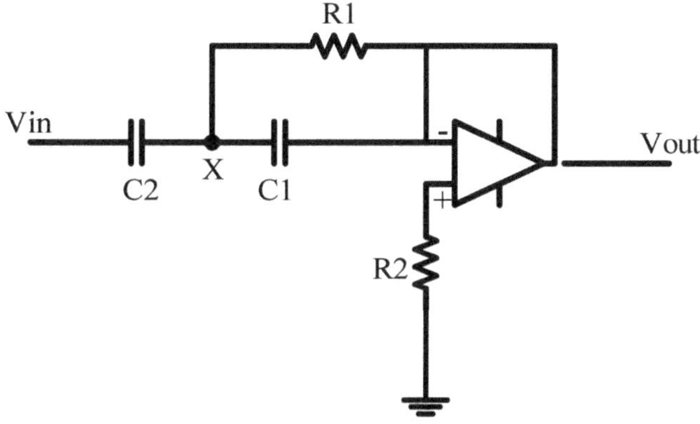

Figure 1. Proposed high pass filter with op-amp.

This proposed circuit involves op-amp devices. Other electronic devices (resistors, capacitors, diode, etc.) with different arrangements can be joint to op-amp, in order to use in various operation and applications. This high pass filter is designed based on op-amp devices and cutoff frequency is set to separate frequencies higher than fundamental component. This circuit has a very accurate operation in the category of electrostatic filters in which it reveals all the harmonics higher than the cutoff frequency with high quality. Also, another prominent property is the adjustment of the phase between input and output signal by correct design. This feature is used to synchronize the reference signal with the network current.

Equations are extracted to drive the transfer function of the proposed filter for Figure 1, follows as:
The KCL in node X can be written as:

$$I_{in} = I_{R1} + I_{C1} \tag{1}$$

That I_{in}, I_{R1}, and I_{C1} are the input current, the current passing through the resistance R_1 and the current passing through the capacitor C_1, respectively.
The currents in the op-amp legs are zero, thus:

$$I_{R2} = I_{C1} \tag{2}$$

When considering the voltage of point X (V_x), the above equation can be rewritten:

$$V_x = \frac{1 + R_2 C_1 s}{R_2 C_1 s} V_{out} \tag{3}$$

It is also possible to write the current of branches according to the voltage of point X:

$$I_{in} = \frac{V_{in} - V_x}{\frac{1}{sC_2}} \tag{4}$$

$$I_{C1} = \frac{V_{Out}}{R_2} \tag{5}$$

$$I_{R1} = \frac{V_x - V_{Out}}{R1} \tag{6}$$

By putting above equations in (1) and solving of the equations in the Laplace domain, the transfer function of the circuit can be calculated, as follows:

$$\frac{V_{Out}}{V_{in}} = \frac{s^2}{s^2 + \frac{C_1 + C_2}{R_2 C_1 C_2} + \frac{1}{R_1 R_2 C_1 C_2}} \tag{7}$$

According to the (7), the transfer function of the circuit is second-order that increases the slop of cutoff frequency and the quality of the output signal as well.

In some harmonic orders, the magnitude of the detected harmonic is low (especially higher-order harmonics), so that it cannot trigger the switching system to remove harmonics by power part. Then, it is necessary to amplify the harmonic orders to increase the accuracy of reference signal. The circuit which is depict in Figure 2 is used as the amplifier circuit [51].

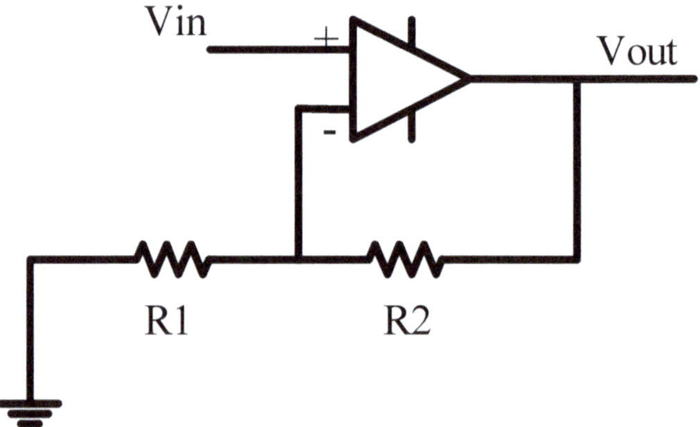

Figure 2. The amplifier circuit with op-amp.

Hysteresis Switching Technique (HYS)

The hysteresis switching technique is more interesting due to some futures such as fast response, less complexity and independence from an additional reference signal (for example triangular wave in PWM Technique) [52]. Grid connection with easy synchronization mood is an outstanding of this Technique. Figure 3 shows the block diagram operation of hysteresis pulse generation. As shown in Figure 3, the reference current which obtained by the detector algorithm (i^*_c) is compared with the output current of the active filter brunch (i_c) and the error due to the difference between these is sent to the fixed hysteresis band. There are two pair group switches for switching, since the active power filter use H-bridge circuit. The group switches work as a cross-pair in the H-bridge. The hysteresis technique is used for both group switches for positive and negative currents. It also has a dead time between the switching.

A constant bandwidth is surrounding the reference signal. If the error value (Δi_c) is higher than the upper band the switch will be off, and if the error value is lower than the lower band, then the switch will be turned on. The operation of the switch between the upper and lower bands for the sinusoidal reference signal is shown in Figure 4.

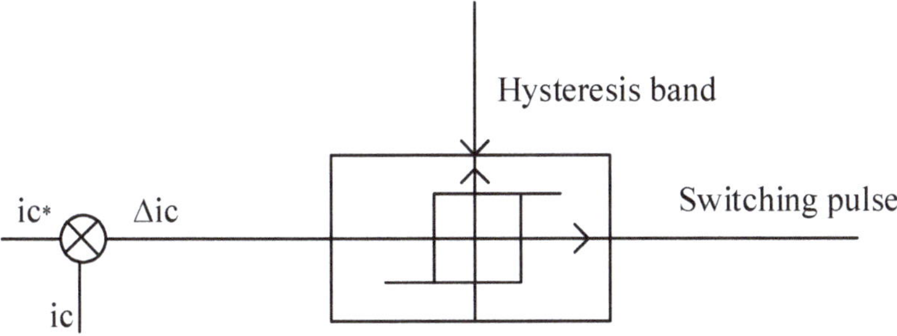

Figure 3. The block diagram operation of hysteresis pulse generation.

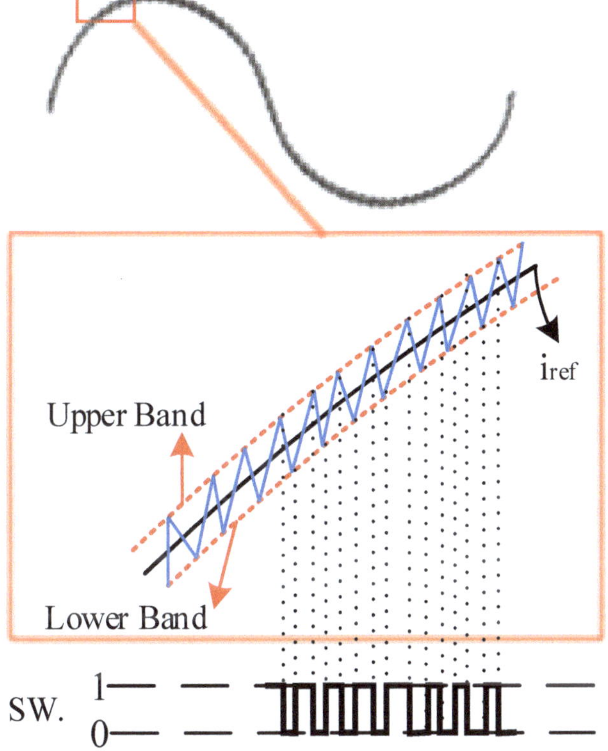

Figure 4. The operation of hysteresis controller to generate switching pulse.

3. Experimental Results

A prototype of active power filter is designed in the laboratory to verify the compensative operation of APHF in order to eliminate the grid harmonics. Figure 5 illustrates the configuration of the study system and the properties and the values of the elements are presented in Table 1.

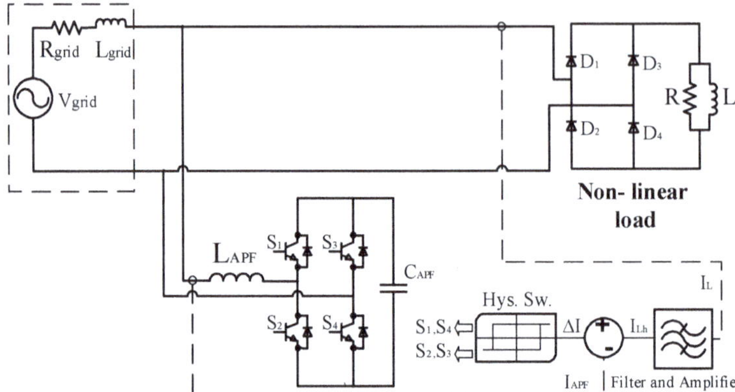

Figure 5. The configuration of study system.

Table 1. The properties and values of the study systems' elements.

Parameters	Magnitude
Power	3 kW
V_{grid}	220 v
R_{grid}	1 Ω
L_{grid}	600 µH
R_{load}	1 Ω
L_{load}	10 mH
C_{APF}	680 µF
L_{APF}	300 mH
$V_{DC\ Link}$	310 v

As shown in Figure 5, a diode-bridge connected with induction and resistor are considered as a non-linear load and they are supplied through the grid. Inductance is used in this system to protect the short circuit between APF and grid as current damper since the APF works according to the current injection. The diode-bridge generates harmonics since it is used as a rectifier, and these harmonics should be supplied through the grid. The APHF is applied in load in parallel to compensate the harmonic and prevent the penetration of it in the grid. IGBT 12n60 and RHR 15120 are used as power electronics switches and diodes in the prototype setup.

The qualification of the extracted harmonics and synchronization switching are investigated. The two sample experimental signals (red colored) are depicted in Figure 6. In order to the accuracy of harmonics extraction of detector circuit (Figure 1). Figure 6a shows the input of semi-square signal and Figure 6b shows semi-triangle ones. Harmonics are exactly extracted from both input signals that reveal harmonic components (blue colored), except the fundamental frequency of 50 Hz. High operation quality of control circuit and accurate extraction of harmonics is obvious in figures.

The amplifier circuit (Figure 2) works properly in which Figure 7 shows the amplified waveform of Figure 6. It is noticeable that the APF injects current to compensate higher harmonics currents too. Thus, controller holds the amplifier in saturated mood, as shown in Figure 7a.

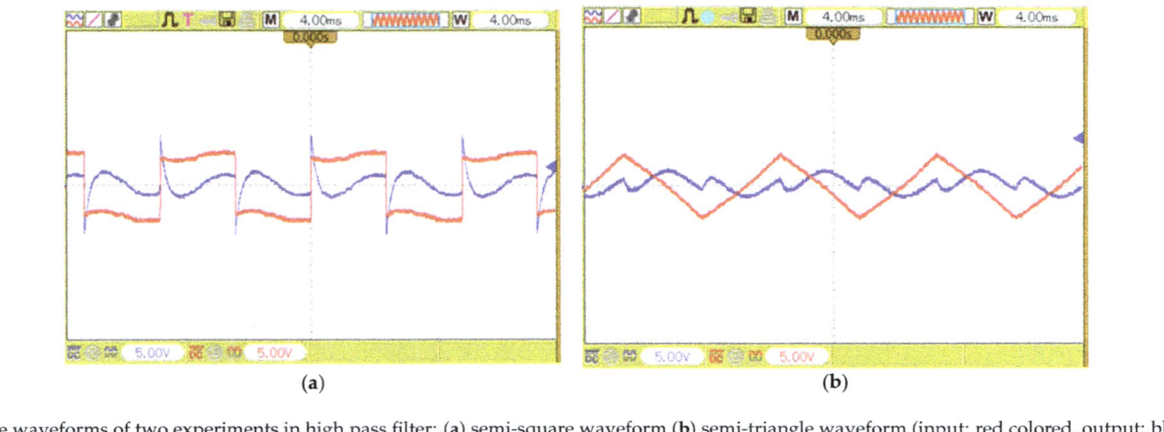

Figure 6. The waveforms of two experiments in high pass filter: (**a**) semi-square waveform (**b**) semi-triangle waveform (input: red colored, output: blue colored).

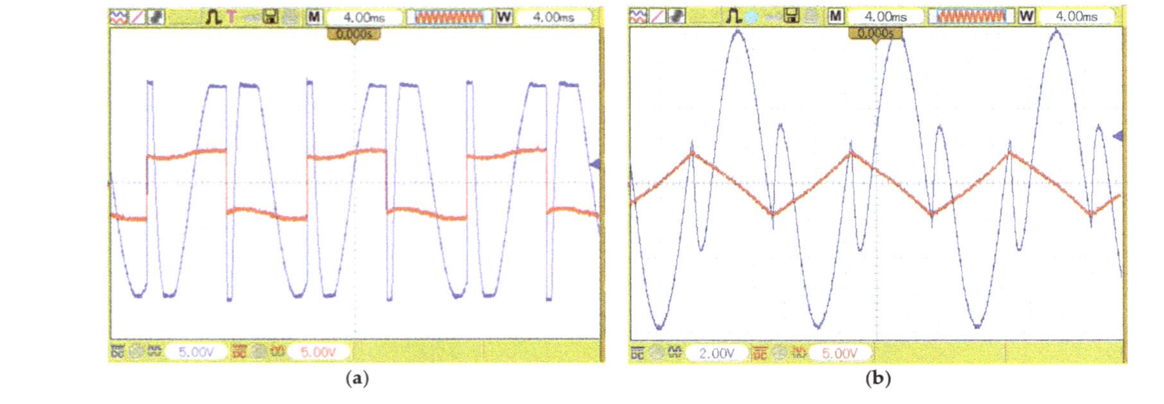

Figure 7. The waveforms of two experiments in amplifier circuit: (**a**) semi-square waveform (**b**) semi-triangle waveform (input: red colored, output: blue colored).

Figure 8 shows the bode and phase diagrams of the proposed control circuit. The cutoff frequency is set in 70 Hz to disappear the fundamental component in the output of the detector circuit for switching.

According to the Figure 8, the amount of phase and magnitude of harmonics are higher than the fundamental component (3rd, 5th, 7th, $\cdots$) passes accurately and unchanged in phase in the proposed controller. This property is very efficient to synchronize the APHF with the grid.

In order to increase the harmonic extraction quality, two series circuits are used to achieve fourth order high pass filter. The extracted and amplified harmonics will be sent to the op-amp comparator circuit to drive and trig power electronic switches. The proposed controller circuit is applied in the prototype system (Figure 9) and the results of evaluations are shown in Figures 10–14 with and without APHF in grid connection.

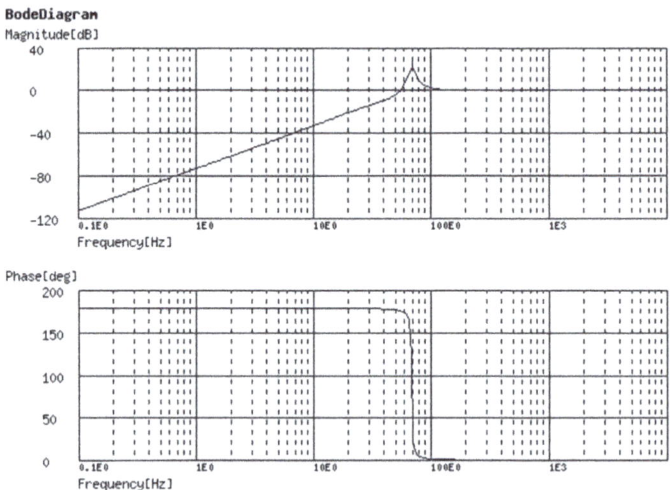

Figure 8. The bod and phase diagrams of proposed high pass filter with op-amp.

Figure 9. The experimental picture of the studied system.

Figure 10 illustrates the current waveform of the grid and Figure 11 shows its harmonic spectrum without APHF. Figures 12 and 13 show the current waveform of the grid and its harmonic spectrum with APHF, respectively. According to the figures, non-sinusoidal waveforms turned to sinusoidal after applying APHF and the harmonics are reduced magnificently from THD% = 24.48 to THD% = 2.86. It is obvious that all orders are decreased under 5%, which satisfy standard IEEE 519.

The injection current of APHF is shown in Figure 14.

Figure 15 also illustrates the smooth voltage waveform of DC link (the capacitor of APHF). It is constant at 310 Volts.

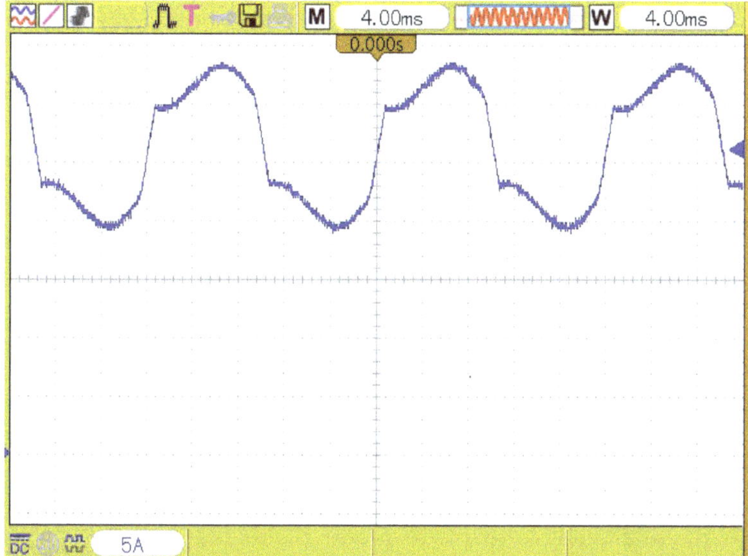

Figure 10. The grid current waveform without active power harmonic filters (APHF).

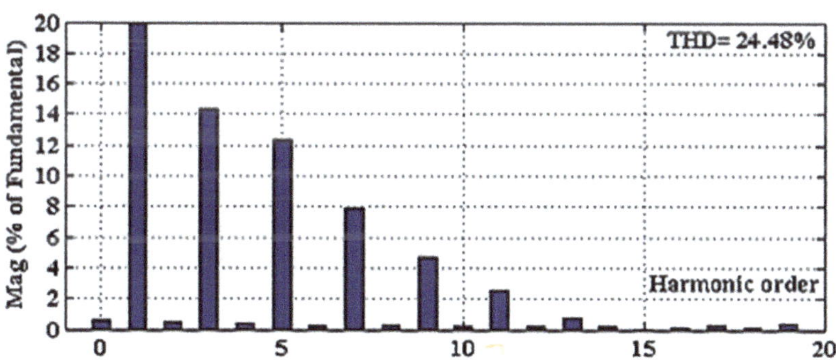

Figure 11. The harmonic spectrum of the current grid without APHF.

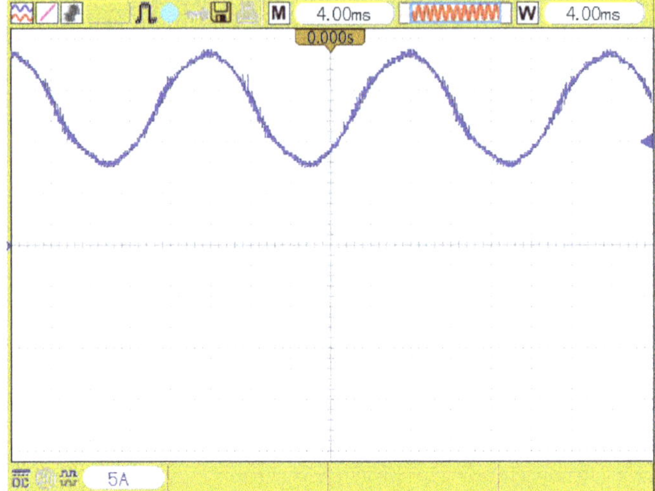

Figure 12. The grid current waveform with APHF.

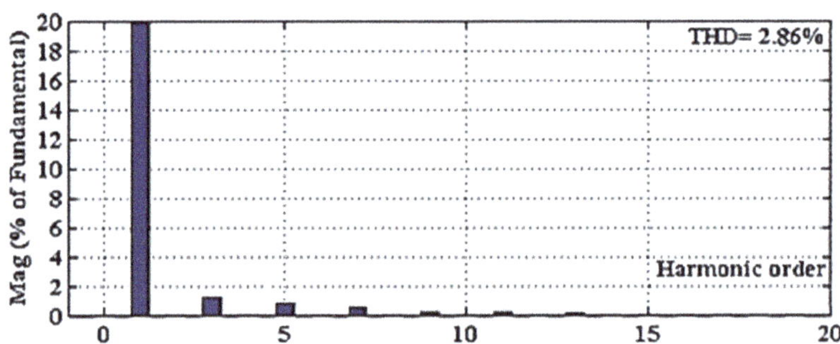

Figure 13. The harmonic spectrum of the current grid with APHF.

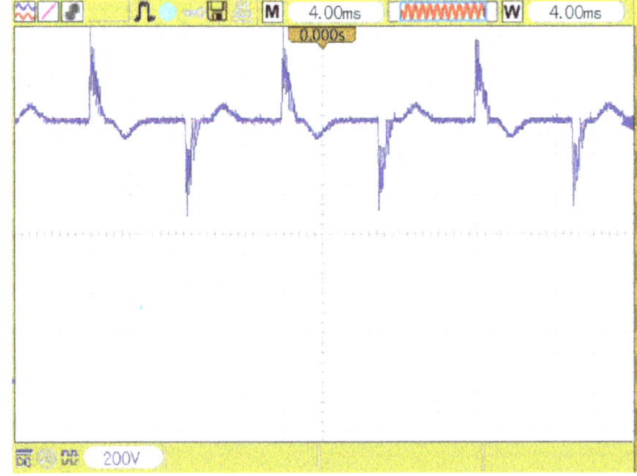

Figure 14. The injection current waveform of APHF brunch.

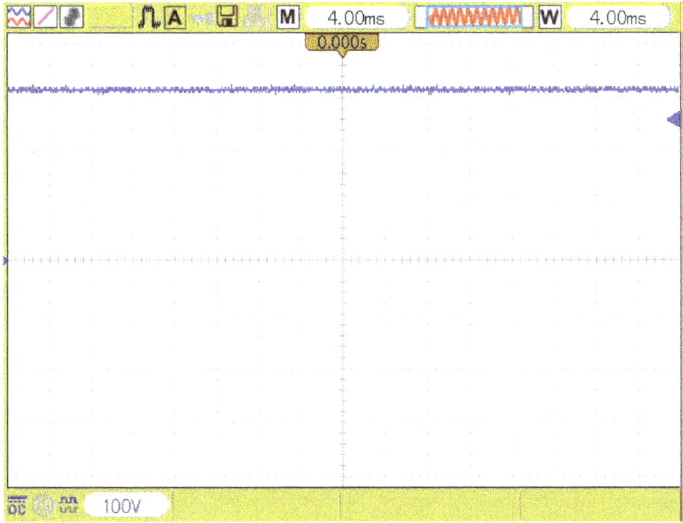

Figure 15. The voltage waveform of APHF's DC link.

4. Conclusions

This paper presented a new controller circuit with op-amp electrostatic circuit for active power harmonic filter. Simplicity, synchronization, and accurate operation are investigated on it. The proposed control system monitors the current of the grid and creates the reference signal and then inject appropriate current to prevent spreading of the load harmonic into the grid. Using the hysteresis switching technique with a precise synchronization made this proposed control system exhibit a fast response with less complexity. A prototype that uses this control circuit is implemented in the laboratory. In study system, the APHF is applied to the non-linear load in parallel with THD% = 24.48 that is supplied from the grid and THD% is reduced to %2.86 in the experimental results. Also, the non-sinusoidal waveform is renovated to sinusoidal waveform by proposed APHF. High operation quality of control circuit and the accurate extraction of harmonics confirm the good performance of the proposed controller.

Author Contributions: All authors contributed equally to this work and all authors have read and approved the final manuscript.

Funding: This research received no external funding.

References

1. Singh, R.; Singh, A. Energy loss due to harmonics in residential campus—A case study. In Proceedings of the 45th International Universities Power Engineering Conference UPEC2010, Cardiff, UK, 31 August–3 September 2010; pp. 1–6.
2. Simpson, R.H. Misapplication of power capacitors in distribution systems with nonlinear loads-three case histories. *IEEE Trans. Ind. Appl.* **2005**, *41*, 134–143. [CrossRef]
3. Yazdani-Asrami, M.; Sadati, S.M.B.; Samadaei, E. Harmonic study for MDF industries: A case study. In Proceedings of the 2011 IEEE Applied Power Electronics Colloquium (IAPEC), Johor Bahru, Malaysia, 18–19 April 2011; pp. 149–154. [CrossRef]
4. Gao, S.; Li, X.; Ma, X.; Hu, H.; He, Z.; Yang, J. Measurement-based compartmental modeling of harmonic sources in traction power-supply system. *IEEE Trans. Power Deliv.* **2017**, *32*, 900–909. [CrossRef]

5. Grady, W.M.; Santoso, S. Understanding power system harmonics. *IEEE Power Eng. Rev.* **2001**, *21*, 8–11. [CrossRef]

6. Samadaei, E.; Khosravi, A.; Sheikholeslami, A. Optimal Allocation of Active Power Filter On real distribution network for improvement of power quality by use of BBO: A case study. *IIUM Eng. J.* **2017**, *18*, 85–99. [CrossRef]

7. Beres, R.N.; Wang, X.; Liserre, M.; Blaabjerg, F.; Bak, C.L. A review of passive power filters for three-phase grid-connected voltage-source converters. *IEEE J. Emerg. Sel. Top. Power Electron.* **2016**, *4*, 54–69. [CrossRef]

8. Chang, G.W.; Chu, S.Y.; Wang, H.L. A new method of passive harmonic filter planning for controlling voltage distortion in a power system. *IEEE Trans. Power Deliv.* **2006**, *21*, 305–312. [CrossRef]

9. Zobaa, A.F. The optimal passive filters to minimize voltage harmonic distortion at a load bus. *IEEE Trans. Power Deliv.* **2005**, *20*, 1592–1597. [CrossRef]

10. Badrzadeh, B.; Smith, K.S.; Wilson, R.C. Designing passive harmonic filters for an aluminum smelting plant. *IEEE Trans. Ind. Appl.* **2011**, *47*, 973–983. [CrossRef]

11. Chang, G.W.; Wang, H.L.; Chu, S.Y. Strategic placement and sizing of passive filters in a power system for controlling voltage distortion. *IEEE Trans. Power Deliv.* **2004**, *19*, 1204–1211. [CrossRef]

12. Zhao, B.; Abramovitz, A.; Smedley, K. Family of bridgeless buck-boost PFC rectifiers. *IEEE Trans. Power Electron.* **2015**, *30*, 6524–6527. [CrossRef]

13. Zhang, J.; Zhao, C.; Zhao, S.; Wu, X. A family of single-phase hybrid step-down PFC converters. *IEEE Trans. Power Electron.* **2017**, *32*, 5271–5281. [CrossRef]

14. Lange, A.D.B.; Soeiro, T.B.; Ortmann, M.S.; Heldwein, M.L. Three-level single-phase bridgeless PFC rectifiers. *IEEE Trans. Power Electron.* **2015**, *30*, 2935–2949. [CrossRef]

15. Liu, Y.; Sun, Y.; Su, M.; Zhou, M.; Zhu, Q.; Li, X. A Single-Phase PFC Rectifier with Wide Output Voltage and Low-Frequency Ripple Power Decoupling. *IEEE Trans. Power Electron.* **2018**, *33*, 5076–5086. [CrossRef]

16. Akagi, H. Modern active filters and traditional passive filters. *Bull. Pol. Acad. Sci. Tech. Sci.* **2006**, *54*, 255–269.

17. Wu, C.J.; Chiang, J.C.; Yen, S.S.; Liao, C.J.; Yang, J.S.; Guo, T.Y. Investigation and mitigation of harmonic amplification problems caused by single-tuned filters. *IEEE Trans. Power Deliv.* **1998**, *13*, 800–806. [CrossRef]

18. Bhattacharya, S.; Cheng, P.T.; Divan, D.M. Hybrid solutions for improving passive filter performance in high power applications. *IEEE Trans. Ind. Appl.* **1997**, *33*, 732–747. [CrossRef]

19. Peng, F.Z. Harmonic sources and filtering approaches. *IEEE Ind. Appl. Mag.* **2001**, *7*, 18–25. [CrossRef]

20. Salam, Z.; Tan, P.C.; Jusoh, A. Harmonic's mitigation using active power filter: A technological review. *Elektr. J. Electr. Eng.* **2006**, *8*, 17–26.

21. Bhattacharya, S.; Frank, T.M.; Divan, D.M.; Banerjee, B. Active filter system implementation. *IEEE Ind. Appl. Mag.* **1998**, *4*, 47–63. [CrossRef]

22. Akagi, H. Active harmonic filters. *Proc. IEEE* **2005**, *93*, 2128–2141. [CrossRef]

23. Akagi, H. New trends in active filters for power conditioning. *IEEE Trans. Ind. Appl.* **1996**, *32*, 1312–1322. [CrossRef]

24. Mohan, N.; Peterson, H.A.; Long, W.F.; Dreifuerst, G.R.; Vithayathil, J.J. Active filters for AC harmonic suppression. In Proceedings of the IEEE Power Engineering Society Winter Meeting, New York, NY, USA, 30 January–4 February 1977.

25. Tuyen, N.D.; Fujita, G. PV-active power filter combination supplies power to nonlinear load and compensates utility current. *IEEE Power Energy Technol. Syst. J.* **2015**, *2*, 32–42. [CrossRef]

26. Javadi, A.; Hamadi, A.; Woodward, L.; Al-Haddad, K. Experimental investigation on a hybrid series active power compensator to improve power quality of typical households. *IEEE Trans. Ind. Electron.* **2016**, *63*, 4849–4859. [CrossRef]

27. Bubshait, A.S.; Mortezaei, A.; Simões, M.G.; Busarello, T.D.C. Power quality enhancement for a grid connected wind turbine energy system. *IEEE Trans. Ind. Appl.* **2017**, *53*, 2495–2505. [CrossRef]

28. Carpinelli, G.; Proto, D.; Russo, A. Optimal Planning of Active Power Filters in a Distribution System Using Trade-off/Risk Method. *IEEE Trans. Power Deliv.* **2017**, *32*, 841–851. [CrossRef]

29. He, J.; Li, Y.W.; Blaabjerg, F.; Wang, X. Active harmonic filtering using current-controlled, grid-connected DG units with closed-loop power control. *IEEE Trans. Power Electron.* **2014**, *29*, 642–653. [CrossRef]

30. Darwish, M.K.; El-Habrouk, M.; Kasikci, I. EMC compliant harmonic and reactive power compensation using passive filter cascaded with shunt active filter. *EPE J.* **2002**, *12*, 43–50. [CrossRef]

31. Javadi, A.; Al-Haddad, K. A single-phase active device for power quality improvement of electrified transportation. *IEEE Trans. Ind. Electron.* **2015**, *62*, 3033–3041. [CrossRef]

32. Antchev, M.H. Classical and Recent Aspects of Active Power Filters for Power Quality Improvement. In *Classical and Recent Aspects of Power System Optimization*; Academic Press: Cambridge, MA, USA, 2018; pp. 219–254.

33. Ko, W.H.; Gu, J.C. Impact of shunt active harmonic filter on harmonic current distortion of voltage source inverter-fed drives. *IEEE Trans. Ind. Appl.* **2016**, *52*, 2816–2825. [CrossRef]

34. Samadaei, E.; Lesan, S.; Cherati, S.M. A new schematic for hybrid active power filter controller. In Proceedings of the 2011 IEEE Applied Power Electronics Colloquium (IAPEC), Johor Bahru, Malaysia, 18–19 April 2011; pp. 143–148. [CrossRef]

35. Samedaei, E.; Vahedi, H.; Sheikholeslami, A.; Lesan, S. Using "STF-PQ" algorithm and hysteresis current control in hybrid active power filter to eliminate source current harmonic. In Proceedings of the 2010 First Power Quality Conferance (PQC), Tehran, Iran, 14–15 September 2010; pp. 1–6.

36. Singh, B.; Solanki, J. An implementation of an adaptive control algorithm for a three-phase shunt active filter. *IEEE Trans. Ind. Electron.* **2009**, *56*, 2811–2820. [CrossRef]

37. Wang, Z.; Wang, Q.; Yao, W.; Liu, J. A series active power filter adopting hybrid control approach. *IEEE Trans. Power Electron.* **2001**, *16*, 301–310. [CrossRef]

38. Kanjiya, P.; Khadkikar, V.; Zeineldin, H.H. Optimal control of shunt active power filter to meet IEEE Std. 519 current harmonic constraints under nonideal supply condition. *IEEE Trans. Ind. Electron.* **2015**, *62*, 724–734. [CrossRef]

39. Saribulut, L.; Teke, A.; Tümay, M. Artificial neural network-based discrete-fuzzy logic controlled active power filter. *IET Power Electron.* **2014**, *7*, 1536–1546. [CrossRef]

40. Suresh, Y.; Panda, A.K.; Suresh, M. Real-time implementation of adaptive fuzzy hysteresis-band current control technique for shunt active power filter. *IET Power Electron.* **2012**, *5*, 1188–1195. [CrossRef]

41. Antoniewicz, K.; Jasinski, M. Experimental comparison of hysteresis based control and finite control state set Model Predictive Control of Shunt Active Power Filter. In Proceedings of the 2015 Selected Problems of Electrical Engineering and Electronics (WZEE), Kielce, Poland, 17–19 September 2015; pp. 1–6. [CrossRef]

42. Zhivich, M.; Cunningham, R.K. The Real Cost of Software Errors. *IEEE Secur. Priv.* **2009**, *7*, 87–90. [CrossRef]

43. Williamson, G.F. Software safety and reliability. *IEEE Potentials* **1997**, *16*, 32–36. [CrossRef]

44. De Almeida, J.R.; Camargo, J.B.; Cugnasca, P.S. Software Safety in Subway and Air Traffic Control Applications. *IEEE Latin Am. Trans.* **2008**, *6*, 106–113. [CrossRef]

45. Mosalikanti, P.; Kurd, N.; Mozak, C.; Oshita, T. Low power analog circuit techniques in the 5th generation intel core TM microprocessor (broadwell). In Proceedings of the 2015 IEEE Custom Integrated Circuits Conference (CICC), San Jose, CA, USA, 28–30 September 2015; pp. 1–4. [CrossRef]

46. Caloz, C.; Gupta, S.; Zhang, Q.; Nikfal, B. Analog signal processing: A possible alternative or complement to dominantly digital radio schemes. *IEEE Microw. Mag.* **2013**, *14*, 87–103. [CrossRef]

47. Gonzalez-Diaz, V.R.; Peña-Perez, A.; Maloberti, F. Opamp gain compensation technique for continuous-time $\Sigma\Delta$ modulators. *Electron. Lett.* **2014**, *50*, 355–356. [CrossRef]

48. Millman, J. *Microelectronics: Digital and Analog Circuits and Systems*; Entire Document; McGraw-Hill: New York, NY, USA, 1979; pp. 523–527, ISBN 0-07-042327-X.

49. Mao, H.; Yang, X.; Chen, Z.; Wang, Z. A hysteresis current controller for single-phase three-level voltage source inverters. *IEEE Trans. Power Electron.* **2012**, *27*, 3330–3339. [CrossRef]

50. Antchev, M.; Petkova, M.; Petkov, M. Single-phase shunt active power filter using frequency limitation and hysteresis current control. In Proceedings of the Power Conversion Conference (PCC'07), Nagoya, Japan, 2–5 April 2007; pp. 97–102. [CrossRef]

51. Lipiansky, E. Ch5. In *The Operational Amplifier as a Circuit Element*; Electrical, Electronics and Digital Hardware Essentials for Scientists and Engineers; Wiley-IEEE Press: Hoboken, NJ, USA, 2013; Volume 1, pp. 287–353, ISBN 9781118414552.

52. Kazmierkowski, M.P.; Malesani, L. Current control techniques for three-phase voltage source PWM converters: A survey. *IEEE Trans. Ind. Electron.* **1998**, *45*, 691–703. [CrossRef]

 *sustainability*

Article

A Countermeasure for Preventing Flexibility Deficit under High-Level Penetration of Renewable Energies: A Robust Optimization Approach

Jinwoo Jeong [1], Heewon Shin [1], Hwachang Song [2] and Byongjun Lee [1,*]

[1] School of Electrical Engineering, Anam Campus, Korea University, 145 Anam-ro, Seongbuk-gu, Seoul 02841, Korea; jinwoo8709@korea.ac.kr (J.J.); redcore451@korea.ac.kr (H.S.)
[2] Department of Electrical and Information Engineering, Seoul National University of Science and Technology, Seoul 01811, Korea; hcsong@seoultech.ac.kr
* Correspondence: leeb@korea.ac.kr; Tel.: +82-10-9245-3242

Received: 23 October 2018; Accepted: 7 November 2018; Published: 12 November 2018

Abstract: An energy paradigm shift has rapidly occurred around the globe. One change has been an increase in the penetration of sustainable energy. However, this can affect the reliability of power systems by increasing variability and uncertainty from the use of renewable resources. To improve the reliability of an energy supply, a power system must have a sufficient amount of flexible resources to prevent a flexibility deficit. This paper proposes a countermeasure for protecting nonnegative flexibility under high-level penetration of renewable energy with robust optimization. The proposed method is divided into three steps: (i) constructing an uncertainty set with the capacity factor of renewable energy, (ii) searching for the initial point of a flexibility deficit, and (iii) calculating the capacity of the energy storage system to avoid such a deficit. In this study, robust optimization is applied to consider the uncertainty of renewable energy, and the results are compared between deterministic and robust approaches. The proposed method is demonstrated on a power system in the Republic of Korea.

Keywords: robust optimization; renewable energy; flexibility; deficit; uncertainty; flexible resource; energy storage systems

---

1. Introduction

A paradigm shift in energy generation has rapidly taken place around the world. The traditional energy industry was aimed at providing energy at a low price. However, the focus is changing to provide safer, cleaner, and more sustainable energy in certain countries. In particular, China has been reinforcing the competitiveness of its sustainable energy industry by supporting a strong policy and developing its technologies. In keeping with this trend, the Republic of Korea has also tried to meet this paradigm shift by establishing an energy policy, which was launched by the government in 2017. As one aspect of this policy, the government announced its Renewable Energy 2030 implementation plan, in which the share of renewable energy sources in the energy mix will increase from its current level of 7% to 20% by 2030. Korea's major energy administration and industry are making an effort to achieve this goal [1,2]. However, such a sudden shift in the energy mix can worsen the conditions of the power system, because renewable energy sources are quite volatile [3,4]. Therefore, some countermeasures are required, including strengthening the grid through investments in the facilities and preparing strategies for the effective operation of renewable energies [5–8].

To achieve stable operation, the power system under high renewable penetration should respond to the variation and uncertainty of renewable energy sources to secure sufficient flexibility. If a power system achieves sufficient flexibility, it can respond rapidly to events such as a sudden decrease in energy output, and ensure stability and superior quality. Owing to its increased importance, studies related to flexibility have been conducted [9–14]. Electric Power Research Institute (EPRI) conducted a study looking at the impact of transmission on system flexibility [9] and developed a multilevel flexibility assessment tool [10]. In [11], the authors clarified flexibility by summarizing the analytic frameworks that recently emerged to measure operational flexibility. The Danish Energy Agency carried out an assessment of flexibility in Denmark and China [12]. Poncela et al. [13] proposed a stepwise methodology based on a set of indicators for future power system flexibility applied to a European case. In [14], flexibility metrics were compared between insufficient ramp resources and the number of periods of flexibility deficit. In particular, the California Independent System Operator carries out annual technical studies to determine the required capacity [15] and has developed a flexible ramping product to handle increasing amounts of variable renewable generation [16].

Representative flexible resources include ramp rates, energy storage systems (ESS), and demand response (DR). Among such flexible resources, an ESS can play an important role in supplying balance to the grid by providing a backup to intermittent renewable energy sources and generating a low-carbon power system [17]. In addition, a decrease in renewable energy curtailment can occur [18]. Therefore, owing to these merits, an ESS was chosen to meet supply and demand against the variations in renewable energy and to ensure nonnegative flexibility under high-level penetration of sustainable energy.

Traditional optimization methodology is aimed at finding a deterministic result by assuming a parameter and variable in a specific state without considering the uncertainty. However, in this case, it is difficult to guarantee a reliable solution unless the data uncertainty is dealt with. For example, if a parameter with uncertainty is estimated to have a certain value, it can be unclear whether the value is correct. In addition, this can make the solution infeasible owing to the possibility of an error. Therefore, many studies on optimization techniques that can apply uncertainty have been carried out. In particular, in the power system industry, studies related to planning and operation have been considered based on an increase in uncertain resources such as renewable energy sources [19–30].

Optimization methods that are able to handle uncertainties have been developed, such as stochastic programming (SP) and robust optimization (RO). Optimization techniques have long been used to deal with uncertainty. Several studies related to SP in power systems have been conducted [19–22]. Jirutitijaroen et al. [19] proposed a mixed-integer stochastic programming approach to find a solution to the generation and transmission line expansion planning problem, including consideration of the system reliability. In [20], SP based on a Monte Carlo approach was introduced to cope with uncertainties, and a new approach to modeling the operational constraints of an ESS was applied to the capacity expansion planning of a wind–diesel isolated grid. In addition, in [21], the authors proposed a multistage decision-dependent stochastic optimization model for long-term and large-scale generation expansion planning. The authors in [22] proposed a novel stochastic planning framework to determine the optimal battery energy storage system (BESS) capacity and the year of installation in an isolated microgrid using a new representation of the BESS energy diagram.

Studies on power system operation and planning using RO have been carried out to consider uncertainties such as renewable energy sources [23–30]. Ruiz and Conejo [23] presented a transmission expansion planning (TNEP) method by constructing the load and RES output into uncertainty sets. In [24], transmission and ESS expansion planning was carried out by characterizing the uncertainty sources pertaining to load demand and wind power production through uncertainty sets. In addition, in [25], energy generation and ESS expansion planning was implemented by handling the net load as an uncertainty set. The authors in [26] used variation in the net load as an uncertainty set, and proposed an economic dispatch to cope with variation in the use of ramp rates. In [27], the authors examined the effectiveness of RO in maximizing the economic benefit for owners of home battery storage systems in the presence of uncertainty in dynamic electricity prices. In [28], the authors proposed an adaptive robust optimization model for multiperiod economic dispatch, and introduced the concept of dynamic uncertainty sets and the methods to construct such sets for modeling the temporal and spatial correlations of uncertainty. Yi et al. [29] presented ESS scheduling by constructing the RES output, load, and real-time thermal rating (RTTR) of transmission lines into an uncertainty set. In [30], algorithms to minimize total cost under Korea's commercial and industrial tariff system based on robust optimization were proposed.

Stochastic Programming assumes that uncertain data have a probability distribution function (PDF), although this method has difficulty in accurately constructing a PDF for the uncertainty. This is based on the generation of scenarios that describe uncertain parameters, the size of which grows with the number of scenarios, which may result in intractability. However, the RO represents an uncertainty parameter set, which can contain any number of scenarios without specific knowledge of the PDF. As its methodology, it also minimizes the objective value under the worst-case scenario. Scenarios do not need to be generated, which makes the RO computationally tractable. Therefore, owing to such advantages, the RO is more appropriate than the SP for solving the optimization problem with uncertainties [31–37]. In this paper, the reasons for using the RO are that it allows for treating uncertainties in the optimization problem and can lead to a robust solution, which is immunized against uncertainty.

Many studies on power system operation and planning with renewable energy have mostly considered its outputs as uncertainties. However, many factors affect the output of renewable energy, including the weather and installation locations, which make it hard to forecast. Thus, this paper proposes using the capacity factor as the output of the RES. Applying the capacity factor can make it simpler to consider the output of renewable energy by using the ratio of the rated capacity to the real outputs of renewable energy without taking the factors into account.

The nameplate capacity of renewable energy is known from the installation planning, while the capacity factor is unknown due to its characteristics, including variable and unpredictable outputs. So, the capacity factor of renewable energy has uncertainty and affects planning because it has difficulty making decisions on how the system will be reinforced. In addition, it needs many scenarios about renewable energy sources. This paper calculates the required capacity of flexible resources like ESS to secure sufficient flexibility without generating scenarios regarding renewable energy resources by constructing its uncertainty set based on the RO.

This paper presents a countermeasure to ensure nonnegative flexibility using flexible resources including the ramp rate and ESS by considering the capacity factor of renewable energy as an uncertainty set. It can be divided into three steps: (i) The range of the capacity factor of renewable energy is predicted in the construction of the uncertainty set. (ii) The initial point where the flexibility deficit occurs within an uncertainty set is detected using the RO. (iii) The capacity of the ESS is estimated to prevent negative flexibility from a variation in renewable energy with the RO. The effectiveness of the proposed method is demonstrated using the Korean Power System for the year 2030.

2. Materials and Methods

2.1. Uncertain Parameter

In the power system planning stage, the output of renewable energy is a typical parameter of uncertainty because it is unpredictable and variable. It is necessary for the output of renewable energy to be expressed as its capacity factor because it is less likely to generate electricity to the nameplate capacity under the influence of many factors, including the installation site and climate. The capacity factor can be expressed based on the ratio of energy generated over a period of time divided by the installed capacity.

$$\text{Capacity Factor} = \frac{Actual\ Energy\ Generated\ (MWh)}{Time\ Period\ (h) \times Installed\ Capacity\ (MW)} \tag{1}$$

The uncertainty set of the capacity factor can be described as follows:

$$\mathrm{CF} = \left\{ CF : \sum_{i \in \mathrm{N}^{RE}} \frac{\left| CF_{re,i} - \overline{CF}_{re,i} \right|}{\hat{CF}_{re,i}} \leq \Gamma_{RE} \sqrt{\mathrm{N}^{RE}}, \right.$$
$$\left. CF_i \in \left[\overline{CF}_{re,i} - \Gamma_{RE} \hat{CF}_{re,i}, \overline{CF}_{re,i} + \Gamma_{RE} \hat{CF}_{re,i} \right] \ \forall i \in \mathrm{N}^{RE} \right\} \tag{2}$$

where N^{RE} denotes the number of renewable energy sources and CF_i is the capacity factor of renewable energy i. In Equation (2), CF_i is located within the range of the upper and lower capacity, and its width is determined based on the deviation $\hat{CF}$. Although the robust optimization has a disadvantage in that its result is usually too conservative, it can overcome such conservativeness by using the budget of uncertainty proposed in [37]. This can be applied using Γ_{RE} in Equation (2), which can control the size of the uncertainty set and lies within the range $0 \leq \Gamma_{RE} \leq 1$. If Γ_{RE} is 1, the capacity factor can have any value within the interval of the uncertainty set. On the contrary, $\Gamma_{RE} = 0$ implies $\mathrm{CF} = CF^{ref}$, which means the uncertainty is not considered.

The capacity factor of renewable energy can be used to construct the uncertainty set by calculating the upper and lower limits through the use of the historical and predicted outputs of renewable energy, shown in Figure 1.

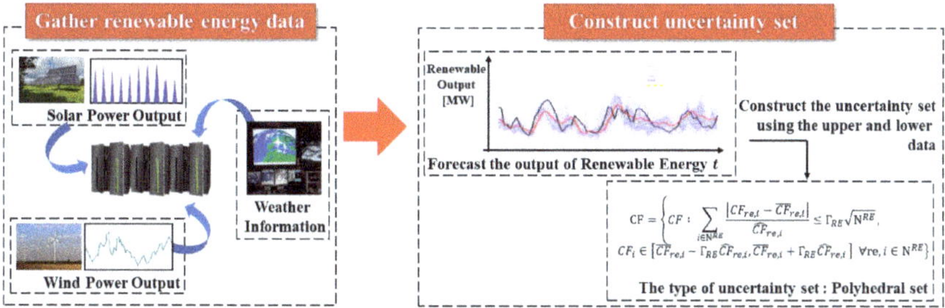

Figure 1. Process of constructing the uncertainty set.

2.2. Mathematical Formulation

This section proposes a way to prevent a flexibility deficit under high-level penetration of renewable energy using a robust optimization methodology. The more the penetration level of the RES increases, the more variation and uncertainty arise. Thus, a measure is needed to keep supply and demand from experiencing an increase in variation and uncertainty, which can be solved by securing a sufficient amount of flexible resources. If the power system can achieve sufficient flexible resources to balance supply and demand from variation and uncertainty, it will not be necessary to install additional flexible resources. However, as an opposite case, a power system needs to invest in additional flexible resources to maintain supply and demand. As a proper process for the installation of flexible resources, the power system is examined to determine how many it would require after looking at whether it can ensure nonnegative flexibility within the uncertainty set. In this paper, the process is divided into 2 steps: (i) finding the initial point of the flexibility deficit within the uncertainty, and (ii) determining the capacity of the flexible resources to ensure nonnegative flexibility within the interval of the capacity factor. The details are presented below.

2.2.1. Searching for the Initial Point of Flexibility Deficit within the Uncertainty Set

Objective Function

The objective function consists of the sum of the cost of the generation and penalty (e.g., load shedding and curtailment). The generation cost is considered based on a quadratic function of the fuel cost of the thermal units, because renewable energy generation has a comparatively low cost.

The penalty cost is expressed based on the amount of flexibility deficit multiplied by its cost coefficient and is a sufficiently large positive constant because it is related to a flexibility deficit causing load shedding and curtailment to balance supply and demand. The objective function can be formulated as follows:

$$\max_{CF \in U_{y \in \Omega(CF, P_{gi}^{pre})}} \min \quad -Cost(P_{gi}^{pre}, P^{FD}) \tag{3}$$

$$Cost(P_{gi}^{pre}, P^{FD}) = \sum_{gi \in NG} \left(\alpha_{gi} + \beta_{gi} \cdot P_{gi}^{pre} + \gamma_{gi} \cdot (P_{gi}^{pre})^2 \right) + c^{FD} \cdot P^{FD} \quad \forall i \in N^G \tag{4}$$

where N^G denotes the number of generators and α_{gi}, β_{gi}, and γ_{gi} denote the cost coefficients of the ith thermal unit; c^{FD} denotes the cost coefficient of a flexibility deficit; and P_{gi}^{pre} and P^{FD} denote the output of the ith thermal unit and the total amount of flexibility deficit, respectively.

An increase in the capacity factor affects the decrease in total net load and generation, which can decrease the total cost of power generation if the power system has sufficient flexible resources to take action against the variation in renewable energy. However, from the perspective of a flexibility deficit, the value of the objective function increases owing to the penalty cost. Thus, just before the point of flexibility deficit is reached, Equation (4) reaches its smallest value. The use of a negative sign in Equation (4) changes it to the largest value just before the point when a flexibility deficit occurs. This allows searching for the initial point of the flexibility deficit within an uncertainty set, because robust optimization considers the worst-case scenario within the set. Accordingly, using this objective function, the initial point of the flexibility deficit can be found, and whether the power system has sufficient flexible resources to ensure nonnegative flexibility within the uncertainty set can be confirmed before determining whether to invest in flexible resources. In this step, it is assumed that the power system has only ramp rates as flexible resources.

Constraints

The constraints are composed of 3 parts: (i) conventional generators and renewable energy output, (ii) power balance, and (iii) power system flexibility.

(i) Output of Conventional Generators and Renewable Energy Sources

The output of a conventional generator is determined within the range of minimum and maximum limits of the generator. In addition, the output of renewable energy changes in accordance with the capacity factor:

$$\underline{P_{gi}^{pre}} \leq P_{gi}^{pre} \leq \overline{P_{gi}^{pre}} \quad \forall i \in N^G \tag{5}$$

$$P_{re,i}^{pre} = CF \times P_{re,i}^{rated} \quad \forall i \in N^{RE} \tag{6}$$

where $\underline{P_{gi}^{pre}}$ and $\overline{P_{gi}^{pre}}$ are the minimum and maximum outputs of the ith generator, respectively, and $P_{re,i}^{rated}$ is the rated capacity of renewable energy sources.

(ii) Power Balance

The output of a conventional generator is determined within the range of minimum and maximum limits of the generator. In addition, the output of renewable energy changes in accordance with the capacity factor:

$$\sum_{i \in N^d} P_{di} = \sum_{i \in NG} P_{gi}^{pre} + \sum_{i \in N^{RE}} P_{re,i}^{pre} + P^{FD} \tag{7}$$

where P_{di} is the demand at load i. The left-hand side of Equation (7) indicates the total load, and the right-hand side indicates the total sum of power generation and the amount of flexibility deficit. When a flexibility deficit occurs owing to a lack of flexible resources, it can meet the power balance using P_{FD}^{pre}.

(iii) Power System Flexibility

In [14], the method for securing power system flexibility is intended to keep the amount of available flexibility higher than the flexibility requirement. The available flexibility means the total amount of flexible resources required to respond to variations in the net load, and the flexibility requirement means the net load ramp. Eventually, to ensure flexibility, the power system should secure flexible resources in advance to prevent a flexibility deficit. In this paper, the flexibility requirement is a variation in renewable energy by applying its variability rate, and the available flexibility simply considers the total sum of the ramp rates, which can be expressed as follows:

$$\Delta P_{re,i}^{pre} = P_{re,i}^{pre} \times Variability \quad Rate \ \forall i \in N^{RE}. \tag{8}$$

$$R_{gi}^{pre_{up}} = Min(Ramprate_{gi}^{up}, \overline{P_{gi}^{pre}} - P_{gi}^{pre}) \ \forall i \in N^G \tag{9}$$

$$R_{gi}^{pre_{down}} = Min(Ramprate_{gi}^{down}, P_{gi}^{pre} - \underline{P_{gi}^{pre}}) \ \forall i \in N^G \tag{10}$$

$$\sum_{i \in N^{RE}} \Delta P_{re,i}^{pre} - \sum_{i \in N^G} R_{gi}^{pre_{up}} \leq P^{FD} \tag{11}$$

$$\sum_{i \in N^{RE}} \Delta P_{re,i}^{pre} - \sum_{i \in N^G} R_{gi}^{pre_{down}} \leq P^{FD} \tag{12}$$

where $\Delta P_{re,i}^{pre}$ is the ramp of renewable energy as the requirement of flexibility, and $R_{gi}^{pre_{up}}$ and $R_{gi}^{pre_{down}}$ are the upward and downward reserves, respectively, at the ith generator that are able to offer active power over a certain period of time depending on the ramp rate. Therefore, Equation (8) defines the flexibility requirement, and Equations (9) and (10) are the available flexibility at the ith generator, enabling an increase and decrease in output within the time interval. Equations (11) and (12) determine the amount of flexibility deficit. When the left-hand side is lower than zero, the flexible resources are adequate to ensure nonnegative flexibility. However, when the left-hand side is higher than zero, the power system has inadequate flexible resources to respond to the net load ramp.

2.2.2. Determining the Capacity of Flexible Resources to Ensure Nonnegative Flexibility

Objective Function

The objective function is composed of the sum of the cost of thermal generation and ESS installation. The cost of thermal generation is the same as the quadratic equation given above. The cost of ESS installation can be expressed based on its capacity multiplied by its cost coefficient. The magnitude of the cost coefficient is sufficiently large to minimize the required capacity of the ESS to prevent a flexibility deficit from the net load ramp. The objective function is formulated as follows:

$$\max_{CF \in U} \ \min_{y \in \Omega(CF, P_{gi}^{post})} Cost(P_{gi}^{post}, P^{ESS}) \tag{13}$$

$$Cost(P_{gi}^{post}, P^{ESS}) = \sum_{gi \in N^G} \left(\alpha_{gi} + \beta_{gi} \cdot P_{gi}^{post} + \gamma_{gi} \cdot (P_{gi}^{post})^2 \right) + c^{ESS} \cdot P^{ESS} \ \forall i \in N^G \tag{14}$$

where c^{ESS} denotes the cost efficiency of installing the ESS, and P_{gi}^{post} and P^{ESS} denote the output of the ith thermal unit and the amount of the required ESS, respectively.

The cost of charging and discharging an ESS is neglected in the objective function because it is much lower than that of installing the ESS. In addition, the cost of installation may considerably affect the value of the objective function compared with charging and discharging the ESS. When the initial flexibility deficit is found in the previous stage, installing an ESS is needed to secure sufficient flexible resources. In this stage, the minimum capacity of the ESS needed to ensure nonnegative flexibility within the uncertainty set is shown.

Constraints

The constraints consist of four parts: (i) conventional generators and renewable energy output, (ii) power balance, (iii) power system flexibility, and (iv) an ESS.

(i) Energy Output of Conventional Generators and Renewable Energy Sources

The constraints on the output of a conventional generator and renewable energy sources are the same as in Equations (5) and (6), which can be expressed as follows:

$$\underline{P_{gi}^{post}} \leq P_{gi}^{post} \leq \overline{P_{gi}^{post}} \quad \forall i \in N^{G} \tag{15}$$

$$P_{re,i}^{pre} = CF \times P_{re,i}^{rated} \quad \forall i \in N^{RE} \tag{16}$$

(ii) Power Balance

The constraint of a power balance is almost the same as in Equation (7), except that the flexibility deficit is substituted with the charge and discharge of the ESS, which can be represented as follows:

$$\sum_{i \in N^{d}} P_{di} = \sum_{i \in N^{G}} P_{gi}^{pre} + \sum_{i \in N^{RE}} P_{re,i}^{pre} + \left(P_{ess}^{dis} - P_{ess}^{cha} \right) \tag{17}$$

where P_{ess}^{dis} and P_{ess}^{cha} are the magnitude of discharging and charging the ESS to protect the flexibility deficit within the uncertainty, respectively. Here, P_{ess}^{dis} can respond to an upward flexibility deficit, and P_{ess}^{cha} is able to cope with a downward flexibility deficit. Therefore, the power system is reinforced by the ESS, avoiding a flexibility deficit.

(iii) Power System Flexibility

The constraint of the power system flexibility is almost the same as in Equations (8)–(12) except for an additional part, the charging and discharging ESS in Equations (21) and (22), which is formulated as follows:

$$\Delta P_{re,i}^{post} = P_{re,i}^{post} \times Variability \quad Rate \; \forall i \in N^{RE}. \tag{18}$$

$$R_{gi}^{pre_{up}} = Min \left(Ramprate_{gi}^{up}, \overline{P_{gi}^{post}} - P_{gi}^{post} \right) \quad \forall i \in N^{G} \tag{19}$$

$$R_{gi}^{pre_{down}} = Min \left(Ramprate_{gi}^{down}, P_{gi}^{post} - \underline{P_{gi}^{post}} \right) \quad \forall i \in N^{G} \tag{20}$$

$$\sum_{i \in N^{RE}} \Delta P_{re,i}^{post} - \left(\sum_{i \in N^{G}} R_{gi}^{post_{up}} + P_{ess}^{dis} \right) \leq P^{Flex} \tag{21}$$

$$\sum_{i \in N^{RE}} \Delta P_{re,i}^{post} - \left(\sum_{i \in N^{G}} R_{gi}^{post_{down}} + P_{ess}^{cha} \right) \leq P^{Flex} \tag{22}$$

where P^{Flex} denotes the additional required flexibility. If $P^{Flex} = 0$, minimum flexibility is ensured within the uncertainty set by installing the minimum ESS. In this paper, P^{Flex} is set to zero because the minimum capacity of the ESS is calculated to secure adequate flexible resources with the uncertainty set.

(iv) Energy Storage System (ESS)

When a net load ramp caused by variability in renewable energy sources occurs, it may be necessary to supply or absorb electricity. Thus, the capacity of an ESS can be composed of the sum of the charge and discharge required to avoid a flexibility deficit unless sufficient ramp rates exist as a flexible resource in the previous stage, which can be expressed as follows:

$$P^{ESS} = \eta_{ess}^{cha} \cdot P_{ess}^{cha} + \frac{1}{\eta_{ess}^{dis}} \cdot P_{ess}^{cha} \tag{24}$$

$$0 \leq P_{ess}^{cha}, P_{ess}^{dis} \tag{25}$$

where η_{ess}^{cha} and η_{ess}^{dis} are the efficiency of the charging and discharging ESS. Indeed, losses occur when the ESS is charging and discharging. To calculate the capacity of the ESS, the efficiency should be considered in the problem.

2.3. Description of the Proposed Method

This paper presents a counterplan to prevent a flexibility deficit with flexible resources including ramp rates and ESS, which is coordinated with robust optimization. The process is composed of three parts: (i) constructing the uncertainty set with the capacity factor of renewable energy, (ii) searching the initial point of the flexibility deficit within the uncertainty, and (iii) calculating the capacity of the ESS to secure flexibility within the uncertainty set. First, based on historical or predicted data about renewable energy sources, the uncertainty set of the capacity factor is constructed. Whether the initial point of the flexibility deficit is found by the robust optimization within the uncertainty set is then examined. If an initial point exists, the power system needs additional flexible resources to secure flexibility under high-level penetration of renewable energy against the net load ramp. After searching for the initial point, based on the robust optimization, how much ESS capacity is needed to ensure nonnegative flexibility from the net load ramp can be calculated. The proposed method is shown in Figure 2.

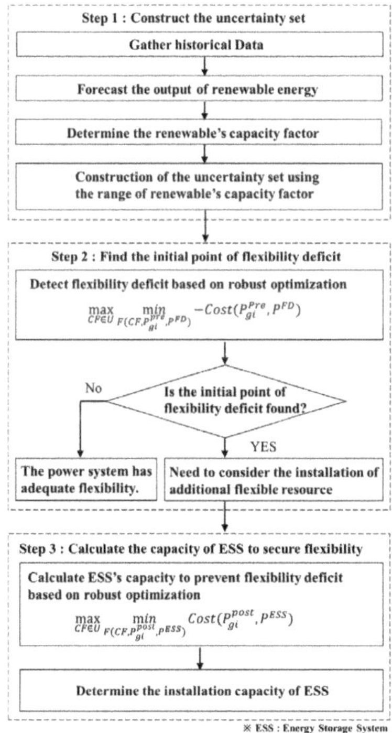

Figure 2. Proposed method.

3. Simulation and Results

This section describes verification of the proposed method using a power system in the Republic of Korea. After determining the interval of the capacity factor as the uncertainty set, the system checks whether a flexibility deficit occurs or whether sufficient ramp rates exist to provide flexible resources, which is considered a ramp rate in this step. Next, the capacity of flexible resources required by the power system to prevent a flexibility deficit is determined. This study was implemented using MATLAB 2017a, YALMIP20180612 as the optimization model language [38], and CPLEX 12.7 as the optimization solver. YALMIP can solve the robust optimization problem based on MATLAB with a variable solver including CPLEX and GROUBI.

3.1. Data Description

This study was conducted on a power system in the Republic of Korea. The offline generators were not considered because the flexibility was provided using in-service generators within a short period of time. There are 143 conventional generators in service, with a total capacity of approximately 93.175 GW. The system is composed of a gas turbine, hydropower, coal-fuel, liquefied natural gas (LNG), and nuclear units. The generation cost depends on the cost coefficient of the generators. Table 1 shows the generator data including number of generators, average ramp rates, maximum and minimum outputs, and cost coefficients.

Table 1. Data of in-service generators. Liquefied natural gas (LNG)

	Total Number	Ramp Rate (MW/h)	P_{min} (MW)	P_{max} (MW)	a_i (₩/h)	b_i (₩/MWh)	c_i (₩/MW²h)
Gas turbine	18	26.669	259.500	627.044	370.5158	1.29171	0.000976
Hydro	22	119.154	90.227	280.727	16.54394	1.70669	0.006959
Coal-fuel	60	15.702	357.353	649.480	164.3439	1.82554	0.000317
LNG	25	25.564	230.800	614.139	94.9775	1.43732	0.000270
Nuclear	18	1.766	973.111	1188.333	473.3557	1.69536	0.000296

According to [1,2], the total capacity of renewable energy will be from 11.3 GW in 2017 to 58.5 GW in 2030, when solar and wind power will be the main renewable resources and make up more than 88% of the total capacity of renewable energy. In this study, the output of renewable energy source is related to the capacity factor, which is considered an uncertainty set by assuming a range of 20–40%. In addition, the peak load predicted for 2030 is considered and is predicted to be 100.5 GW, assuming that the load will increase by an average of 1.3% per year [2].

3.2. Searching for the Initial Point of Flexibility Deficit with Robust Optimization

The proposed robust optimization in step 2 is applied to find the initial point of the flexibility deficit within the uncertainty set. In this simulation, for a flexible resource, only the ramp rates are considered by assuming that there are no other flexible resources, such as an ESS. The result is shown in Table 2. The initial point of the flexibility deficit occurs at a capacity factor of 0.3051, which means the power system requires additional flexible resources such as ramp rates, an ESS, and the demand response to prevent a flexibility deficit. In addition, it has the highest value within the range of the capacity factor because the objective function has a negative sign.

Table 2. Initial point of flexibility deficit.

	Capacity Factor	Value of Objective Function
Robust	0.3051	−179,890

The above results were analyzed in more detail by increasing the capacity factor by 0.005, from 0.2 to 0.4, using a deterministic method, which is represented in Figures 3 and 4. Figure 3a is divided into three parts: a section ensuring a flexibility deficit (A), a section with an increase in the flexibility deficit (B), and a section with a nearly uniform flexibility deficit (C). Unless a flexibility deficit exists, the more the capacity factor of renewable energy increases, the greater the total decrease in power generation, which can affect the cost of power generation or the absolute value of the objective function. So, in section A, the value of the objective function gradually increases by increasing the capacity factor of renewable energy, since the objective function has a negative sign, shown in Figure 3b. However, after the initial point of the flexibility deficit, the absolute value of the objective function can increase, owing to the penalty of the deficit. Thus, section A has a higher cost than sections B and C, because not only is there no flexibility deficit, but the objective function also has a negative sign. The initial point of the flexibility deficit is also located at the capacity factor of renewable energy = 0.3 between section A and section B from the deterministic approach. Through this result, the robust and deterministic approach is almost the same.

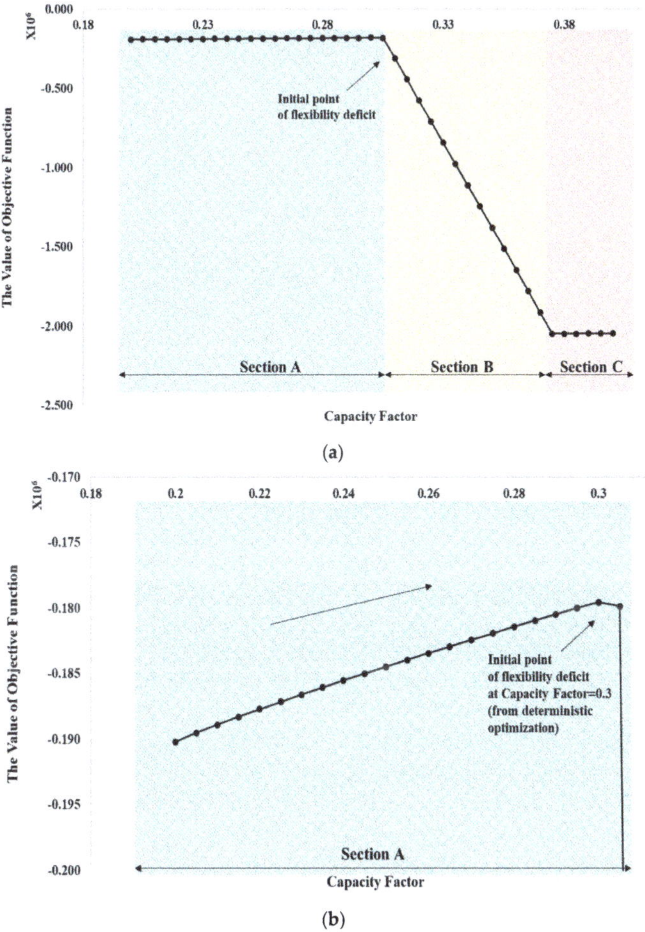

Figure 3. Results of deterministic optimization per 0.005 increase in capacity factor for step 2: values of objective function (**a**) within the interval of the capacity factor and (**b**) of section A.

Owing to the flexibility deficit, sections B and C need additional flexible resources to respond to the net load ramp. Section B shows an increase in the flexibility deficit because of insufficient ramp rates as flexible resources. Although section C is also an interval with a flexibility deficit, its interval seldom has an increase in flexibility deficit owing to the increase in ramp rates caused by a reduction in power generation. An increase in the generation of renewables can increase variation and uncertainty, while it can affect the decrease in net load and total power generation and the increase in ramp rates.

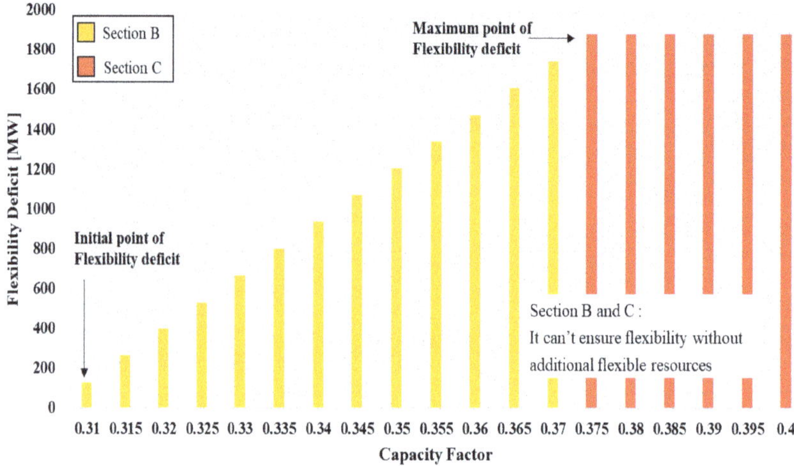

Figure 4. Results of deterministic optimization per 0.005 increase of capacity factor for step 2.

3.3. Determining the Capacity of Eenergy Storage System(ESS) to Ensure Nonnegative Flexibility

The capacity of an ESS to ensure nonnegative flexibility through the proposed robust optimization applied in step 3 was determined. The results are compared between deterministic and robust optimizations, shown in Table 3. Using a deterministic approach, the results show that the cost is small when the capacity factor is 0.2. The cost of the capacity factor, 0.4, is about 42.06 times higher than that of the capacity factor, 0.2, which occurs from the ESS installation to ensure nonnegative flexibility. Indeed, when the capacity factor is 0.4, many more flexible resources are needed than with a capacity factor of 0.2. It is reasonable that there is a need for flexible resources caused by an increase in the variation of renewable energy when the capacity factor is higher. Reviewing the results of robust optimization, when the capacity factor is 0.3749 within the uncertainty set, the necessary capacity of the ESS has a maximum value of 1875.7 MW. This means the system requires the largest capacity of the ESS installation within the uncertainty set.

Table 3. Comparison between deterministic and robust results.

	Capacity Factor	Value of Objective Function	Installation of ESS (MW)
Deterministic	0.2000	215,279	0
Deterministic	0.4000	9,056,140	1777.4
Robust	0.3749	9,545,556	1875.7

To analyze this result in detail, the previous deterministic approach is used, which is a method for increasing the capacity factor from 0.2 to 0.4 by steps of 0.05. The results are shown in Figure 5. Within the range of the capacity factor, the results based on a deterministic approach show that the value of the objective function is the smallest when the capacity factor is 0.300 and the highest when the capacity factor is 0.375, as shown in Figure 5a. In section A, the cost of power generation decreases through an increase in the capacity factor of renewable energy and a decrease in total generation and net load, shown in Figure 5b. The point at a capacity factor of 0.300 is the smallest value of the objective function and is located near the initial point of the flexibility deficit. In section C, the value of the objective function gradually decreases because the ramp rates increase by reducing the power generation. The maximum point of the value of objective function is placed at a capacity factor of renewable energy = 0.375 between sections B and C from the deterministic approach. Also, the capacity factor is almost the same as the result of robust optimization. Sections B and C both require additional flexible resources to prevent a flexibility deficit. However, section B increases the flexibility deficit and section C no longer increases the flexibility deficit by increasing the capacity factor of renewable energy. Due to this fact, in section C, the value of the objective function decreases by increasing the capacity factor of renewable energy, shown in Figure 5c. The details of this are introduced in Figure 6.

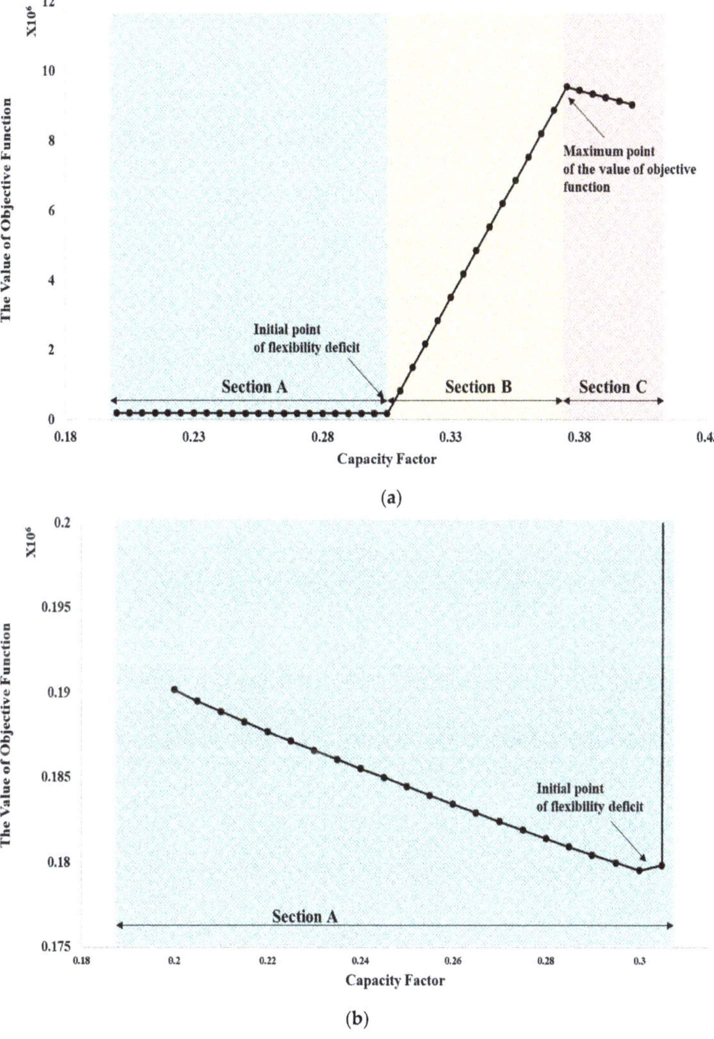

Figure 5. *Cont.*

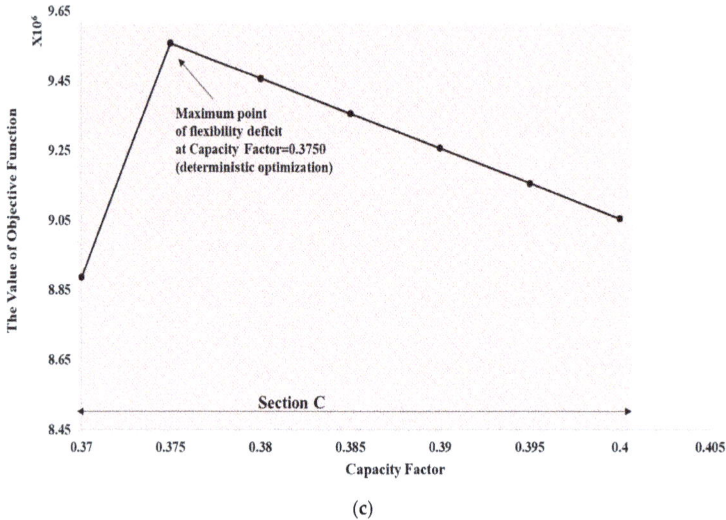

(**c**)

Figure 5. Results of deterministic optimization per 0.005 increase in capacity factor for step 3: values of objective function (**a**) within the interval of the capacity factor, (**b**) of section A, and (**c**) of section C.

In section A, the total sum of ramp rates increases owing to a decrease in the net load through an increase in the renewable energy output. In section B, the total sum of ramp rates gradually decreases because of a decline in the number of generators required to maintain supply and demand. In addition, in section C, the total sum of ramp rates increases again because the number of generators out of service no longer increases. This may be why the necessary capacity of an ESS is not the highest at the maximum capacity factor despite the increase in variability and uncertainty from renewable energy.

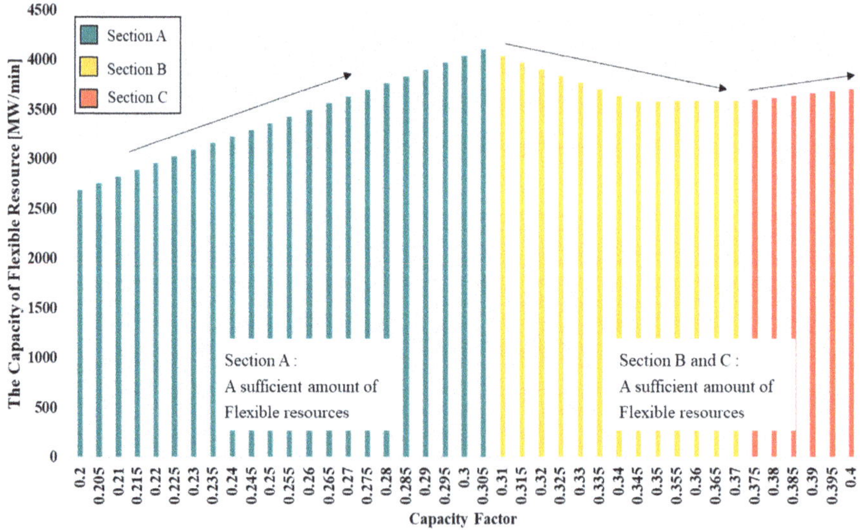

Figure 6. Total capacity of flexible resources by capacity factor.

4. Conclusions

A deterministic optimization cannot consider uncertainty, which can undermine the reliability of the solution owing to an inability to reflect the uncertainty. In previous studies using deterministic optimization and stochastic programming, in order to consider uncertainty such as that found with renewable energy, it was necessary to make scenarios. For example, all capacity factors of renewable energy constructed by a planner are reviewed to decide how the system will be reinforced for stable operation despite variation and uncertainty, which may require much effort to do. However, robust optimization does not require creating scenarios or using much effort because it needs the uncertainty set. Therefore, it is proper to use robust optimization to include uncertainty. In actuality, in power system planning and operation, because it is extremely difficult to take into account all possible scenarios, it is reasonable to prepare a countermeasure for the worst case. Therefore, robust optimization is a suitable model in power system planning and operation.

This paper presents a robust optimization model to secure flexible resources and prevent the occurrence of a flexibility deficit from the variability and uncertainty of renewable energy. This model considers the capacity factor of renewable energy as the uncertainty set and is divided into two steps: (i) searching for the initial point of the flexibility deficit and (ii) determining the capacity of the ESS to ensure nonnegative flexibility. In the first step, it is determined whether a flexibility deficit point occurs within the interval of the capacity factor when only considering ramp rates as flexible resources. This step takes place before determining whether to invest in flexible resources. In the next step, the necessary capacity of the ESS is calculated, which can ensure nonnegative flexibility within the uncertainty set. Through this study, the results of the worst case using a deterministic approach and robust optimization are similar. Indeed, searching the worst case using a deterministic approach may require many things, from making to studying scenarios, but robust optimization may be able to reduce the effort of considering the worst case without creating scenarios.

Future work will include more detailed modeling, including power flow limits of transmission lines and unit commitment to improve the quality of the solution. It will also be necessary to contain realistic conditions to guarantee a solution.

Author Contributions: J.J. conceived and designed the research methodology, performed the system simulations, and wrote this paper. B.L. supervised the research, improved the system simulation, and made suggestions regarding this research. The other authors discussed and contributed to the writing of the paper.

Funding: This research was supported by the Korea Electric Power Corporation (grant number: R17XA05-4) and the Human Resource Program in Energy Technology of the Korea Institute of Energy Technology Evaluation and Planning (KETEP) granted financial resource from the Ministry of Trade, Industry & Energy, Republic of Korea (No. 20174030201820).

Conflicts of Interest: No conflicts of interest relevant to this article are reported.

References

1. The Ministry of Trade, Industry and Energy. *New and Renewable Energy 3020 Implementation Plan*; The Ministry of Trade, Industry and Energy: Sejong-si, Korea, 2017. (In Korean)
2. The Ministry of Trade, Industry and Energy. *The 8th Basic Plan for Long-Term Electricity Supply and Demand (2017–2031)*; The Ministry of Trade, Industry and Energy: Sejong-si, Korea, 2017. (In Korean)
3. Stram, B.N. Key challenges to expanding renewable energy. *Energy Policy* **2016**, *96*, 728–734. [CrossRef]
4. Kunz, H.; Hagens, N.; Balogh, S. The Influence of Output Variability from Renewable Electricity Generation on Net Energy Calculations. *Energies* **2014**, *7*, 150–172. [CrossRef]
5. Haas, J.; Cebulla, F.; Cao, K.; Nowak, W.; Palma-Behnke, R.; Rahmann, C.; Mancarella, P. Challenges and trends of energy storage expansion planning for flexibility provision in low-carbon power systems—A review. *Renew. Sustain. Energy Rev.* **2017**, *80*, 603–619. [CrossRef]
6. Bae, M.; Lee, H.; Lee, B. An Approach to Improve the Penetration of Sustainable Energy Using Optimal Transformer Tap Control. *Sustainability* **2017**, *9*, 1536.

7. Liu, Z.; Chen, Y.; Luo, Y.; Zhao, G.; Jin, X. Optimized Planning of Power Source Capacity in Microgrid, Considering Combinations of Energy Storage Devices. *Appl. Sci.* **2016**, *6*, 416. [CrossRef]
8. Shigenobu, R.; Noorzad, A.; Muarapaz, C.; Yona, A.; Senjyu, T. Optimal Operation and Management of Smart Grid System with LPC and BESS in Fault Conditions. *Sustainability* **2016**, *8*, 1282. [CrossRef]
9. Tuohy, A. *Flexibility in Transmission and Resource Planning*; EPRI: Palo Alto, CA, USA, 2014.
10. EPRI. *Metrics for Quantifying Flexibility in Power System Planning*; EPRI: Palo Alto, CA, USA, 2014.
11. Cochran, J.; Miller, M.; Zinaman, O.; Milligan, M.; Arent, D.; Palmintier, B.; O'Malley, M.; Mueller, S.; Lannoye, E.; Tuohy, A.; et al. *Flexibility in 21st Century Power Systems*; National Renewable Energy Lab. (NREL): Golden, CO, USA, 2014.
12. Danish Energy Agency. Flexiblity in the Power System—Danish and European Experience, 2015. Available online: https://ens.dk/sites/ens.dk/files/Globalcooperation/flexibility_in_the_power_system_v23-lri.pdf (accessed on 9 November 2018).
13. Poncela, M.; Purvins, A.; Chondrogiannis, S. Pan-European Analysis on Power System Flexibility. *Energies* **2018**, *11*, 1765. [CrossRef]
14. Lannoye, E.; Daly, P.; Tuohy, A.; Flynn, D.; O'Malley, M. Assessing Power System Flexibility for Variable Renewable Integration: A Flexibility Metric for Long-Term System Planning. *CIGRE Sci. Eng. J.* **2015**, *3*, 26–39.
15. California ISO. Final Flexible Capacity Needs Assessment for 2019, 2018. Available online: http://www.caiso.com/informed/Pages/StakeholderProcesses/FlexibleCapacityNeedsAssessmentProcess.aspx (accessed on 9 November 2018).
16. California ISO. Flexible Ramping Product Uncertainty Calculation and Implementation Issues, 2018. Available online: https://www.caiso.com/Documents/FlexibleRampingProductUncertaintyCalculationImplementationIssues.pdf (accessed on 9 November 2018).
17. Rodrigues, E.M.G.; Godina, R.; Santos, S.F.; Bizuayehu, A.W.; Contreras, J.; Catalão, J.P.S. Energy storage systems supporting increased penetration of renewables in islanded systems. *Energy* **2014**, *75*, 265–280. [CrossRef]
18. Denholm, P. Energy storage to reduce renewable energy curtailment. In Proceedings of the 2012 IEEE Power and Energy Society General Meeting, San Diego, CA, USA, 22–26 July 2012; pp. 1–4.
19. Jirutitijaroen, P.; Singh, C. Reliability Constrained Multi-Area Adequacy Planning Using Stochastic Programming with Sample-Average Approximations. *IEEE Trans. Power Syst.* **2008**, *23*, 504–513. [CrossRef]
20. Hajipour, E.; Bozorg, M.; Fotuhi-Firuzabad, M. Stochastic Capacity Expansion Planning of Remote Microgrids with Wind Farms and Energy Storage. *IEEE Trans. Sustain. Energy* **2015**, *6*, 491–498. [CrossRef]
21. Zhan, Y.; Zheng, Q.P.; Wang, J.; Pinson, P. Generation Expansion Planning with Large Amounts of Wind Power via Decision-Dependent Stochastic Programming. *IEEE Trans. Power Syst.* **2017**, *32*, 3015–3026. [CrossRef]
22. Alharbi, H.; Bhattacharya, K. Stochastic Optimal Planning of Battery Energy Storage Systems for Isolated Microgrids. *IEEE Trans. Sustain. Energy* **2018**, *9*, 211–227. [CrossRef]
23. Ruiz, C.; Conejo, A.J. Robust transmission expansion planning. *Eur. J. Oper. Res.* **2015**, *242*, 390–401. [CrossRef]
24. Dehghan, S.; Amjady, N. Robust Transmission and Energy Storage Expansion Planning in Wind Farm-Integrated Power Systems Considering Transmission Switching. *IEEE Trans. Sustain. Energy* **2016**, *7*, 765–774. [CrossRef]
25. Liu, D.; Shang, C.; Cheng, H. A two-stage robust optimization for coordinated planning of generation and energy storage systems. In Proceedings of the 2017 IEEE Conference on Energy Internet and Energy System Integration (EI2), Beijing, China, 26–28 November 2017; pp. 1–5.
26. Thatte, A.A.; Sun, X.A.; Xie, L. Robust Optimization Based Economic Dispatch for Managing System Ramp Requirement. In Proceedings of the 2014 47th Hawaii International Conference on System Sciences, Waikoloa, HI, USA, 6–9 January 2014; pp. 2344–2352.
27. Ampatzis, M.; Nguyen, P.H.; Kamphuis, I.R.; van Zwam, A. Robust optimisation for deciding on real-time flexibility of storage-integrated photovoltaic units controlled by intelligent software agents. *IET Renew. Power Gener.* **2017**, *11*, 1527–1533. [CrossRef]
28. Lorca, A.; Sun, X.A. Adaptive Robust Optimization with Dynamic Uncertainty Sets for Multi-Period Economic Dispatch Under Significant Wind. *IEEE Trans. Power Syst.* **2015**, *30*, 1702–1713. [CrossRef]

29. Yi, J.; Lyons, P.F.; Davison, P.J.; Wang, P.; Taylor, P.C. Robust Scheduling Scheme for Energy Storage to Facilitate High Penetration of Renewables. *IEEE Trans. Sustain. Energy* **2016**, *7*, 797–807. [CrossRef]
30. Kim, J.; Choi, Y.; Ryu, S.; Kim, H. Robust Operation of Energy Storage System with Uncertain Load Profiles. *Energies* **2017**, *10*, 416. [CrossRef]
31. Ben-Tal, A.; El Ghaoui, L.; Nemirovski, A. *Robust Optimization*; Princeton University Press: Princeton, NJ, USA, 2009; Volume 28.
32. Ben-Tal, A.; Nemirovski, A. Robust optimization—Methodology and applications. *Math. Program.* **2002**, *92*, 453–480. [CrossRef]
33. Steuben, J.C.; Turner, C.J. Robust optimization of mixed-integer problems using NURBs-based metamodels. *J. Comput. Inf. Sci. Eng.* **2012**, *12*, 041010. [CrossRef]
34. Bertsimas, D.; Sim, M. The price of robustness. *Oper. Res.* **2004**, *52*, 35–53. [CrossRef]
35. Bertsimas, D.; Dunning, I.; Lubin, M. Reformulation versus cutting-planes for robust optimization. *Comput. Manag. Sci.* **2016**, *13*, 195–217. [CrossRef]
36. Yuan, Y.; Li, Z.; Huang, B. Robust optimization under correlated uncertainty: Formulations and computational study. *Comput. Chem. Eng.* **2016**, *85*, 58–71. [CrossRef]
37. Bertsimas, D.; Brown, D.B.; Caramanis, C. Theory and Applications of Robust Optimization. *SIAM Rev.* **2011**, *53*, 464–501. [CrossRef]
38. Löfberg, J. Automatic robust convex programming. *Optim. Methods Softw.* **2012**, *27*, 115–129. [CrossRef]

 *sustainability*

Article

Transient Impact Analysis of High Renewable Energy Sources Penetration According to the Future Korean Power Grid Scenario

Seungchan Oh, Heewon Shin, Hwanhee Cho and Byongjun Lee *

School of Electrical Engineering, Korea University, Anam Campus, 145 Anam-ro, Seongbuk-gu,
Seoul 02841, Korea; tmdckssla@korea.ac.kr (S.O.); redcore451@korea.ac.kr (H.S.); whee88@korea.ac.kr (H.C.)
* Correspondence: leeb@korea.ac.kr; Tel.: +82-2-3290-3246; Fax: +82-2-3290-3692

Received: 11 October 2018; Accepted: 8 November 2018; Published: 11 November 2018

Abstract: Efforts to reduce greenhouse gas emissions constitute a worldwide trend. According to this trend, there are many plans in place for the replacement of conventional electric power plants operating using fossil fuels with renewable energy sources (RESs). Owing to current needs to expand the RES penetration in accordance to a new National power system plan, the importance of RESs is increasing. The RES penetration imposes various impacts on the power system, including transient stability. Furthermore, the fact that they are distributed at multiple locations in the power system is also a factor which makes the transient impact analysis of RESs difficult. In this study, the transient impacts attributed to the penetration of RESs are analyzed and compared with the conventional Korean electric power system. To confirm the impact of the penetration of RESs on transient stability, the effect was analyzed based on a single machine equivalent (SIME) configuration. Simulations were conducted in accordance to the Korean power system by considering the anticipated RES penetration in 2030. The impact of RES on transient stability was provided by a change in CCT by increasing of the RES penetration.

Keywords: transient impact; renewable energy source penetration; power system stability

---

1. Introduction

As part of the expended efforts to reduce greenhouse gas emissions in recent years, the installation of renewable power sources (RESs) is continually increasing. Countries around the world are formulating plans to replace existing fossil-fuel-based power stations with RES [1–4]. The Korean power system is also planning to increase its RES penetration. According to the recently announced the 8th basic plan for the supply and demand of electricity, 60 GW of renewable power will be installed in the Korean power system by 2030. The total installed capacity of wind and photovoltaic power is planned to be 53 GW, and the ratio of solar power to wind power is expected to be 2:1. The peak load of the Korean power system in 2030 is expected to be approximately 100 GW. In effect, the scale of the wind power and photovoltaic power generation capacities will be 53% of the peak load.

Many parts of the RESs have characteristics that are different from those of conventional generators. One of the most important differences is fluctuation of the electrical power output. In the case of the conventional generator, the output can be controlled. Therefore, when the stability of the conventional generator is examined, no major problems are anticipated, even if it is operated at the rated output power. However, in the case of RES, such as the wind power generations and photovoltaics, the output cannot be arbitrarily adjusted. Based on these characteristics, the study of the cooperation of RESs must be able to reflect their output variations. To cope with such RES fluctuations, multiple research studies have been conducted on the effects of the mixing of various RESs and the utilization of energy storage system (ESS) [5–8]. Based on the penetration of RESs, the scale of the impact of their output fluctuations has changed. Correspondingly, studies of the impact of RES penetration on the individual power grid have been conducted [9–13], in view of the increased importance of the penetration of RESs and their influences on the stability of the electric power system.

Changes in the static characteristics of the power system due to the penetration of RESs can be easily calculated. However, in the case of transient stability, it is not easy to analyze the effects of the penetration of RESs. In the MIGRATE report in Europe, the decrease in the transient stability margin owing to the increase of the penetration of RESs is specified as an important observation point [14]. According to the results of this study, it can be observed that transient stability is improved initially according to the expansion of the penetration of RESs. However, according to the expansion of penetration of RESs, it has been documented that the transient stability has continually deteriorated [9,11]. Generally, when the specific inertia of the renewable power source is small, the RES output is replaced with the output of the conventional generators, thereby improving the transient stability response. However, when the ratio of the renewable power source was increased, the existing generator had to be stopped, and the transient stability deteriorated owing to the decrease of the system inertia [15].

Various studies have been conducted to analyze the impact of expansion of RESs on transient stability [16–19]. Most of the studies have been based on simulations whereby the penetration of RESs affect the transient stability. However, only a limited number of prior studies exist which have investigated the main factor that affects the transient stability in the situation where the penetration of RESs has expanded. The study by Liu et al. explained the influence of RESs on the transient stability based on effective equal area criterion (EEAC) [19]. In this study, it was shown that the main factor that affected the transient stability were the locations of the RESs and the adjustment of the output of conventional generators. However, this study was considered only when RESs directly replaced the output of the conventional generator.

This paper investigated the impact of the distributed penetration of RESs on transient stability. The major factors of change at transient stability are derived by configuration of hybrid method reflecting RES penetration. The simulation was conducted in the Korean power system in 2030. The penetration of RESs was distributed to individual buses based on the planned capacity in accordance to geographical region. The impact of RES penetration on the transient was provided as the change of the CCT in each area. Based on simulation, the impact of the penetration of RESs in the Korean power system was represented.

2. Dynamic Impact Analysis for RES with Hybrid Method

2.1. Equal Area Criterion (EAC) and Single Machine Equivalent (SIME) Configuration

The equal area criterion (EAC) is a method used to evaluate the transient stability of the power system. When a contingency occurs, the electric power of the generator decreases owing to a voltage drop, thus resulting in an acceleration of the generator. After the elimination of the fault, the generator is decelerated because the electric power is increased after the phase angle induced by the fault is increased. If the acceleration area is larger than the deceleration area, the generator is desynchronized.

To analyze the transient stability of the power system with EAC, the system must be modeled in a one machine infinite bus (OMIB) [20]. However, the actual power system consists many generators and complex transmission system. Therefore, all generators are reduced as the one generator by SIME configuration. For this reduction, all system generators are classified into critical or non-critical group. The equivalent phase angles of these two groups are defined according to their central of angle (COA) values. The definition of angle difference and COA for each group is given as follows:

$$\delta \triangleq \delta_C - \delta_N \tag{1}$$

$$\begin{cases} \delta_C \triangleq M_C^{-1} \sum_{i \in C} M_i \delta_i \\ \delta_N \triangleq M_N^{-1} \sum_{k \in N} M_k \delta_k \end{cases} \tag{2}$$

$$\left(M_C = \sum_{i \in C} M_i \,,\; M_N = \sum_{k \in N} M_k \right) \tag{3}$$

where:

δ = the rotor angle of the equivalent machine
δ_C = COA of the critical group
δ_N = COA of the non-critical group
δ_i, δ_k = the rotor angle of the machine i, k
M_i, M_k = the inertia of the machine i, k
M_C = COA of the critical group
M_N = COA of the non-critical group

Other parameters, including the angular velocity and acceleration, mechanical input, and electrical output, can be calculated based on the above definitions:

$$\begin{cases} \omega_C = \dot{\delta}_C = M_C^{-1} \sum_{i \in C} M_i \omega_i \\ \omega_N = \dot{\delta}_N = M_N^{-1} \sum_{k \in N} M_k \omega_k \end{cases} , \quad \begin{cases} \ddot{\delta}_C = M_C^{-1} \sum_{i \in C} M_i \ddot{\delta}_i \\ \ddot{\delta}_N = M_N^{-1} \sum_{k \in N} M_k \ddot{\delta}_k \end{cases}$$
$$\omega = \omega_C - \omega_N, \quad \ddot{\delta} = \ddot{\delta}_C - \ddot{\delta}_N \tag{4}$$

$$P_m = M \left(M_C^{-1} \sum_{i \in C} P_{mi} - M_N^{-1} \sum_{k \in N} P_{mk} \right)$$
$$P_e = M \left(M_C^{-1} \sum_{i \in C} P_{ei} - M_N^{-1} \sum_{k \in N} P_{ek} \right) \tag{5}$$
$$(M = M_C M_N / (M_C + M_N))$$

where:

ω = the rotor angle of the equivalent machine
ω_C = the angular velocity of the critical group
ω_N = the angular velocity of the non-critical group
ω_i, ω_k = the rotor angle of the machine i, k
P_m = COA of the critical group
P_e = COA of the non-critical group
P_{mi}, P_{mk} = mechanical input of the machine i, k
P_{ei}, P_{ek} = electrical output of the machine i, k

2.2. Transient Impact Analysis of RES Penetration with the Hybrid Method

In this section, the impact of the RES penetration on the transient stability is derived from the SIME configuration. If the generator included in the critical group is stopped, it can affect to the transient stability. However, if the generator included in the non-critical group is stopped, impact of stopping generator can ignore. Therefore, only the situation where the stop of the generator included in the critical group is considered. In the case of the conventional generator that included in critical group with the RESs, mechanical input, and electrical output of the equivalent generator is changed as follows:

$$\begin{aligned} P'_m &= \frac{(M_C - M_d) \cdot M_N}{(M_T - M_d)} \cdot \left((M_C - M_d)^{-1} \left(\left(\sum_{i \in C} P_{mi} \right) - P_{md} \right) - M_N^{-1} \sum_{k \in N} P_{mk} \right) \\ &= P_m + \frac{M_N \cdot M_d}{M_T \cdot (M_T - M_d)} \cdot \left(\sum_{i \in C} P_{mi} + \sum_{k \in N} P_{mk} \right) - \frac{M_N}{(M_T - M_d)} \cdot P_{md} \end{aligned} \tag{6}$$

$$\begin{aligned} P'_e &= \frac{(M_C - M_d) \cdot M_N}{(M_T - M_d)} \cdot \left((M_C - M_d)^{-1} \left(\left(\sum_{i \in C} P_{ei} \right) - \alpha \cdot P_{RES} \right) - M_N^{-1} \sum_{k \in N} P_{ek} \right) \\ &= P_e + \frac{M_N \cdot M_d}{M_T \cdot (M_T - M_d)} \cdot \left(\sum_{i \in C} P_{ei} + \sum_{k \in N} P_{ek} \right) - \frac{M_N}{(M_T - M_d)} \cdot \alpha \cdot P_{RES} \end{aligned} \tag{7}$$

$$P'_a = P_a + \frac{M_N}{(M_T - M_d)} \cdot \left(\frac{M_d}{M_T} \cdot \left(\sum_{j \in T} P_{mj} - P_{ej} \right) - (P_{md} - \alpha \cdot P_{RES}) \right) \tag{8}$$

$$(M_T = M_C + M_N)$$

where:

P'_m = new mechanical input of the equivalent machine reflecting RES penetration

P'_e = new electrical output of the equivalent machine reflecting RES penetration

P_{md}, M_d = mechanical input and inertia of the stropped generator

P_{RES} = amount of the RES penetration

α = the ratio about RES output and electrical output reduction of the equivalent machine

Equations (6)–(8) represent the cases where the generator is stopped and the RES has been installed. Equation (6) represents the mechanical input of the equivalent generator reflecting stop of the conventional generator. Mechanical input of the equivalent generator decreases by the stop of the generator which is included in the critical group. In Equations (7) and (8), α denotes the ratio of the RES power generation which contributes to the reduction of the electrical output. The value of α is unity if the RES is installed in the same location as the generator of the critical group. Conversely, α is zero when the installation position of the RES is electrically far away from the critical group. Equation (7) is the electrical output of the equivalent generator with RES penetration. Though, some type of the RES has a mechanical input and inertia, it cannot be applied to the mechanical input because it is connected to the power system through the inverter device and asynchronous characteristic. Equation (8) shows acceleration energy of the equivalent generator.

In Equation (8), change of the stop of the generator and RES expansion is canceled. However, impact of the contingency affect differently in mechanical input and electrical output. Mechanical input is not affected by the contingency. On the other hand, electrical output reduced by the power system voltage, especially during the contingency condition. Most of RESs have little risk about loss of synchronism and storing acceleration energy in the inertia. That is the reason why the impact of RES is applied to the electrical output of the equivalent generator.

To apply RES penetration in SIME as above Equations (6)–(8), RES should not be included in SIME configuration. It means that RES only applied in dynamic time domain simulation to affect the simulation result. By the RES penetration, the electrical output of the equivalent generator is changed. However, the mechanical input is affected by the stop of the generator only. Proposed SIME configuration reflecting RES penetration can provide the numerical margin to analyze the transient impact.

2.3. Change of the Power-Angle Curve by the RES Penetration

To understand the impact of the RES penetration on the transient stability, transient impact of the RESs derived from Equations (6)–(8) is applied in the power-angle curve. Power-angle curve is provided to explain the impact of the RES penetration with four scenarios. Table 1 summarizes these scenarios.

Table 1. Four scenarios used to investigate the impact of RES penetration.

	Scenario
Case 0	Base case
Case A	Reduction of mechanical input in critical group
Case B	Increment of RES power generation in the proximity of the critical group
Case C	Replacement of conventional generator with RES
Case D	Decrease of generator inertia in critical group

2.3.1. Case A—Reduction of Mechanical Input in Critical Group

Case A is the case where only the mechanical input decreases owing to the termination of the operation of the generator. Since there is no RES input, P_{RES} is equal to zero. For convenience, inertia effects are ignored in this analysis. Figure 1 depicts the power–angle curve in the case where a reduction of mechanical input has occurred. Points a–e respectively represent the operating points before the fault, at the instant at which the fault occurs, after the fault, after the fault clears, and the case at which the operating point has an angular velocity of zero. The subscript indicates the relevant scenario. The colored area shows the acceleration/deceleration area of the base case, and the diagonally highlighted area represents the area within which the scenario has been applied. As is known, the reduction in the mechanical input improves the transient stability.

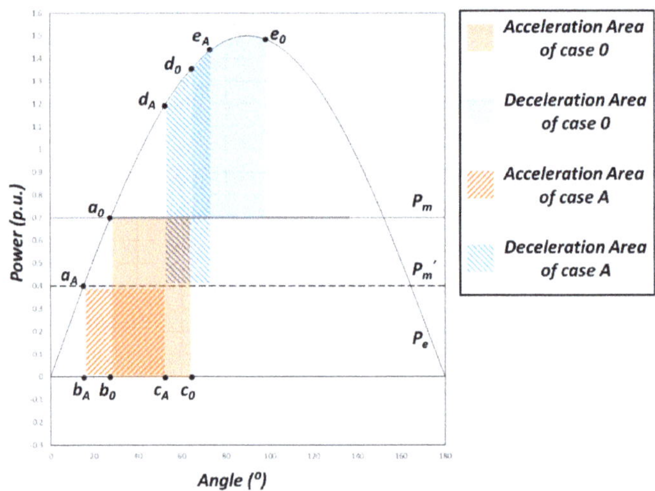

Figure 1. Impact of the reduction of mechanical input in the critical group.

2.3.2. Case B—Increment of RES Power Generation in the Proximity of the Critical Group

Case B is the case where only the output of RES is increased. Since the generator is still in operation, P_{md} is zero. For convenience, inertia effects are ignored. However, unlike the mechanical input, the output of the RES becomes zero when the fault occurs. At the fault state, the RES cannot maintain the required electrical output. Figure 2 shows the power–angle curve to which the installation of the RES is applied. The points displayed on the graph have the same meanings as those listed in Figure 1. The effect of the penetration of the RES is exactly the opposite of the effect induced when the generator stopped. In reference to cases A and B, the influences seem to have the same magnitude but opposing trends. However, at the fault condition, it is very important that the output of the RES is zero. Correspondingly, the size of the acceleration area of Case B is the same as the base case. In case A, the size of the acceleration area decreased because the mechanical input decreased. However, owing to the change of the normal operating point and the reduction of the deceleration area, transient stability deteriorated.

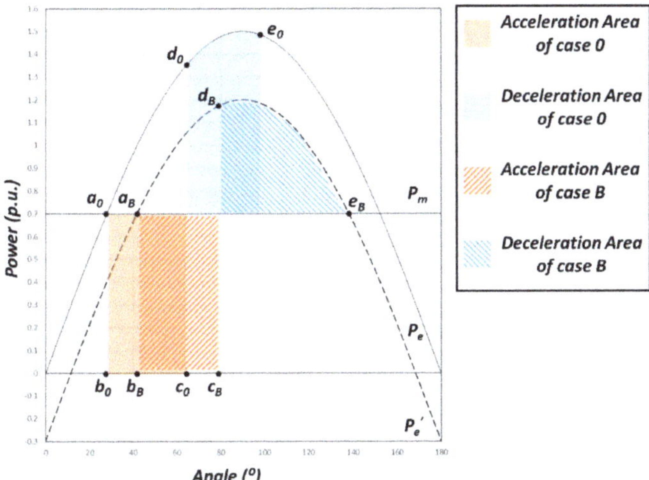

Figure 2. Impact of the power generation increment induced by the RES in the proximity of the critical group.

2.3.3. Case C—Replacement of Conventional Generator with RES

Case C is the case where the conventional generator is replaced by the RES. In this case, α is equal to unity and P_{RES} is equal to P_{md}. For convenience, inertia effects are ignored here. Figure 3 shows the power–angle curve for the case where the conventional generator is replaced by the RES. Since the output of the RES is not affected during the fault condition, only the change of the mechanical input is reflected in acceleration area changes. However, after the clearing of the fault, the deceleration area is the same as the base case. It means that transient stability is improved when the conventional generator replaced with RES like case C. In the general operating condition, this effect is greater because the value of α is less than one.

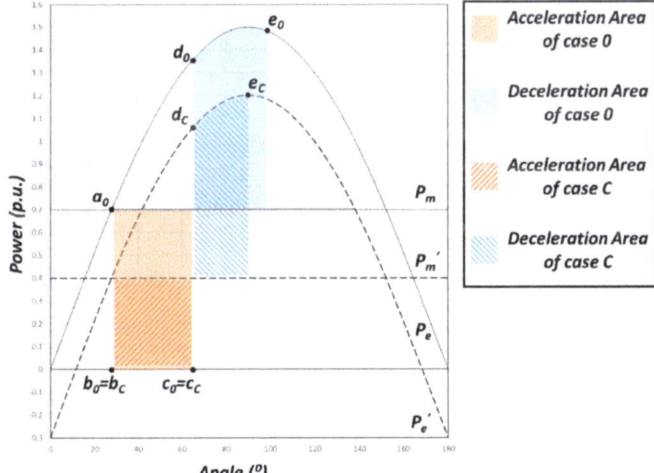

Figure 3. Impact of conventional generator replacement with RES.

2.3.4. Case D—Decrease of the Inertia of the Generator in the Critical Group

Case D shows the effect of the change of the inertia of the generator on the power–angle curve. The change in the inertia does not impose distinct changes in the power–angle curve. The main change elicited by the inertia variations relates to the increase in the phase angle during the fault, i.e., the distance between points b and c. As the overall inertia of the critical group decreases, the distance between points b and c increases. Figure 4 shows the change of the acceleration/deceleration areas induced by the reduction of the inertia. Since the change in the inertia does not cause a change in the mechanical input/electrical output, the power–angle curve is the same. When the same contingency occurs, the phase angle increases more than the base case. Such a change can cause transient instability.

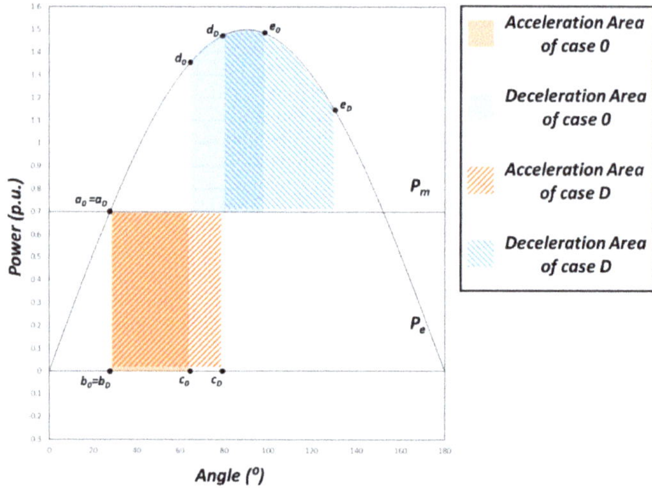

Figure 4. Impact of the decrease of the inertia of the generator in the critical group.

2.3.5. Other Characteristics of RES Affecting Transient Stability

Based on prior considerations, the RES has dynamic characteristics that are different from conventional generation. In addition, the RES has a different transient characteristic than conventional generators. The part that greatly affects the transient stability is the voltage/reactive power characteristic. The RES is generally connected to an inverter, or is installed together with a reactive power compensator. Generally, reactive power compensation contributes to system transient stabilization. This effect does not apply to the SIME configuration, however, it does affect the simulation result and electrical output of the equivalent generator. The effect of the reactive power support is also reflected in the transient stability margin.

2.4. Impact of RES Penetration in Transient Characteristics

Cases A–D show the influences of the RES penetration on the transient stability. Case A shows that the termination of the operation of the conventional generator by the penetration of the RES improves transient stability. Conversely, Case B shows that the increase in the generation of RES deteriorates transient stability. Case C represents the concurrent effects of Cases A and B. The result of case C shows that the transient stability improves, even if the effects of cases A and B are concurrently reflected. Case D shows the influence of the change of the inertia. When the inertia of the generator decreases considerably, the transient stability can deteriorate. However, assuming that the inertia values of the generators in the power system do not differ considerably, it can be assumed that the impact of inertia can be neglected.

If the power system changes owing to the penetration of RES are the same as those elicited in case C, the transient stability of the power system will be improved. However, this response is different from that elicited based on the research results. The existing research results show that the transient stability is improved when the penetration of the RES is small. However, as the scale of RES penetration increases, the transient stability deteriorates considerably. The reason for this discrepancy is attributed to the increased imbalance of regional supply and demand by the penetration of the RES. The area associated with power generation increases owing to the penetration of the RES is not same as the area associated with power generation decreases owing to the penetration of the RES. Therefore, the general change of the electrical grid by the penetration of the RES is denoted by cases A and B, and not by case C. Additionally, the installation location of the RES is separated by a distinct distance to the critical group. This means that the adverse effects of the penetration of the RES are not observed when the scale is small. Therefore, when the scale of the RES penetration is small, the improvement effect elicited in case A is conspicuous. However, as the penetration scale of the RES increases, adverse effects attributed to case B start to become evident.

3. Future Korean Grid Scenario with Increased RES Penetration

3.1. Penetration of RESs in the Future Korean Power Grid

According to the plan, the scale of RES will increase to approximately 53 GW by 2030. At 2030, the peak load of the Korean electric power system will become 100 GW. Unlike the conventional generators, it is not possible to accurately ascertain the scale and locations of the RESs based on this plan. Therefore, in consideration of the generally known size, the capacity of RES is distributed in accordance to geographical area. The RES capacity ratio in accordance to geographical region was set at 0% for the capital area, 25% for the Gangwon area, 5% for the Chungcheong area, 35% for Honam area, 35% for the Yeongnam area. The composition of the RES considers a 2:1 distribution ratio of photovoltaic and wind powers. The scenario reflects the distribution of the future RESs of the Korean power system. Figure 5 represents the outline of the RES penetration scenario.

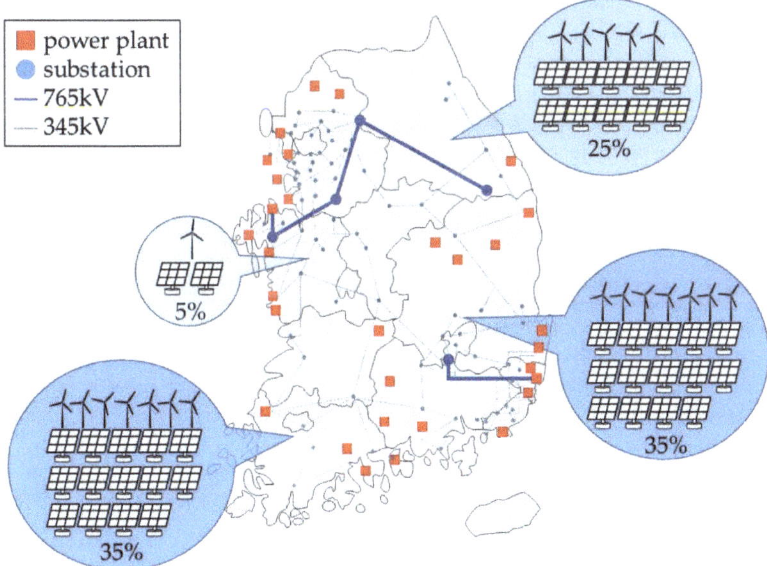

Figure 5. The outline of transmission system and power plant in the Korea power grid, and RES penetration scenario.

3.2. Reactive Power Capability of RES

The characteristics of the RES vary depending on the energy source and design. For accurate power system analyses, the differences of RESs should be identified. In order to reflect these, the type, design, capacity, and location of the RESs that need to be installed must be known. However, it is impossible to accurately determine the RES capacity of each bus at the system planning stage. Specifically, in the case of RES, there is an increased possibility that many plants will be installed by many businesses. As a result, the installation plan of RESs is very inaccurate at the planning stage of the long-term power system. Therefore RESs are modeled with several assumptions.

The first assumption is that most of the RESs consist of wind and photovoltaic sources. 8th basic plan for the supply and demand of electricity will install RESs with a total capacity of 60 GW of which 53 GW is consisted of wind and photovoltaic sources. The second assumption is that the wind source consists mostly of type 3 or type 4 generators. According to the grid code of the Korean electric power grid, in the case of wind power generators, it is necessary to be able to output reactive power that corresponds to 0.95 p.f. of the rated power output [3]. Therefore, the wind source was modeled as a type 3 or type 4 wind power generator capable of reactive power control. Furthermore, assuming that the operating characteristics of type 3 generators do not differ considerably from those for type 4 generators, all wind generator models are of type 4.

In the case of photovoltaics, it is assumed that the system is driven with a unity power factor because there is no grid code. There is no standard related to the voltage and reactive power of the photovoltaic source in the grid code of the Korean power system. Therefore, photovoltaic sources are not expected to produce reactive power for power system stability. The dynamic model of each RES utilizes the 2nd-generation model of WECC. In this model, each element is defined as a module so that the characteristics of the RESs are properly described, and the modules are connected to each other as needed [20,21].

3.3. Allocation of the Capacity of RESs

Conventional power plants have large capacities and their positions are determined at the power system planning stage. Additionally, the construction period is long and it is mainly connected to the high-voltage bus of the system. Conversely, RES has a smaller capacity than conventional plants, and is distributed over a relatively large area. Given these characteristics, the modeling and evaluation methods for RESs differ from those used for conventional sources. Nevertheless, to analyze the effect of the RES penetration, a RES installation scenario is required. Nevertheless, to analyze the effect of the RES penetration, a RES installation scenario is required. In this study, the capacity of the RES distributed by the limits of individual buses. In the planning stage, if the limits of individual buses are violated, the bus cannot accept the RES installation. Therefore, it can be inferred that more RESs will be installed on the bus that have more margin. In this paper, RES penetration scenario is formed as follow process.

Step 1. Perform individual evaluations on all buses that are candidates for the RES installation. Apply RES to each bus, apply contingencies within two-levels, and verify the overload and load flow convergence. The maximum allowable capacity that does not cause overload and load-flow divergence is the RES capacity limit of each bus. At this time, the upper limit of the RES capacity can be arbitrarily set. In that case, if a problem does not occur when the upper limit is set in place, the limit capacity of the RES is maintained to the upper limit.

Step 2. Determine the total capacity of the RES and the ratio of each region. The regional capacity is distributed to the individual buses at the rate of the RES limit for each bus, as calculated in Step 1. The relevant equations are listed below:

$$P_a = P_t \cdot r_a \tag{13}$$

$$P_i = (P_{li} \cdot P_a) \Big/ \sum_{k \in a} P_{lk} \tag{14}$$

where P_a is the RES capacity to be allocated to region a, P_t is the total capacity of the RES, r_a is the capacity ratio of region a, P_i is the RES capacity of the bus I included in region a, and P_{li} is the limit capacity of the bus i calculated in Step 1.

Step 3. Set the step of the penetration of the RES and increase the RES penetration step-by-step until the load flow calculation diverges. In this case, the dispatch of the supply and demand is based on the merit order.

4. Case Study

The voltage levels of the transmission facilities of KEPCO are 765 kV, 345 kV, and 154 kV. The contingency list includes double faults of transmission lines at the voltage level of 345 kV. However, transmission lines directly connected to the generator are excluded. The maximum RES penetration applied in this simulation was 30 GW. The composition of the RES considered a 2:1 ratio for the generated photovoltaic and wind powers. The scenario reflects the composition of the future RES of the Korean power system. Simulations were conducted with PSS/E to analyze the dynamic impact of the RES penetration.

In conventional studies, a change in the transient stability was observed based on a number of contingencies with CCT that were less than the reference value. The number of contingencies smaller than the reference CCT value decreased initially, but increased abruptly subsequently. To observe the changes of the transient characteristics of the Korean electric power system based on the penetration of the RES, the number of contingencies with CCT values smaller than the reference CCT was estimated. The applied reference CCT value was 200 ms.

Figures 6 and 7 shows the proportion of contingencies with CCT values less than 200 ms according to the penetration of RES and the studied geographical areas. Similar to the existing research results, the proportion of contingency tends to decrease in the section where the penetration of RES is small. As shown, this tendency is reverses when the RES penetration increases more than 20 GW. Subsequently, the proportion of contingencies with CCT less than 200 ms increases abruptly as the RES penetration increases. Figure 8 shows the decreases of the amounts of power generation in each geographical area according to the penetration of the RES. As shown, the amount of power generation decreases in the cases of the Capital and Chungcheong areas. Conversely, the amount of power generation increases in the cases of the areas of Gangwon, Honam, and Yeongnam. These outcomes considered the power generation amounts of the RES. In view of these results, it can be confirmed that imbalances of regional supply and demand occur based on the penetration of RES.

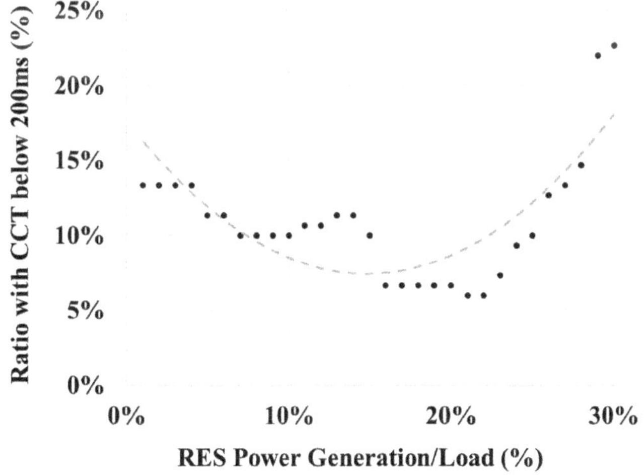

Figure 6. Change of ratio with CCT below 200 ms by RES penetration.

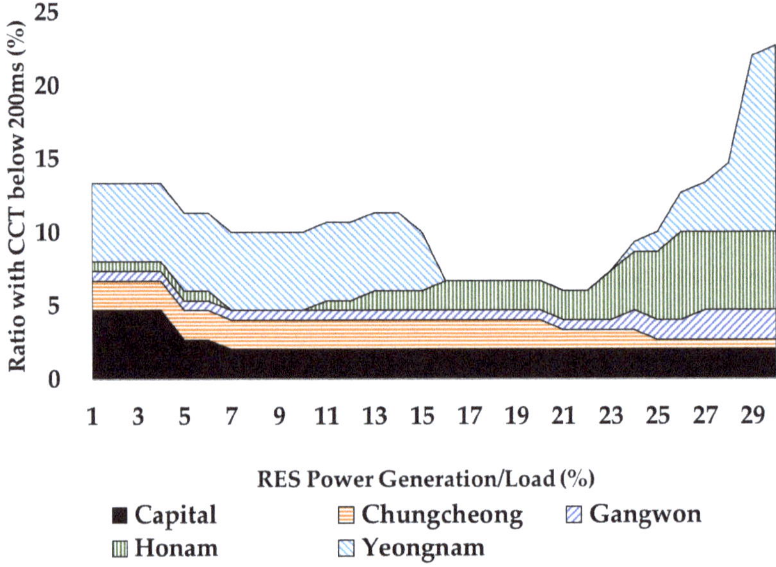

Figure 7. Classification of ratio with CCT below 200 ms for each studied area.

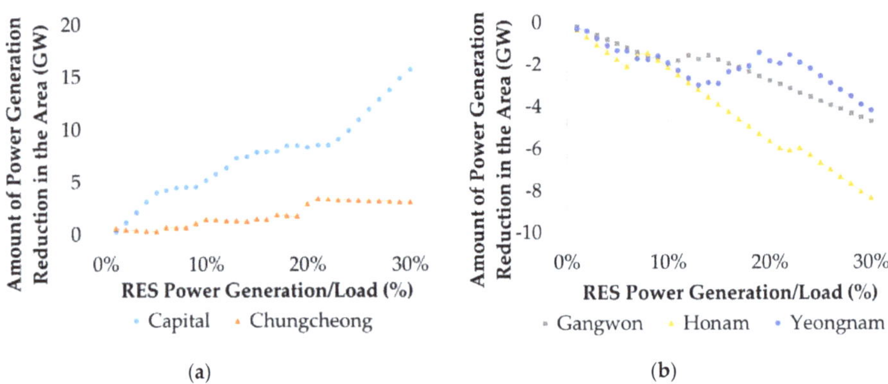

Figure 8. Power generation decreases as a function of the penetration of RES in each of the studied areas. (**a**) Capital, Chungcheong; and (**b**) Gangwon, Honam, Yeongnam.

In the cases of the Capital and Chungcheong areas, the rates of the contingencies were sustainably reduced. This is consistent with the decreasing tendency in the power generation within the area. Conversely, in the case of the areas which is power generation decreased, and with the exception of the Gangwon area, the ratio tends to increase after the temporary decrease. This is similar to the decreasing trend in power generation within the area. An area that elicits a significant change as a function of the RES penetration is the Yeongnam area, whereby the change of power generation is large. It can be observed that the rate rapidly increases in the 15 to 20 GW section where the power generation in the respective areas decreases. In order to observe these trends in detail, changes in CCT and power generation in some areas are observed. These areas are Chungcheong, Honam, and Yeongnam. The Chungcheong area has large-scale power generation complexes, but the area's power generation is reduced due to RES penetration. The Honam is the area where the generator is less than the load in the area. The Yeongnam area is a place where supply and demand are balanced, and there are generators with high generation cost.

Figures 9–11 show the respective changes in the power generation amounts and CCT values in the areas of Chungcheong, Honam, and Yeongnam. It can be confirmed that the decreasing tendency of the power generation in the respective areas and the changes in the CCT tend to be similar based on these graphs. This indicates that the imbalance in the local supply and demand owing to the penetration of RES is the main cause of the change in the transient characteristics. In the case of the Chungcheong area, the amount of power generation in this area sharply decreases at 20 GW. As a result, the CCT value in the vicinity of this point also tends to increase. A similar phenomenon is observed at a power of 7 GW, even in the Honam area. In the case of the area of Yeongnam, this kind of influence is also observed. In the case of the Yeongnam area, the amount of power generation in the area increases for an RES penetration in the range of 0 to 15 GW. For RES penetrations in the range of 15 to 20 GW, the amount of power generation in the area decreases. Subsequently, the amount of power generation in the area increases again. As the power generation in the area increases, the CCT values in that area exhibits a continually decreasing trend, despite their small magnitude. However, the CCT value shows a rapidly increasing tendency for RES penetrations in the range of 15 to 20 GW. Subsequently, the CCT decreases according to the increase of the RES penetration.

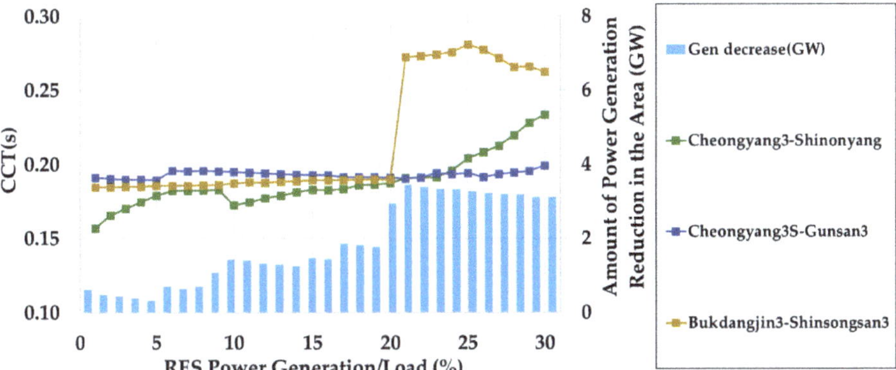

Figure 9. Changes in power generation and CCT variations in the Chungcheong area.

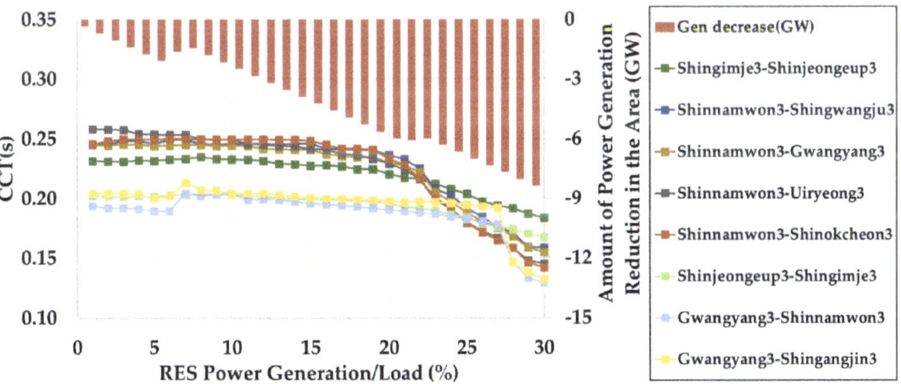

Figure 10. Changes in power generation and CCT variations in the area of Honam.

It can be confirmed based on the conducted case study that the changes in the regional power generation owing to the RES penetration and transient stability are closely related. Generally, if the RES output within the area increases, the CCT tends to decline overall. Conversely, stopping the operation of the conventional generator is a factor that causes the CCT value to increase. However, the effect of the termination of the operation of the conventional generator is limited by the contingency that is affected by that generator. In addition, in the case of the RES penetration, its influences on the CCT values are observed within the entire area since it is constructed in a form that allows even distribution within the area. Analysis of the transient impact of the RES penetration based on the conducted case study confirmed that the variations in the power generation within the area can constitute distinct observation points.

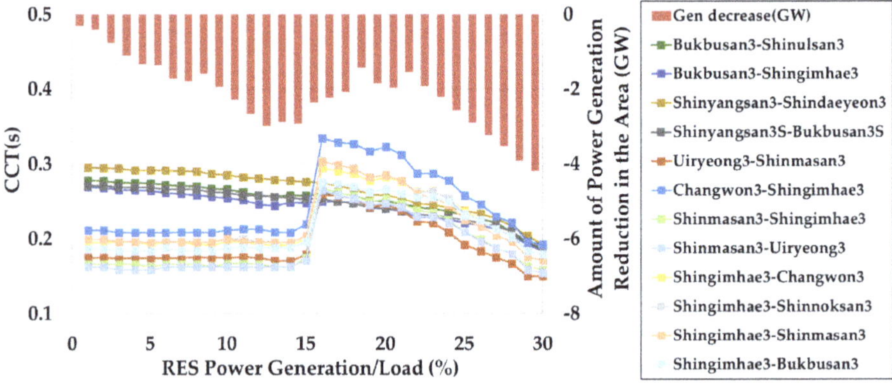

Figure 11. Changes in power generation and CCT variations in the area of Yeongnam.

5. Conclusions

The expansion of RES in future power systems constitutes an inevitable trend. To this present day, the capacity of the RES has been relatively small, which has not disturbed the stability of the power system. In addition, if the project is a large scale, the input location is determined in a manner similar to that for conventional generators. However, it is expected that in the future, RESs will be installed in the low-voltage grid as distributed power sources. Therefore, an analysis is necessary on the impact of the RES penetration in the future planning of the power system.

This study analyzed the impact of the expansion of RES penetration in the future Korean electric power system based on the transient characteristics, and analyzed the causes of such changes. To understand the changes of the transient characteristics elicited by the penetration of the RESs, the method to reflect RES to the SIME configuration is provided. Based on the modified method, the impact of the RES penetration was explained. It can be confirmed that the reduction of the conventional generation in the SIME configuration was contributed to the transient stabilization of the system. Conversely, the penetration of RESs adversely affected the transient stability of the system. Especially, it is confirmed that transient stability is improved when the conventional generator replaced with RES. Based on these facts, it was estimated that the transient characteristics worsened in the case where the regional power generation increased in accordance to the RES penetration. Conversely, in the area where the power generation within the area decreased, it was estimated that the transient characteristics improved.

The simulation was conducted in the Korean electric power system for the year 2030. A number of the contingency which has smaller CCT than 200 ms decrease when the RES penetration is small. At some point, however, the number of contingencies increases with increasing RES penetration. This phenomenon also observed in the previous Irish case. To confirm the major factors of the transient impact, a number of the contingencies and the change of power generation were observed by each area. By a result, it is confirmed that the number of contingencies and the regional power generation are related closely.

Author Contributions: S.O. conceived and designed the research, conducted the simulations, and wrote the paper; H.S. contributed analysis tools and analyzed the data; H.C. analyzed the data and contributed to visualization; and B.L. improved the theoretical part and guided the research.

Funding: This research received no external funding.

Acknowledgments: This work was supported by "Human Resources program in Energy Technology" of the Korea Institute of Energy Technology Evaluation and Planning (KETEP) granted financial resource from the Ministry of Trade, Industry and Energy, Republic of Korea (no. 20174030201820).

Conflicts of Interest: The authors declare no conflict of interest.

References

1. Li, J.; Geng, X.; Li, J. A comparison of electricity generation system sustainability among g20 countries. *Sustainability* **2016**, *8*, 1276. [CrossRef]

2. Brouwer, A.S.; Van Den Broek, M.; Seebregts, A.; Faaij, A. Impacts of large-scale intermittent renewable energy sources on electricity systems, and how these can be modeled. *Renew. Sustain. Energy Rev.* **2014**, *33*, 443–466. [CrossRef]

3. Bae, M.; Lee, H.; Lee, B. An approach to improve the penetration of sustainable energy using optimal transformer tap control. *Sustainability* **2017**, *9*, 1536.

4. Lee, H.; Bae, M.; Lee, B. Advanced reactive power reserve management scheme to enhance LVRT capability. *Energies* **2017**, *10*, 1540. [CrossRef]

5. de Oliveira Costa Souza Rosa, C.; Costa, K.A.; da Silva Christo, E.; Braga Bertahone, P. Complementarity of hydro, photovoltaic, and wind power in Rio de Janeiro state. *Sustainability* **2017**, *9*, 1130. [CrossRef]

6. Mo, J.Y.; Jeon, W. How does energy storage increase the efficiency of an electricity market with integrated wind and solar power generation?—A case study of Korea. *Sustainability* **2017**, *9*, 1797. [CrossRef]

7. Palmer, D.; Koubli, E.; Betts, T.; Gottschalg, R. The UK solar farm fleet: A challenge for the national grid? *Energies* **2017**, *10*, 1220. [CrossRef]

8. Graabak, I.; Korpås, M. Variability characteristics of European wind and solar power resources—A review. *Energies* **2016**, *9*, 449. [CrossRef]

9. Siemens PTI Study Team. PREPA Renewable Generation Integration Study. Available online: https://www.aeepr.com/Docs/Siemens%20PTI%20Final%20Report%20-%20PREPA%20Renewable%20-%20final-11.pdf (accessed on 23 February 2018).

10. Tamimi, B.; Cañizares, C.; Bhattacharya, K. System stability impact of large-scale and distributed solar photovoltaic generation: The case of Ontario, Canada. *IEEE Trans. Sustain. Energy* **2013**, *4*, 680–688. [CrossRef]

11. TransGrid, New South Wales Transmission Annual Planning Report 2017. Available online: https://www.transgrid.com.au/news-views/publications/transmission-annual-planning-report/Documents/Transmission%20Annual%20Planning%20Report%202017.pdf (accessed on 23 February 2018).

12. EirGrid. *All Island TSO Facilitation of Renewable Studies*; EirGrid: Dublin, Ireland, 2010.

13. Rodriguez, R.A.; Becker, S.; Andresen, G.B.; Heide, D.; Greiner, M. Transmission needs across a fully renewable European power system. *Renew. Energy* **2014**, *63*, 467–476. [CrossRef]

14. Deliverable D1.1 Report on Systemic Issues. Available online: https://www.h2020-migrate.eu/_Resources/Persistent/9bf78fc978e534f6393afb1f8510db86e56a1177/MIGRATE_D1.1_final_TenneT.pdf (accessed on 19 February 2018).

15. Dudurych, I.; Burke, M.; Fisher, L.; Eager, M.; Kelly, K. Operational security challenges and tools for a synchronous power system with high penetration of non-conventional sources. In Proceedings of the CIGRE 2016 Session, Paris, France, 21–26 August 2016. C2-116.

16. Liu, Z.; Liu, C.; Li, G.; Liu, Y.; Liu, Y. Impact study of PMSG-based wind power penetration on power system transient stability using EEAC theory. *Energies* **2015**, *8*, 13419–13441. [CrossRef]

17. Kabouris, J.; Kanellos, F. Impacts of large-scale wind penetration on designing and operation of electric power systems. *IEEE Trans. Sustain. Energy* **2010**, *1*, 107–114. [CrossRef]

18. Hossain, M.J.; Pota, H.R.; Mahmud, M.A.; Ramos, R.A. Investigation of the impacts of large-scale wind power penetration on the angle and voltage stability of power systems. *IEEE Syst. J.* **2012**, *6*, 76–84. [CrossRef]

19. Eftekharnejad, S.; Vittal, V.; Heydt, G.T.; Keel, B.; Loehr, J. Impact of increased penetration of photovoltaic generation on power systems. *IEEE Trans. Power Syst.* **2013**, *28*, 893–901. [CrossRef]
20. Pavella, M.; Ernst, D.; Ruiz-Vega, D. *Transient Stability of Power Systems—A Unified Approach to Assessment and Control*; Springer: New York, NY, USA, 2000; ISBN 2196-3185.
21. WECC Renewable Energy Modeling Task Force. WECC Wind Plant Dynamic Modeling Guidelines. Available online: http://www.wecc.biz/committees/StandingCommittees/PCC/TSS/MVWG/Shared%20Documents/MVWG%20Approved%20Documents/WECC%20Wind%20Plant%20Dynamic%20Modeling%20Guidelines.pdf (accessed on 27 April 2018).

Article

A Scenario Analysis of Solar Photovoltaic Grid Parity in the Maldives: The Case of Malahini Resort

Tae Yong Jung [1], Donghun Kim [1,*], Jongwoo Moon [2] and SeoKyung Lim [1]

[1] Graduate School of International Studies, Yonsei University, Seoul 03722, Korea;
    tyjung00@yonsei.ac.kr (T.Y.J.); seokyung.lim@gmail.com (S.L.)
[2] Research Center for Global Sustainability, Institute for Global Engagement & Empowerment,
    Yonsei University, Seoul 03722, Korea; slide1234@yonsei.ac.kr
* Correspondence: dhkim2@yonsei.ac.kr; Tel.: +82-2-2123-6287

Received: 10 October 2018; Accepted: 2 November 2018; Published: 5 November 2018

Abstract: The Maldives, one of the Small Island Developing States (SIDS) with great solar potential, is keen to promote renewable energy systems to reduce its heavy reliance on imported diesel for power generation. However, adopting renewable energy systems is still burdensome for the Maldives not only because of its high initial costs and insufficient financial resources but also because of a lack of understanding about whether the deployment of a renewable system is economically feasible. Therefore, the concept of grid parity is explored as an important concept in this paper to examine the possible timeframe for reaching it. A distinctive feature of the paper is that the paper used actual cost and technical information to analyze the levelized cost of energy (LCOEs) of the independent renewable system in a remote island and examined its timeframe for reaching the grid parity condition. Based on economic and technical information from a project for replacing existing diesel generator to photovoltaic (PV) with energy storage system (ESS) in Kuda Bandos Island in the Maldives, the paper considers three different system configurations and evaluates which configuration could result in the most optimal off-grid energy systems in this remote island. With sensitivity analysis on various uncertainties, the paper shows the range of the levelized costs of energy and the periods required for reaching grid parity for deploying solar photovoltaics and ESSs in Kuda Bandos Island, Maldives. The result indicates that the photovoltaic system is an economically feasible option for the resort, and that grid parity can be reached within the project lifetime. However, the result shows that the use of advanced ESSs is still an expensive option and would not be economically reasonable.

Keywords: levelized cost of energy; photovoltaic with energy storage system; HOMER simulation; LCOE comparison; sensitivity analysis

---

1. Introduction

In most remote islands, centralized generation of electricity is nearly impossible due to their unique conditions, and it makes diesel generation the primary source for power generation. High dependence on diesel price puts a huge economic and environmental burden on Small Island Developing States (SIDS). Reducing heavy reliance on imported diesel for power generation has been an important agenda for SIDS. This challenge has been encouraging SIDS to increase the deployment of renewable energy systems [1]. The recent technology development provides opportunities for the Maldives to deploy more renewable energy systems by reducing the costs of solar photovoltaics (PV) with or without energy storage systems (ESS) as well as increasing their efficiencies [2]. However, adopting renewable energy systems is still burdensome for the Maldives because of its initial costs and insufficient financial resources. Also, the countries lack understanding about whether a renewable system is an economically feasible option compared to conventional energy systems. Since the countries have limited financial resources and capacity, the economic feasibility is inevitably an important factor for the selection of energy systems. Therefore, achieving grid parity where the cost of renewables becomes the same or less than that of conventional sources is an important milestone for the successful diffusion of renewables in SIDS.

Reducing high diesel dependence and adopting renewable systems are also an important national agenda for the Maldives, one of the SIDS. The Maldives is in the Indian Ocean and is composed of more than 1000 islands. Due to its location in tropical zones, the Maldives has a high potential for solar systems. The country reached 100% electrification rate, but the dominant system is still off-grid diesel generation. The Maldives has established an ambitious target of achieving 100% renewable energy by 2020 in 2009 and encouraged the deployment of renewable systems [3]. However, the pace of the deployment of renewable systems has been slow, and only 1.28% of electricity generated is provided by renewables [4].

The paper examines the possible timeframe for reaching grid parity in a SIDS country from the case project of replacing an existing diesel generator to photovoltaic power generation with ESS in a resort on Kuda Bandos Island, Maldives. Kuda Bandos Island is located near Male, the capital of the Maldives. The main economic activity of the island is tourism, so the resort on the island has demands for establishing a sustainable energy system. Also, this small island is remote from other islands, and the resort is the only demand source for energy. Thus, both the reliability and economic feasibility of the energy system are important. The project is funded by the Korea Institute of Energy Technology Evaluation and Planning and the Ministry of Trade, Industry, and Energy of the Republic of Korea. The purpose of the project is to replace the existing diesel generator to PV with ESS and diesel generator. The project implements the installation of PV 290 kW with Power Conversion System (PCS) 295 kW, 728 kWh Lithium battery and 250 kWh Vanadium Redox Flow Battery (VRFB). The objective of the project is to develop advanced VRFB and implement the independent renewable system on the island. Thus, the paper tests the system's technical feasibility and compares levelized cost of energy (LCOEs) of three different systems, which are existing diesel generation, solar photovoltaic systems with diesel generators, and solar photovoltaic systems with ESSs based on the information given from the resort and the companies providing system components. The study uses costs, such as logistics, technology, construction, and operations costs, and system specifications from the case project. Since the project site is a remote island and requires a self-sufficient energy system, the project contains additional issues, such as including logistics costs and operations and maintenance issues. Thus, the concept of grid parity would be applied to the comparison among LCOEs of feasible options. The study examines the timeframe for reaching grid parity for those renewable systems compared to the existing diesel generation. Examining LCOEs of different system settings, including ESS, and their timeframe for reaching grid parity, can contribute to accelerating deployment of renewable systems in the Maldives as well as other SIDS countries. Furthermore, the results can provide further implications to other islands in the Maldives, those have similar characteristics, such as remote island and a self-sufficient diesel generation, to the case projects.

Literature Review

In the literature, various papers studied the economic and technical feasibility of sustainable energy systems in developing countries and SIDS. The objectives of the studies were to find the optimal independent renewable system and demonstrate the economic and financial feasibility of the selected system. (Jung et al. (2017) [5]; Ali et al. (2018) [6]; and Kaldellis et al. (2009) [7]) Jung et al. (2017) and Ali et al. (2018) both examined technical and economic feasibilities of independent PV systems on the islands of the Maldives by Hybrid Optimization of Multiple Electric Renewables (HOMER) simulations and demonstrated that the PV system is a feasible option to replace the diesel generation of the small islands in the Maldives. In addition, Kalidellis et al. (2009) found the optimal PV-ESS system configuration, as Jung et al. (2017) did, for small remote islands and conducted a cost-benefit analysis. The study also confirmed that the optimal renewable system with ESS can be considered as an effective option for solving electrification issues on remote islands, especially located in high and medium-high solar potential regions [7]. Though those studies demonstrated the technical and economic feasibility of the independent renewable systems, those analyses limitedly consider the timeframe for reaching grid parity condition of renewable systems. Therefore, the analysis of the paper can contribute to providing additional information when the grid parity condition of independent renewable systems can be reached in remote islands.

Moreover, the optimal selection and the management of renewable systems are essential parts for understanding economic feasibility of the renewable systems. In particular, the recent progress in computational simulation capacity allowed the introduction of the advanced methodologies in achieving optimal management of renewable systems with ESSs. Though this paper analyzes based on a project with a pre-determined renewable system configuration, Rangel et al. (2018) presented a methodology for choosing an optimal size, type, and site of ESS. The purpose of the study was to obtain the optimal charge and discharge strategy and find the best cost option [8]. Ghanaatian and Lotfifard (2018) proposed a method for the optimal control of the flywheel ESS by using a tube-based Model Predictive Control (MPC) model. The study included uncertainties and external disturbances in the model and conducted simulations to show the effectiveness of the model [9]. Li and Hennessy (2013) conducted a study to reduce grid dependence of midsize European town by adopting a large-scale PV, Wind, and VRFB. The study targeted to reduce grid dependence, grid purchase, and spilled wind lost and achieve shorter payback periods for the ESS. By considering different size of VRB-ESS, the study found the optimal size and technical specification of ESS that satisfies the objectives [10]. These studies were to find the optimal selection and management of the battery system that satisfy the optimal operation and the reliability of the renewable system. In addition, the advanced methodologies are presented for the management of multiple micro grid systems. Tavakoli et al. (2018) proposed a way to find an optimal management system of battery energy for the commercial building microgrid using a linear optimization programming with the Conditional Value at Risk (CVaR). This approach considers maximizing the management of microgrid photovoltaic system and battery energy while reducing key uncertainties, such as electricity price and power generation, in operating photovoltaic energy system with battery. It not only improves the efficient use of renewable energy system but also enhances the resilience of the commercial microgrid [11]. Marzband et al. (2018a) and Marzband et al. (2018b) demonstrated a smart Transactive Energy (TE) framework that established a coalition of multiple home microgrids and found an optimal resource management and profit sharing among the participants [12,13]. Marzband et al. (2018a) showed a smart TE framework that builds a coalition of multiple home microgrids and found an optimal way to achieve fair allocation of coalition profit. The study used a multi-stage stochastic programming based on artificial bee colony (MSSP-ABC) algorithm to simulate the formation of possible coalitions and find a "fair" distribution of profits. The simulation result demonstrated that the coalition encourages the participation of home microgrid owners and consumers in the deregulated market and contributes to mitigating electric load fluctuations [12]. Marzband et al. (2018b) used improved optimization techniques to solve the non-linear and non-convex Market Operator Transactive Energy (MO-TE) structure issue. The method proposed by the study was used to find the home microgrids' optimal electricity and thermal resources and maximize home microgrids owners' profits. This approach enables to find the optimal timing for home microgrids owners to exchange electricity and provides a chance to reduce the exploitation cost of home microgrids owners [13]. However, the case project used in this paper is a resort in a remote island with no grid connections to other islands, and the resort operates its self-sufficient energy system. Also, the pre-determined renewable system design was implemented in the resort. Thus, these advanced methodologies for optimal management of multiple home-grid systems as well as the selection of ESS are hard to be implemented in this case.

Along with feasibility studies of on-grid and off-grid renewable energy systems, various research has been undertaken on grid parity analysis. To compare the different types of energy technologies, LCOE calculation method is widely used in the grid party analysis. In the early 2010s, the studies showed that a solar PV system showed a great potential to be a feasible option in the future but reaching grid parity condition needed much more time. (Branker et al. (2011) [14]; Yang (2010) [15]; Lund (2011) [16]) However, the continued technological development made the on-grid and off-grid solar PV systems more attractive options, and the recent studies expected shorter timeframes for reaching grid parity of PV systems. Though the recent studies found that both off-grid and on-grid solar systems are still expensive options for developing countries (Baurzhan and Jenkins (2016) [17]; Zou et al. (2017) [18]), many studies found the grid parity of solar PV systems can be reached in countries within the next few decades. For instance, Zou et al. (2017) and Mundada et al. (2016) studied whether the off-grid PV systems can reach the grid parity condition in China and the US Zou et al. (2017) found optimal solar PV systems in five different Chinese cities and applied learning curves for calculating LCOEs. It found that the grid parity condition of off-grid PV systems in those Chinese cities can be reached in between about 2026 and 2050 [18]. Mundada et al. (2016) chose a residential sector in Houston, MI with an unfavorable environment for solar systems as a case study and calculated LCOE of off-grid PV with ESSs and combined heat and power (CHP) systems. The study showed that LCOEs of the hybrid system are lower than grid costs in many cases and argued that the hybrid system can be developed further in the US in the near future [19]. Moreover, there has been in-depth studies which included uncertainties in calculating LCOEs of PV systems. (Biondi and Moretto (2015) [20]; and Said et al. (2015) [21]) Biondi and Moretto (2015) examined solar grid parity in Italy by using a real option approach. The study evaluated uncertainties, such as energy prices and PV module costs, and demonstrated how those uncertainties affect the PV market in Italy. It showed that PV systems seem to achieve grid parity soon thanks to relatively high electricity tariffs and good solar radiation, but the uncertainties may delay grid parity condition in Italy for several years [20]. Said et al. (2015) presented an improved modeling of LCOE that used the effective lifetime of different types of PV technologies and different environmental or technical conditions. The study included the input data's uncertainties in the analysis and examined whether PV technologies reached grid parity in Egypt. It found the use of the effective lifetime affects LCOE and the total electricity generation during the lifetime [21].

Various studies on the levelized cost of energy are conducted on comparing LCOEs of different technologies in national level and on comparing LCOEs of grid-connected renewable systems to other conventional energy systems. However, in the countries composed of many islands, such as the Maldives, the grid connection among systems is nearly impossible, and the deployment of independent renewable systems have additional risk factors. This paper uses actual technology, logistics, construction, and operations costs from the project. Also, the paper applies the concept of the grid parity to the project in a remote island in the Maldives where the central grid system cannot be established. The findings from the paper can be further applied to other remote islands. Moreover, the paper conducts sensitivity analysis on key risk factors, such as changes in discount rate and diesel price, to examine how the external risk factors can affect the economic feasibilities and the grid parity conditions of renewable energy systems in the Maldives.

2. Materials and Methods

2.1. System Design and Optimization

Prior to examining the LCOEs of energy systems and the grid parity timeframe, HOMER software version 3.12 (HOMER Energy, Boulder, CO, USA) is used to find the optimal system configurations. HOMER software is a powerful tool developed by US National Renewable Energy Laboratory that simulates a viable system for all possible combinations of different energy systems based on their technical and economic merit [22]. HOMER can model grid-connected as well as off-grid power systems comprising any combination of wind turbines, PV arrays, fuel cells, small hydropower, biomass, converter, batteries, and conventional generators [23]. The simulated systems are sorted and filtered based on the criteria the user determines. Next, HOMER allows analysts to conduct sensitivity analyses and to examine the impacts of uncertainty and changes in input variables. This energy modeling software simulates multiple system combinations based on input variables such as monthly load, discount rate, and system costs and finds the optimal system designs, satisfying the technical constraints at the lowest net present cost [24]. The main objective of the HOMER simulation is to evaluate the economic and technical feasibility of renewable energy systems to be installed in a specific location based on given information of equipment and energy resource availability.

The simulation results show that three system configurations were found to be viable designs based on the input variables we entered, as shown in Figure 1. The system I represents a system design that uses only the existing diesel generator. System II represents a system configuration in which PV and diesel generators are used simultaneously, and System III is composed of PV, ESS and diesel generators.

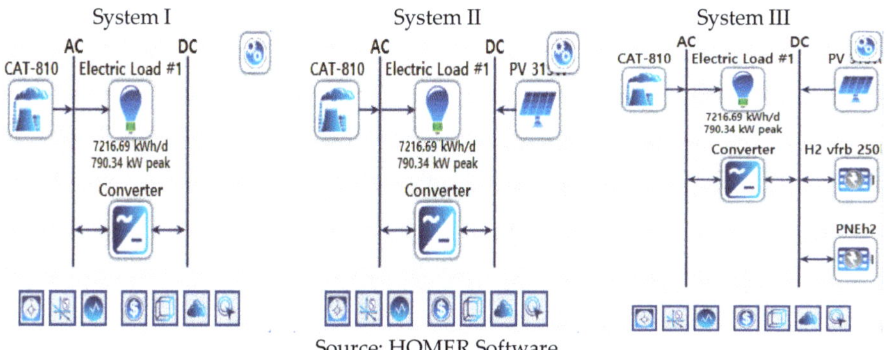

Figure 1. System Configuration.

2.2. Site Specification and Load Patterns

This study is conducted to replace existing diesel generators with either PV-only or PV with ESS in the Malahini Kuda Bandos resort located on Kuda Bandos Island, Maldives. The Malahini Kuda Bandos resort is the only resort in this small island, and the resort is self-satisfying its electric loads from a diesel generator. The resort has three diesel generators. However, operating one diesel generator at a time meets the electric loads of the resort. One of the two other diesel generators has to be operated manually if the electric load exceeds 500 kW. This PV-ESS project was implemented to eliminate the inconvenience of turning on diesel generators manually.

The Maldives is hot and sunny all throughout the year, with average temperatures of 23 to 31 degrees Celsius. The high season falls between December and March, and the monsoon runs from May to October [25]. Despite fluctuations in weather condition, the Malahini Kuda Bandos resort has a constant and stable electric load throughout the year. Its monthly average of electric loads falls between 303.76 kW and 338.57 kW.

2.3. Assumptions and System Components

This paper assesses the feasibility of two new technical options for replacing diesel generators operating at Kuda Bandos Island with either PV-only or PV-ESS. The capacity of PV is 290 kW. The PCS that stores the power produced by Solar PV in the ESS system is 295 kW. There are also two types of ESSs included in this project: a 728 kW Lithium-Ion Battery (LIB) type and a 250 kW VRFB type.

In addition to the assumptions of system components, various financial assumptions are considered. The debt-to-equity ratio is assumed to be 70:30 [26]. The rate of return for equity is assumed to be nominal at 13.5%, and the debt rate is assumed to be 8.5%. The inflation rate is set as 2.2%, which is an average rate of the Maldives between 2013 and 2017 [27]. Based on these assumptions, the real discount rate is set as 7.8%. The debt repayment period is set at 7 years, and the grace period is given as 1 year during the construction period. Operation and Maintenance (O&M) costs are assumed to be 2% of the total system cost, and the O&M costs are expected to rise at the same rate as the annual inflation rate. In terms of electricity generation, electricity generation from solar is assumed to decrease by 0.5% per year as solar panel performance declines, and diesel consumption per liter is calculated as electricity generated by diesel (kWh) times L/kWh. We also assume that diesel prices are set at $1.06/L [28], which is equal to the world average, and we assume that annual diesel price rises every year at 2.2%, which is equivalent to the inflation rate of the Maldives.

Based on this input data, the technical simulation shows the three possible system configurations at the Kuda Bandos Island. The possible system configurations are shown in Table 1. The base system configuration is a diesel-only system that uses the existing diesel generator already in operation at the resort (System I). Another possibility is to use the existing diesel generator and 290 kW PV together (System II), and the other option is to supply electricity to the resort using the existing diesel generator, 290 kW PV, and two types –LIB and VRFB of ESS (System III).

Table 1. System Configuration.

System	PV	PCS	ESS	Diesel
System I	-	-	-	648 kW
System II	290 kW	295 kW	-	648 kW
System III	290 kW	295 kW	978 kWh	648 kW

According to optimizations and simulations, in both System II and System III, PV provides approximately 472,428 kWh for the first year and accounts for 17.1% of electricity generated from the entire system. As solar performance degrades 0.5% annually, the remaining amount of electricity required will be covered by additionally using diesel generator. The technical information indicates that the ESS is used to avoid the occurrence of capacity shortage and unmet electric load and reduce excess electricity. Without an ESS, 80,525 kWh (2.92% of total electricity generation) of electricity is considered to be excess electricity every year.

3. Results

3.1. LCOE and Grid Parity

To find the grid parity, this study calculated yearly LCOEs based on the technical information as well as financial assumptions. The LCOE methodology, which is the unit cost of the energy generated by a system over its lifetime is abstracted from reality and is used as a benchmarking or ranking tool to evaluate the cost-effectiveness of other energy generation technologies [29]. The LCOE method is useful in that it allows fair and direct comparison of different electricity generation technologies [30]. Because photovoltaic and ESSs require huge initial capital investments, both System II and System III start with very high LCOEs of 1.08 and 0.49, respectively, as shown in Figure 2. Also, the debt interests and repayments occurred in the first 7 years make their LCOEs higher. As the system continues to generate more electricity, LCOEs continues to decrease. The study assumed the real diesel price to be fixed, LCOE of System I stays at $0.3208 per kWh. According to the study, LCOE of System II (PV + Diesel) reaches grid parity in 5.77 years (between 2022 and 2023). Afterward, it continues to decrease and has LCOE of $0.3056/kWh at the end of the project period.

However, this study finds that System III fails to reach grid parity. Adding ESS requires high system costs and financing costs, and it makes LCOE of the system very high. Though its LCOE decreases rapidly at the early stage, the difference between System III and System I is about $0.11/kWh in 2023 when System I reaches the grid parity condition. At the end of the project period, LCOE of System III reaches $0.3614/kWh, and it is still $0.04/kWh higher than that of System III. This result is consistent with Yang (2010)'s conclusion that solar PV is much less cost effective in a distributed system. Yang attributed the result in the fact that many analysts did not amortized all the end-user costs and incorrectly considered the cost of $1/W as manufacturing costs instead of retail installation costs when calculating grid parity [15].

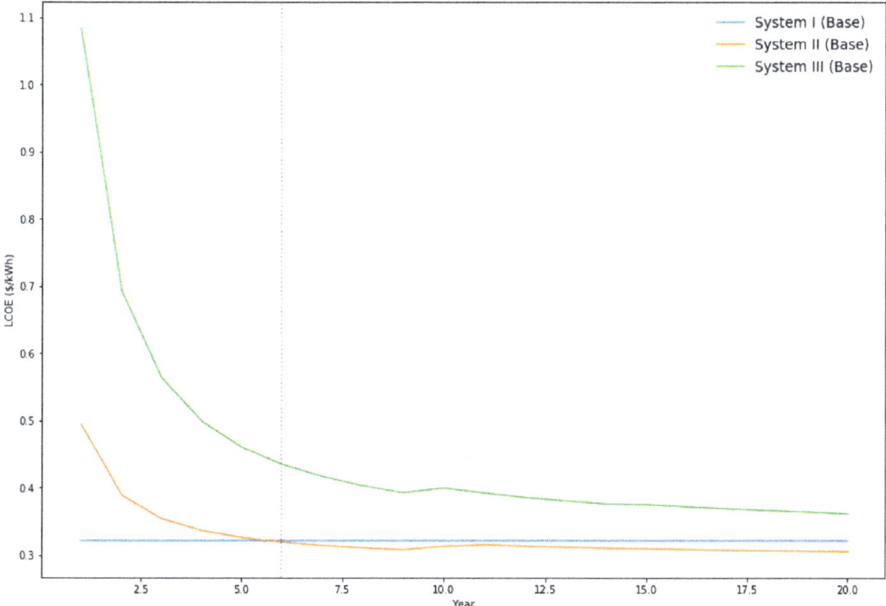

Figure 2. Yearly LCOEs of the Systems (Unit: $/kWh).

3.2. Sensitivity Analysis

Installing off-grid renewable systems in SIDS can face various unexpected circumstances and these issues can increase project costs and threaten project operations. Thus, the study conducts a sensitivity analysis to examine how the changes in diesel prices and discount rates affect the systems' LCOEs. Table 2 shows the sensitivity analysis conditions considered in this study.

Table 2. Sensitivity Analysis Conditions.

Sensitivity Analysis			
Discount Rate	10%		12%
Diesel Price (Real)	$0.954 (−10%)	$1.166 (+10%)	$1.272 (+20%)

3.2.1. Discount Rate

The project investors require the rate of return on their investments based on their perception of the project risk level. Local conditions and risk factors of installing and operating renewable systems in SIDS can make project investors to require a higher rate of return. To reflect it, the study conducts a sensitivity analysis on discount rate by altering the discount rate to 10% and 12%. The value of the discount rate must be carefully assessed because it can influence the invertor decision towards one option or another. The value chosen for the discount rate can be influenced by the investor's choice and should be carefully assessed. Thus, we conduct a sensitivity analysis of discount rates based on the conservative discount rates assumed by the International Energy Agency, which are between 10% and 12% for PV systems [31].

Table 3 and Figure 3 show the LCOEs of the three systems when the discount rate changes, and Figure 4 shows when the grid parity condition can be achieved at the different discount rates.

If the discount rate increases to 10% from 7.8%, LCOEs of the renewable systems increase slightly while LCOE of System I decreases to $0.3206/kWh. Still, System II reaches grid parity, but it is delayed and occurs in 6.33 years (between 2023 and 2024). System III becomes more expensive and the LCOE gap between System III and System I becomes $0.05/kWh at the end of the project periods. If the discount rate increases up to 12%, System II reaches grid parity in 6.95 years (between 2023 and 2024). This result is supported by the research of Simsek et al. (2018) demonstrating that debt fraction and discount rate illustrated significant sensitivities on both LCOE and government cost. By increasing discount rate from 6% to 12%, the research of Simsek et al. (2018) also proves that both LCOE and government cost showed an increasing trend [32]. As the economic feasibility of the project depends highly on feed-in tariffs concessions, the Maldives' unstable political and legislative situations could negatively affect the discount rates of PV and ESS technologies. However, as the technology matures, and the solar PV markets evolve, the significance of discount rates will decline over time.

Table 3. LCOEs of the systems when the discount rate changes (Unit: $/kWh).

Sensitivity Analysis (Discount Rate)	**7.8%**	**10%**	**12%**
LCOEs (System I)	0.3208	0.3206	0.3204
LCOEs (System II)	0.3056	0.3085	0.3113
LCOEs (System III)	0.3614	0.3735	0.3851

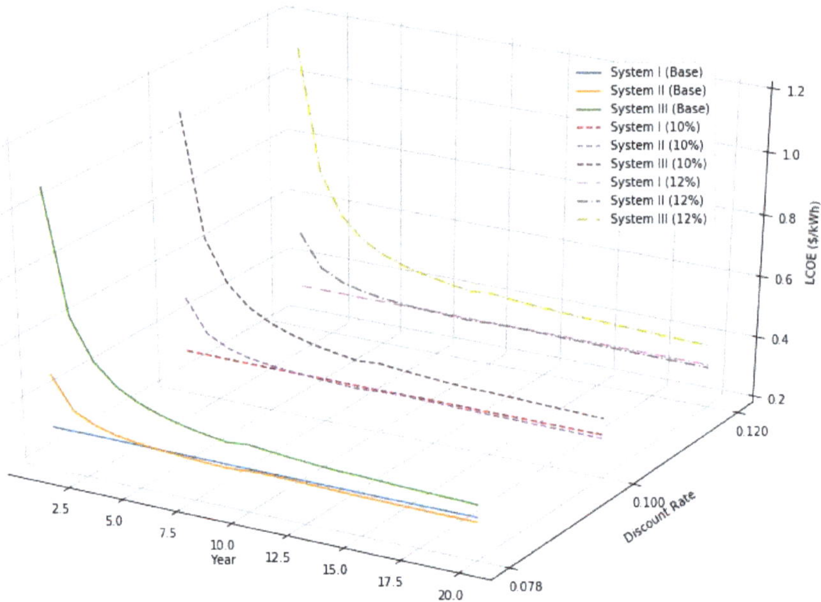

Figure 3. Yearly LCOEs if Discount Rate changes.

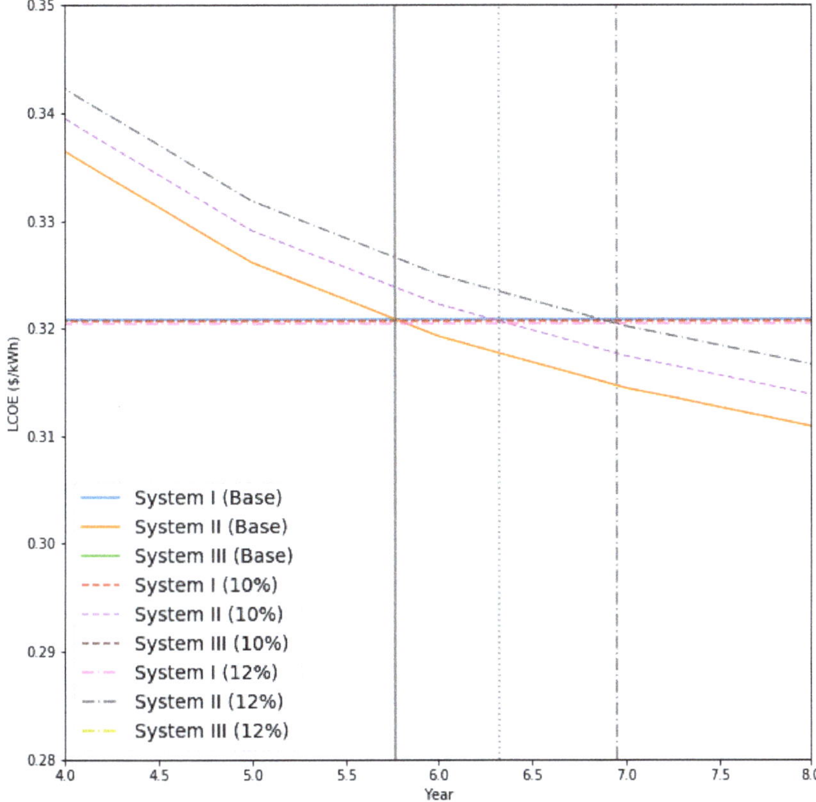

Figure 4. Grid Parity (Sensitivity Analysis on Discount Rate).

3.2.2. Diesel Prices

Due to all systems using a large portion of diesel generation, the changes in diesel prices can affect LCOEs of the systems significantly. In particular, under the circumstances of increasing global diesel prices in recent years, the increase in diesel price can increase LCOEs of the project.

Table 4 and Figure 5 show the LCOEs of the systems when the diesel price changes, and Figure 6 shows when the grid parity condition can be achieved at the different diesel prices. If the diesel price decreases to \$0.954/L, LCOEs of the systems also decrease. Grid parity of System II can be reached in 6.6 years (between 2023 and 2024). However, System III cannot reach grid parity. On the other hand, if the diesel price increases by 10% and becomes \$1.166/L, grid parity of System II is reached in 5.11 years (between 2022 and 2023). Also, the LCOE difference between System III and System I reduces to \$0.035/kWh. If the diesel price further increases to \$1.272/L, the grid parity condition of System II can be achieved earlier. Grid parity is achieved in 4.64 years (between 2021 and 2022). This indicates that the project is likely to benefit if the recent trends of increasing global oil price continue and the diesel price of the Maldives increases. This result is supported by research done by Peerapong and Limmeechokchai who conducted a sensitivity analysis using variations such as solar radiations, cost of diesel prices, real interest rates and load consumptions. The variations of diesel prices set from \$0.9/L to \$1.2/L affect NPC, COE, and renewable shares in the system. The study finally concludes that minimum diesel price should be at least \$0.561/L for the hybrid diesel/PV system with ESS to compete with the diesel-only existing system [33].

Table 4. LCOEs of the systems when the diesel price changes (Unit: (Diesel) $/L; (LCOE) $/kWh).

Sensitivity Analysis (Diesel Price)	$0.954	$1.166	$1.272
LCOEs (System I)	0.2907	0.3509	0.3810
LCOEs (System II)	0.2799	0.3314	0.3571
LCOEs (System III)	0.3365	0.3862	0.4111

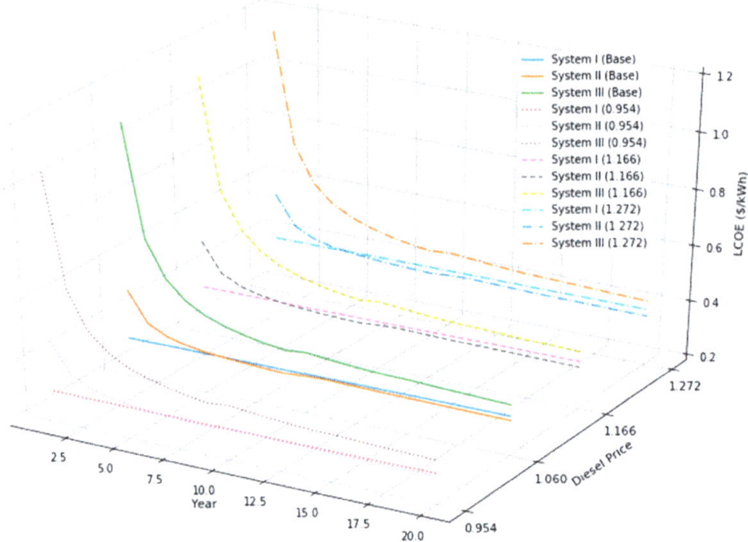

Figure 5. Yearly LCOEs when Diesel Price changes.

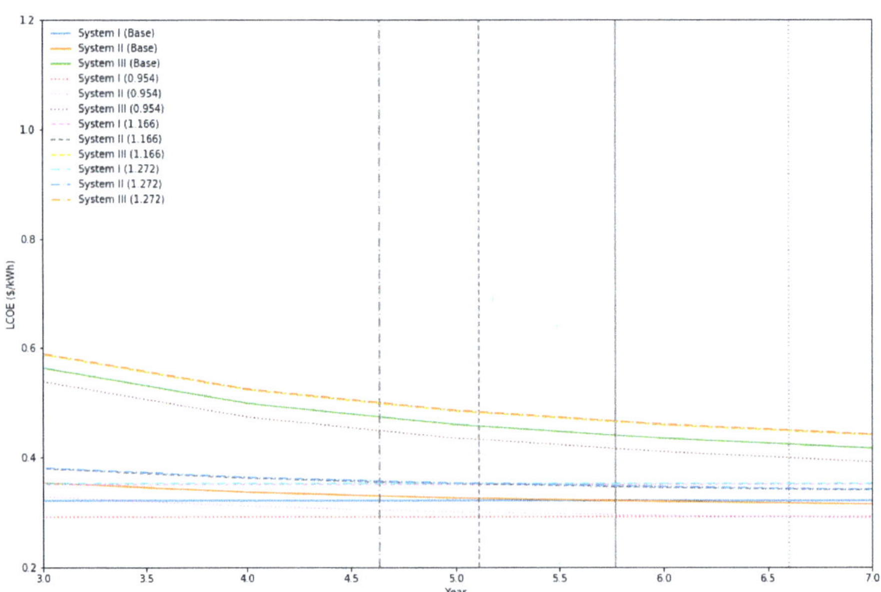

Figure 6. Grid Parity (Sensitivity Analysis on Diesel Price).

3.2.3. Summary of Sensitivity Analysis

The sensitivity analysis shows the time required to reach grid parity if the diesel price and discount rate change, and the detailed results of the sensitivity analysis are shown in Table 5. Though the analysis indicates that the system design with an ESS cannot reach grid parity, System II reaches grid parity mostly within 7 years. If the diesel price goes down and the discount rate increases, System II requires more time to reach the grid parity as shown in Table 6. An increase in diesel price benefits the renewable systems consuming less diesel, and an increase of discount rate, which means the project is perceived as a riskier one, can make the renewable systems less attractive.

Table 5. Summary of Results.

LCOEs		System I	System II	System III
@7.8%	$1.06	0.3208	0.3056	0.3614
@7.8%	$0.954	0.2907	0.2799	0.3365
@7.8%	$1.166	0.3509	0.3314	0.3862
@7.8%	$1.272	0.3810	0.3571	0.4111
@10%	$1.06	0.3206	0.3085	0.3735
@12%	$1.06	0.3204	0.3113	0.3851

Table 6. When the Grid Party can be reached (System II).

Sensitivity Analysis	$0.954	$1.06	$1.166	$1.272
7.8%	6.6 years	5.77 years	5.11 years	4.64 years
10%	7.35 years	6.33 years	5.58 years	4.95 years
12%	8.25 years	6.95 years	6.01 years	5.35 years

4. Discussion

The study examined whether renewable systems are feasible options in SIDS based on the case of the Kuda Bandos Island, Maldives. Also, the study used the LCOE method to find when the grid parity condition can be met. The case project is to replace the existing diesel generator to a solar photovoltaic system with ESS and diesel generator.

Though the project included two types of energy storage systems, the sensitivity analysis showed that the grid parity of PV with ESS is hard to be reached within the project lifetime. However, if the project only uses solar PV and diesel generator, the grid parity can be reached in 5.77 years, and the renewable system is an effective solution for replacing the existing diesel generator and reducing diesel consumption. Moreover, in all cases, the sensitivity analysis showed that solar PV and diesel generator (System II) reaches the grid parity condition. Thus, the paper demonstrates that solar PV already reached the grid parity condition and became an effective solution for achieving the renewable target of the Maldives and reducing the heavy dependence on imported fossil fuel.

The results of the sensitivity analysis indicate that the discount rate and diesel price are major factors affecting the economic feasibilities of renewable systems and the grid parity conditions. The challenges, such as logistics and maintenance issues, of the SIDS countries caused by their unique characteristics often make investors perceive the deployment of the off-grid renewable system in those countries as a risky project. This perception can increase the discount rate required to evaluate a renewable energy project. As shown in the sensitivity analysis on the discount rate, a higher discount rate makes renewable systems which require a huge upfront investment less attractive, and it delays the periods needed for achieving the grid parity condition. Thus, a proper assessment of risk factors of renewable projects and supportive policies for reducing those risk factors should be considered.

Moreover, the result of the sensitivity analysis on diesel price provides information about how the introduction of carbon taxes or environmental taxes can accelerate the grid parity condition in the Maldives. the Maldives is vulnerable to climate change and focuses on fuel switching from diesel to renewable energy options [34]. Also, tourism is one of the key parts of the Maldives' economy. Thus, the country has incentives to reduce diesel generation, which is the dominant energy system in the Maldives, emitting greenhouse gases and air pollutants. If the country introduces environmental taxes or carbon taxes, its impact is likely to be similar to the sensitivity analysis to an increase of diesel price. Thus, it can further accelerate the grid parity condition of PV systems in the Maldives. However, PV with ESS will remain as a very expensive option. As the sensitivity analysis indicates, even a 20% increase of diesel price does not bring the grid parity condition of PV with ESS and diesel generator. This implies that it will require much longer time to fully eliminate the diesel generator from the system.

In recent years, the global oil price has been increasing rapidly, and the Brent Oil price has exceeded $80 per barrel in the late September 2018 [35]. As shown in the sensitivity analysis on diesel price, a continued trend of increasing global oil price is likely to benefit the deployment of renewable systems. Moreover, the system configuration of the project is pre-determined in the case project, and the project included high-performance but expensive batteries in the system configuration. If the trend of increasing global oil price continues and the optimal system configuration is considered, the solar photovoltaic with ESS can become a more attractive option.

Author Contributions: Conceptualization, T.Y.J. and D.K.; Data curation, J.M. and S.L.; Formal analysis, T.Y.J., D.K., J.M. and S.L.; Investigation, T.Y.J., D.K., J.M. and S.L.; Methodology, T.Y.J., D.K., J.M. and S.L.; Project administration, T.Y.J. and D.K.; Resources, T.Y.J. and D.K.; Software, J.M. and S.L.; Supervision, T.Y.J. and D.K.; Validation, T.Y.J. and D.K.; Visualization, J.M.; Writing—original draft, T.Y.J., D.K., J.M. and S.L.; Writing—review & editing, T.Y.J. and D.K.

Funding: This research was funded by the Korea Institute of Energy Technology Evaluation and Planning (KETEP) and the Ministry of Trade, Industry & Energy (MOTIE) of the Republic of Korea (No. 20162010103860).

Acknowledgments: This work was supported by Korea Institute of Energy Technology Evaluation and Planning (KETEP) grant funded by the Korea government (MOTIE) (20162010103860, Development and Demonstration of Multiple Linked ESS for Special Environmental Areas).

Conflicts of Interest: The authors declare no conflict of interest.

Acronyms and Abbreviations

SIDS	Small Island Developing States
PV	Photovoltaic
ESS	Energy Storage System
LCOE	Levelized Cost of Energy
LIB	Lithium-Ion Battery
VRFB	Vanadium Redox Flow Battery
KETEP	The Korea Institute of Energy Technology Evaluation and Planning
MOTIE	The Ministry of Trade, Industry and Energy of the Republic of Korea
HOMER	Hybrid Optimization of Multiple Energy Resources
CVaR	Conditional Value at Risk
MSSP-ABC	Multi-stage Stochastic Programming based on Artificial Bee Colony
MO-TE	Market Operator Transactive Energy
CHP	Combined Heat and Power System
PCS	Power Conversion System
O&M	Operations and Maintenance

References

1. International Renewable Energy Agency. Renewable Islands: Setting for Success. 2014. Available online: http://www.irena.org/-/media/Files/IRENA/Agency/Publication/2017/Oct/GREIN_Settings_for_Success.pdf?la=en&hash=A57CFED57FA49FB8EEF14D3A0BFBB0721D9B731F (accessed on 1 July 2018).

2. International Renewable Energy Agency. Renewable Power Generation Costs in 2017. 2018. Available online: http://www.irena.org/publications/2018/Jan/Renewable-power-generation-costs-in-2017 (accessed on 10 August 2018).

3. International Renewable Energy Agency. Renewable Energy Roadmap: The Republic of Maldives. 2015. Available online: http://www.irena.org/EventDocs/Maldives/Maldivesroadmapbackgroundreport.pdf (accessed on 10 August 2018).

4. World Bank. Renewable Energy Output (% of total Electricity Output). 2018. Available online: https://data.worldbank.org/indicator/EG.ELC.RNEW.ZS?view=chart (accessed on 27 September 2018).

5. Jung, T.Y.; Kim, D. A solar energy system with energy storage system for Kandooma Island, Maldives. *Korean Energy Econ. Rev.* **2017**, *16*, 24.

6. Ali, I.; Shafiullah, G.M.; Urmee, T. A preliminary feasibility of roof-mounted solar PV systems in the Maldives. *Renew. Sustain. Energy Rev.* **2018**, *83*, 18–32. [CrossRef]

7. Kaldellis, J.K.; Zafirakis, D.; Kaldelli, E.L.; Kavadias, K. Cost benefit analysis of a photovoltaic-energy storage electrification solution for remote islands. *Renew. Energy* **2009**, *34*, 1299–1311. [CrossRef]

8. Rangel, C.A.S.; Canha, L.; Sperandio, M.; Severiano, R. Methodology for ESS-type selection and optimal energy management in distribution system with DG considering reverse flow limitations and cost penalties. *IET Gener. Transm. Distrib.* **2018**, *12*, 1164–1170. [CrossRef]

9. Ghanaatian, M.; Lotfifard, S. Control of flywheel energy storage systems in presence of uncertainties. *IEEE Trans. Sustain. Energy* **2018**. [CrossRef]

10. Li, H.; Hennessy, T. European Town Microgrid and Energy Storage Application Study. In Proceedings of the 2013 IEEE PES Innovative Smart Grid Technologies Conference (ISGT), Washington, DC, USA, 24–27 February 2013; pp. 1–6.

11. Tavakoli, M.; Shokridehaki, F.; Akorede, M.F.; Marzband, M.; Vechiu, I.; Pouresmaeil, E. CVaR-based energy management scheme for optimal resilience and operational cost in commercial building microgrids. *Int. J. Electr. Power Energy Syst.* **2018**, *100*, 1–9. [CrossRef]

12. Marzband, M.; Azarinejadian, F.; Savaghebi, M.; Pouresmaeil, E.; Guerrero, J.M.; Lightbody, G. Smart transactive energy framework in grid-connected multiple home microgrids under independent and coalition operations. *Renew. Energy* **2018**, *126*, 95–106. [CrossRef]

13. Marzband, M.; Fouladfar, M.H.; Akorede, M.F.; Lightbody, G.; Pouresmaeil, E. Framework for smart transactive energy in home-microgrids considering coalition formation and demand side management. *Sustain. Cities Soc.* **2018**, *40*, 136–154. [CrossRef]

14. Branker, K.; Pathak, M.J.M.; Pearce, J.M. A review of solar photovoltaic levelized cost of electricity. *Renew. Sustain. Energy Rev.* **2011**, *15*, 4470–4482. [CrossRef]

15. Yang, C.J. Reconsidering solar grid parity. *Energy Policy* **2010**, *38*, 3270–3273. [CrossRef]

16. Lund, P.D. Boosting new renewable technologies towards grid parity—Economic and policy aspects. *Renew. Energy* **2011**, *36*, 2776–2784. [CrossRef]

17. Baurzhan, S.; Jenkins, G.P. Off-grid solar PV: Is it an affordable or appropriate solution for rural electrification in Sub-Saharan African countries? *Renew. Sustain. Energy Rev.* **2016**, *60*, 1405–1418. [CrossRef]

18. Zou, H.; Du, H.; Brown, M.A.; Mao, G. Large-scale PV power generation in China: A grid parity and techno-economic analysis. *Energy* **2017**, *134*, 256–268. [CrossRef]

19. Mundada, A.S.; Shah, K.K.; Pearce, J.M. Levelized cost of electricity for solar photovoltaic, battery and cogen hybrid systems. *Renew. Sustain. Energy Rev.* **2016**, *57*, 692–703. [CrossRef]

20. Biondi, T.; Moretto, M. Solar Grid Parity dynamics in Italy: A real option approach. *Energy* **2015**, *80*, 293–302. [CrossRef]

21. Said, M.; El-Shimy, M.; Abdelraheem, M.A. Photovoltaics energy: Improved modeling and analysis of the levelized cost of energy (LCOE) and grid parity—Egypt case study. *Sustain. Energy Technol. Assess.* **2015**, *9*, 37–48. [CrossRef]

22. Lambert, T.; Gilman, P.; Lilienthal, P. Micropower system modeling with HOMER. *Integr. Altern. Sources Energy* **2006**, *1*, 379–418.

23. HOMER Pro 3.12. Available online: http://www.nrel.gov/homer (accessed on 29 July 2018).

24. Bahramara, S.; Moghaddam, M.P.; Haghifam, M.R. Optimal planning of hybrid renewable energy systems using HOMER: A review. *Renew. Sustain. Energy Rev.* **2016**, *62*, 609–620. [CrossRef]

25. Stojanov, R.; Duzi, B.; Kelman, I.; Nemec, D.; Prochazka, D. Local perceptions of climate change impacts and mitigation patterns in Male, Maldives. *Geogr. J.* **2016**, 370–385. [CrossRef]

26. Maldives Statistics. European Commission. Available online: http://trade.ec.europa.eu/doclib/docs/2006/september/tradoc_115814.pdf (accessed on 26 September 2018).

27. Global Landscape of Renewable Energy Finance 2018. International Renewable Energy Agency (IRENA). Available online: https://www.irena.org/-/media/Files/IRENA/Agency/Publication/2018/Jan/IRENA_Global_landscape_RE_finance_2018.pdf (accessed on 15 August 2018).

28. Global Petrol Prices. Diesel Prices, Liter. Available online: https://www.globalpetrolprices.com/diesel_prices/ (accessed on 20 July 2018).

29. Jacqueline, T.Y.; Anton, F. Moving beyond LCOE: Impact of various financing methods on PV profitability for SIDS. *Energy Policy* **2016**, *98*, 749–758.

30. Benjamin, P.; de Sandro, S.; Joao, D.B. Grid parity analysis of distributed PV generation using Monte Carlo approach: The Brazilian case. *Renew. Energy* **2018**, *127*, 974–988.

31. International Energy Agency (IEA). *Energy Technology Perspectives 2010: Scenarios and Strategies to 2050*; International Energy Agency (IEA/OECD): Paris, France, 2010; pp. 1–706.

32. Yeliz, S.; Carlos, M.; Amador, G.M.; Jose, C.M.; Rodrigo, E. Sensitivity and effectiveness analysis of incentives for concentrated solar power projects in Chile. *Renew. Energy* **2018**, *129*, 214–224.

33. Prachuab, P.; Bundit, L. Optimal electricity development by increasing solar resources in diesel-based micro grid of island society in Thailand. *Energy Rep.* **2017**, *3*, 1–3.

34. Ministry of Environment and Energy, Government of Maldives. *Maldives' Intended Nationally Determined Contribution (INDC)*; Ministry of Environment and Energy, Government of Maldives: Malé, Maldives, 2015.

35. Bloomberg. Crude Oil and Natural Gas. Available online: https://www.bloomberg.com/energy (accessed on 27 September 2018).

Article

Long-Term Decision on Wind Investment with Considering Different Load Ranges of Power Plant for Sustainable Electricity Energy Market

Jaber Valinejad [1], Mousa Marzband [2,3], Mudathir Funsho Akorede [4], Ian D Elliott [2], Radu Godina [5], João Carlos de Oliveira Matias [6,7] and Edris Pouresmaeil [8,*]

[1] Bradley Department of Electrical and Computer Engineering, Virginia Tech, Northern Virginia Center, Falls Church, VA 24043, USA; jabervalinejad@vt.edu

[2] Department of Physics and Electrical Engineering, Faculty of Engineering and Environment, Northumbria University Newcastle, Newcastle upon Tyne NE18ST, UK; mousa.marzband@gmail.com (M.M.); mousa.marzband@northumbria.ac.uk (I.D.E.)

[3] Department of Electrical Engineering, Lahijan Branch, Islamic Azad University, Lahijan 4416939515, Iran

[4] Department of Electrical & Electronics Engineering, Faculty of Engineering and Technology, University of Ilorin, P.M.B. 1515, Ilorin 240003, Nigeria; akorede@unilorin.edu.ng

[5] Centre for Aerospace Science and Technologies—Department of Electromechanical Engineering, University of Beira Interior, 6201-001 Covilhã, Portugal; rd@ubi.pt

[6] DEGEIT—Department of Economics, Management, Industrial Engineering and Tourism, University of Aveiro, 3810-193 Aveiro, Portugal; jmatias@ua.pt

[7] GOVCOPP—Research Unit on Governance, Competitiveness and Public Policies, University of Aveiro, 3810-193 Aveiro, Portugal

[8] Department of Electrical Engineering and Automation, Aalto University, 02150 Espoo, Finland

* Correspondence: edris.pouresmaeil@aalto.fi; Tel.: +358-505-984-479

Received: 23 September 2018; Accepted: 15 October 2018; Published: 22 October 2018

Abstract: The aim of this paper is to provide a bi-level model for the expansion planning on wind investment while considering different load ranges of power plants in power systems at a multi-stage horizon. Different technologies include base load units, such as thermal and water units, and peak load units such as gas turbine. In this model, subsidies are considered as a means to encourage investment in wind turbines. In order that the uncertainties related to demand and the wind turbine can be taken into consideration, these effects are modelled using a variety of scenarios. In addition, the load demand is characterized by a certain number of demand blocks. The first-level relates to the issue of investment in different load ranges of power plants with a view to maximizing the investment profit whilst the second level is related to the market-clearing where the priority is to maximize the social welfare benefits. The bi-level optimization problem is then converted to a dynamic stochastic mathematical algorithm with equilibrium constraint (MPEC) and represented as a mixed integer linear program (MILP) after linearization. The proposed framework is examined on a real transmission network. Simulation results confirm that the proposed framework can be a useful tool for analyzing the investments different load ranges of power plants on long-term strategic decision-making.

Keywords: capacity investment; market power; wind resources; dynamic planning; stochastic approach

---

1. Introduction

Investment in renewable energy sources, including wind power plant, is of particular importance due to the increase in the efficiency of clean energy, and the need to reduce pollution and fuel consumption [1–3]. However, the production of wind resourced units involves an inherent uncertainty and limited capacity, which alone may not be responsible for load growth in a power grid [4–7]. In addition, investment in wind resources may be less profitable than other technologies for investor due to budget constraints [8–12]. Consequently, there is a need to pay special attention to investing in renewable energy, including wind power, when carrying out generation expansion planning. Moreover, the possibility of investing in other technologies are taken into account in the model when generation expansion planning (GEP) is undertaken. However, this issue in previously presented models is often overlooked.

On the other hand, the analysis of the dynamic approach to this problem is of great importance due to the fact that yearly investment decisions (including capacity, location and time of construction) depend on investment decisions carried out in previous years [13–17]. As a result, in this article, we focus on a dynamic approach including renewable energy such as wind power in the grid used in the highest possible value. For this reason, investors will be encouraged to invest in the units while giving consideration to subsidies for new wind units. Whilst it is possible that the above incentives may not result in investment, other technologies that exploit peak load and base load units that consider various aspects (such as budget constraints, uncertainty and the limited capacity of wind turbines, etc.) during the planning stages, may be more profitable and therefore more attractive to investors.

In Figure 1, the effectiveness of the proposed GEP model is shown along with the above studies' shortcomings. The following points outline the important research that has been done in this field:

- The model presents more reliable and better results if those that are intended for the short-term market include the market clearing aspect, whereas those intended for the longer term take into account investment. Therefore, the investment problem in this paper is bi-level; however, those presented in [5,18–24] are not bi-level.
- In those models that have been proposed for market clearing, the supply function model in particular is a more accurate and realistic model when compared to others. Therefore, in this respect, a supply function model is used in the spot market whilst Refs. [25–29] make use of other models;
- In this paper, three different load range technologies, including base, peak and wind plant are considered as the candidate units for investment and each technology has different options for investment capacity. However, candidate units for investment in [5,21,25,27,30] are wind unit and in [14,19,20,22,26,31–39] are other technologies. In addition, in this paper, the subsidy is considered in order to stimulate investment in the wind unit. This approach is more in line with reality, whilst largely overlooked in previous cases;
- The proposed model is a dynamic one, whilst the dynamic nature of investment decisions has not been considered in papers concerning wind energy [3,25,40,41] and other technologies [19,20,22,26,32,34–39]. Therefore, the multi-period stochastic model consisting of transmission network constraints is presented here;
- In this network, consideration is given to the uncertainty in the generation of electricity from wind power along with demand consumption. These uncertainties are modeled by means of a scenario based approach.
- In addition, conventional producers are assumed to be competitive, i.e., they offer their respective productions at their marginal costs.

Considering the context above, contributions of this paper are as following:

- to propose a wind investment model considering different load ranges of power plants so that the best production technology in the optimized location are offered.
- to envelope a wind investment multi-stage model for sustainable electricity energy markets
- to provide a methodology to investigate subsidies in both planning and operating planning in a network-constrained electricity market as a game-theoretic model.
- To implement the proposed model to Mazandaran Regional Electric Company (MREC) as a real power network and comprehensively analyze the related results.

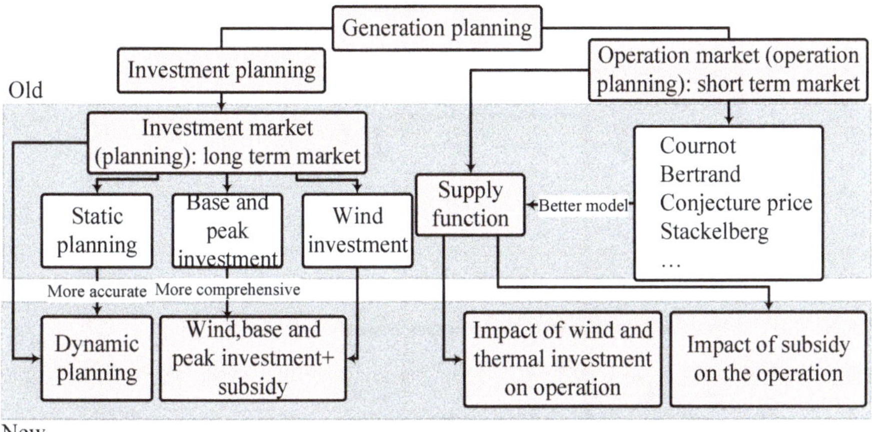

Figure 1. Block diagram illustrating the effectiveness of investment and planning.

2. Planning Features

Dynamic multi-stage property: Planning in this article is done for 10 stages. Each stage can include one to several years depending on the terms planner. The annual discount rate of each stage in the dynamic approach is assumed to be 8.7%. Non-anticipativity issues should be analyzed according to the uncertainty of the demands and wind turbine production, as shown in Figure 2, combined with the dynamic approach of GEP.

Demand model property: the demand blocks are obtained from a stepwise approximation of the load-duration curve, as illustrated in Figure 2a. It is assumed that the per-stage demand is specified with three different demand blocks; namely peak, shoulder and off-peak. Annual growth of demands is assumed to be 6.2%. The weighting factors associated with each demand block (peak, shoulder and off-peak) are assumed to be 20%, 50% and 30%, respectively. In each year of the planning period, the weighting factor of the off-peak and shoulder blocks are considered to be 25% and 60% respectively of the associated forecasted peak demand. For the sake of simplicity, each demand considers one bid per block. For MREC demand, three uncertainties are considered, having the same probabilities as those shown in Figure 2b.

Investment technologies property: In this paper, three different load ranges of power plants including base, peak and wind units are considered for investment. Base technologies have relatively low operation costs but high investment costs due to supplying grid demand during all hours of the day. However, peak technologies, whilst also having low operation costs with high investment costs, only supply the grid during hours of peak demand.

In this article, it is assumed that the grid includes the construction of wind turbines. Construction of these units involves higher investment costs than the peak and base units; however, it is assumed that the operation cost of wind turbines units is zero. Renewable energy such as wind resources are usually characterized by uncertainty. The uncertainty of wind turbines can be modeled using a set of scenarios. To represent the uncertainty related to wind production, only three wind intensity factors are considered, despite the fact that a larger number of scenarios may easily be considered for a typical wind turbine in this model. Limiting the number of intensity factors prevents an increase in the computational complexity and corresponding time to solve the mathematical model by means of software simulation. Therefore, nine uncertainties are obtained with respect to three uncertainties of demand.

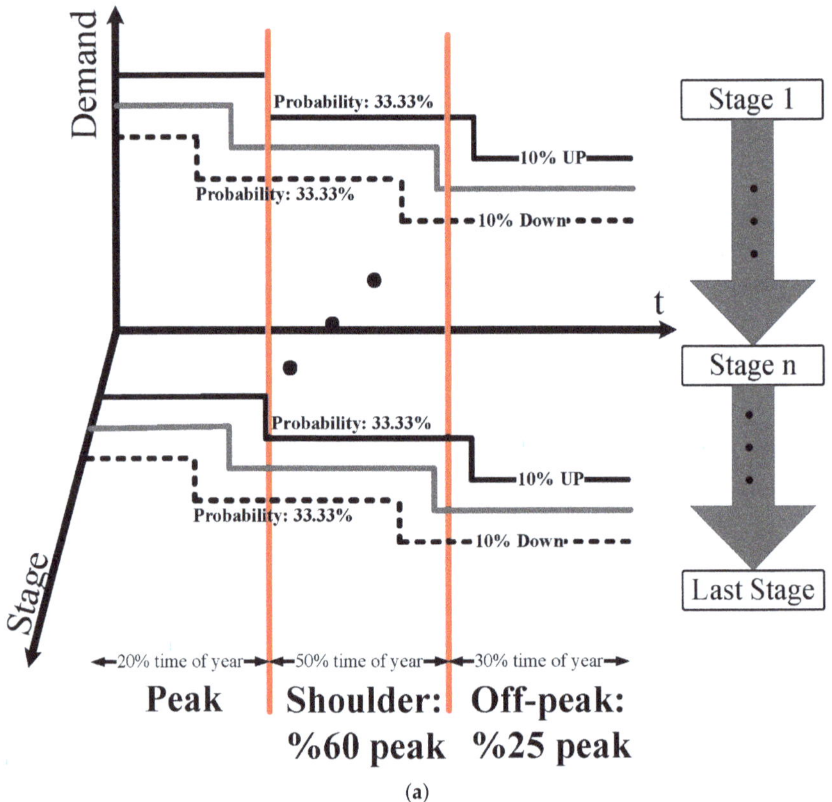

(a)

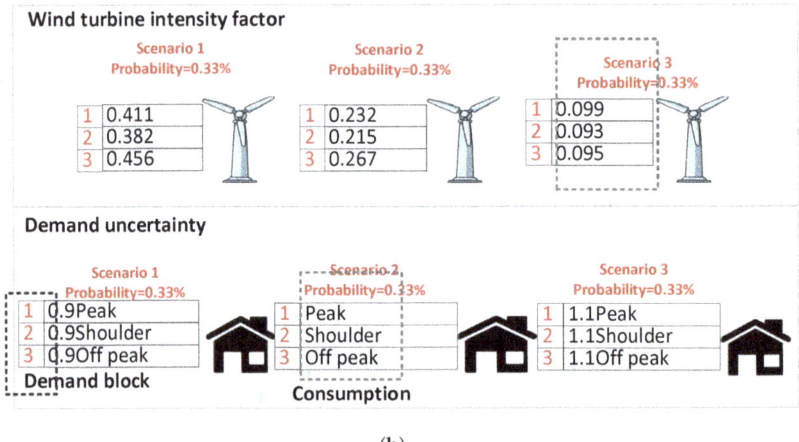

(b)

Figure 2. Features of demand and wind turbine. (**a**) piecewise approximation of the load duration curve for the planning year at a particular bus; (**b**) the uncertainty of demand and wind turbine.

The modeling of wind turbine and converter: the Kinetic energy produced by wind is converted into mechanical power by the wind turbine rotor:

$$P_{wt} = 0.5\rho\pi R^2 V^3 = 0.5\rho\pi R^2 (\frac{R}{\lambda})^3 (W_R)^3,\tag{1}$$

$$\lambda = \frac{W_R R}{V},\tag{2}$$

$$P_{mec} = C_p(\lambda,\beta)P_{wt} = 0.5C_P(\lambda,\beta)\rho\pi R^2(\frac{R}{\lambda})^3 (W_R)^3.\tag{3}$$

Equations (1)–(3) present equations of wind turbine modelling. Equation (1) presents the output power for wind turbine so that ρ and R are the air density (kg/m^3) and the blade radius (m) accordingly. In addition, V, W_R as well as λ present the wind speed (m/s), rotor speed (rad/s) and the turbine tip-speed ratio, respectively [42]. The relation between turbine mechanical power (P_{mec}) and available power (P_{wt}) is represented by Equation (3) while C_P and β are power coefficient and pitch angle, respectively.

Non-anticipativity property: it should be noted that there is an issue of non-anticipativity nature in the problem of stochastic multi-stage modelling [43]. Figure 3 shows the different states relating to this issue. If the scenarios considered in the model are independent and different from one another at all stages, these provisions should not be applied to the programming (Mode 1, Figure 3). The multi-stage problems that include uncertainty have two modes, if uncertainties in the scenarios being considered are the same in the process of multi-stage models. The constraints related to non-anticipativity properties should be applied to the model (Mode 3, Figure 3) if the problem of stochastic multi-stage is solved by optimization algorithms, such as heuristic algorithms, in order that each iteration results in a different optimal solution. In addition, applying non-anticipativity constraints to the program is optional, if GAMS solvers such as MILP, LP, and NLP are used that result in the same optimal solution in repeated results (iterations). As a result of these issues around optimization, the same result in the optimal solution is obtained for scenarios that are similar in some of the steps (Mode 2, Figure 3). Therefore, applying the non-anticipativity conditions to the program tends to increase the complexity of the model and the corresponding calculations' volume; however, it has no effect on providing the optimal solution. This issue should be considered in the simulation results only to ensure the correct program process; it can be ignored in a model for simplifying purposes and to achieve less computation and complexity. In this article, the provisions of the non-anticipitivity properties can be omitted due to the fact that the defined scenarios are independent of each other and they do not have any similarities with cases spanning a 10-year planning period.

The aforementioned is similar to Mode 1 from Figure 3. For example, the dynamic stochastic [20,44] is similar to the work of this paper and case 1 of Figure 3 and the expressed model in [20] is similar to Mode 3 of Figure 3. The first-level variables, which correspond to the installed generation capacity of the investing agent, will not be stochastic variables, however, since a generation company can only make one investment decision due to the fact that it is impossible to know which scenario is going to occur in reality.

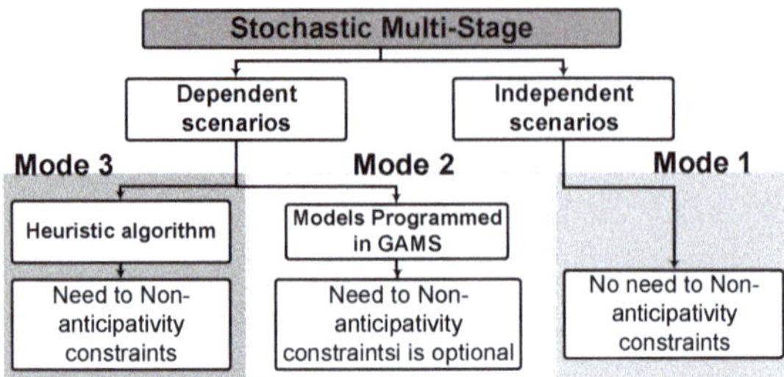

Figure 3. Three different cases for the non-anticipativity condition in a multi-stage stochastic problem.

3. The Proposed Algorithm

This idea is solved by the algorithm presented in Figure 4. The planning carried out with the aim of maximizing the profits of the investment in the wind unit along with different load ranges, power plants including peak and base technologies, and investment in wind units that includes subsidies and grants to encourage investment is invested in the first year. New units are operated in subsequent years as existing units in the network. In the next stage, the intention is to minimize the operation cost so that the output at this stage is the market-clearing results, which includes market-clearing prices, unit production and consumption. This procedure is carried out for the next blocks in the first year. Overall, the benefits of investment and operation are maximized as a result of this planning. In Figure 4, N_r, N_w, N_t, N_{wu} and N_{thu} are number of years, scenarios, time block, new wind units and new thermal units in the planning year, respectively. In addition, more information about bi-level modelling can be figured out in [19,43–45]. In addition, to perceive multi-stage planning, Ref. [14] is useful.

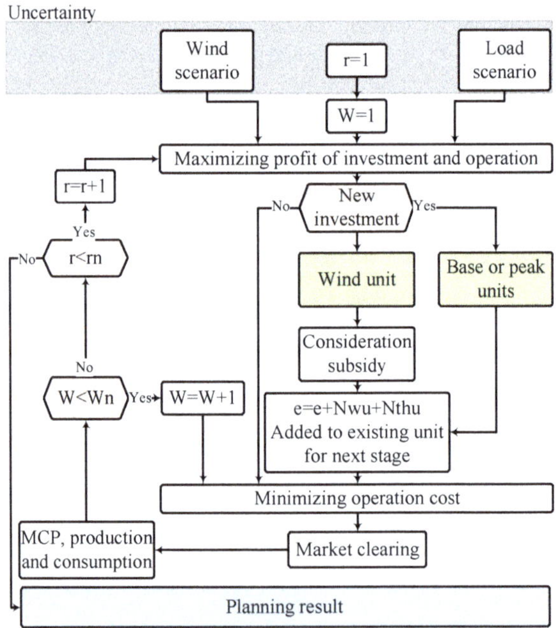

Figure 4. The algorithm for solving the proposal idea.

3.1. Inputs and Outputs

Figure 5 shows the input and output of each stage and stage to stage. The first-level includes 10 stages in which each stage includes a set of market-clearing issues for different scenarios and time demand blocks. The input of the first level for each stage includes input related to investment in different load ranges of power plants k_s^{TH}, CS_s^{TH}, c_n^{invyear}, c_n^{inv}, X_n^{Wmax}, S and the general input N_t^h, γ_{tw}, P_i^{ESmax}, CS_i^{ES}, f.

First-level variables include $X_{\text{ys}}^{\text{TH}} / X_{\text{yn}}^{W}$, $u_{\text{ysh}}^{\text{TH}} / u_{\text{ynl}}^{W}$, and these have a certain effect in the second-level and are therefore considered to be second-level inputs. In addition, the second level includes the general inputs P_i^{ESmax}, d_{yjtw}, B_{nm}, and $F_{nm}^{\max}$. The output of the second-level are variables $P_{\text{ystw}}^{\text{TH}} / P_{\text{yitw}}^{\text{Es}} / P_{\text{yn,tw}}^{W}$, $P_{\text{yy'stw}}^{\text{THY}} / P_{\text{yy'n,tw}}^{\text{THY}} / P_{\text{yy'n,tw}}^{\text{WY}}$, and θ_{ytnw}. These variables are also the variables of first-level. First-level input of each stage to the next stage includes variables P_i^{Esmax}, d_{yjtw}, B_{nm}, $F_{nm}^{\max}$ and the second-level input of the previous stage includes variables $\overline{G}^{\text{theu}}$.

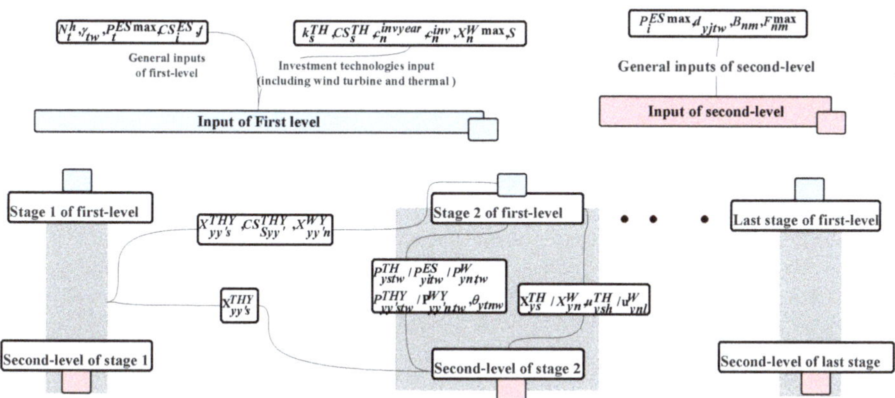

Figure 5. Input and output of each stage and stage to stage.

3.2. Converting Two Level Model to One Level Model

The market and the proposed algorithm are expressed using the bi-level model to implement the proposed ideas in this article. This bi-level model can be solved by heuristic algorithms, GAMS solvers and so forth. The GAMS solvers are used to solve the model proposed in this paper. For this purpose, the bi-level model is converted to a one-level problem and then the optimal solution is obtained using the available solvers, as shown in Figure 6. The second-level problem has constraints including DC power flow, limitations of unit production, and balance between production and consumption. Therefore, the second-level problem is a linear problem and therefore convex. Thus, karush kuhn tucker (KKT) conditions can be used to convert the bi-level problem into a one-level problem. Furthermore, complimentary constraints obtained from KKT conditions are linear when using the theory of big M. The resulting mathematical model becomes a problem of MILP after linearization of the nonlinear relationships and therefore it is solvable by MILP solvers in GAMS.

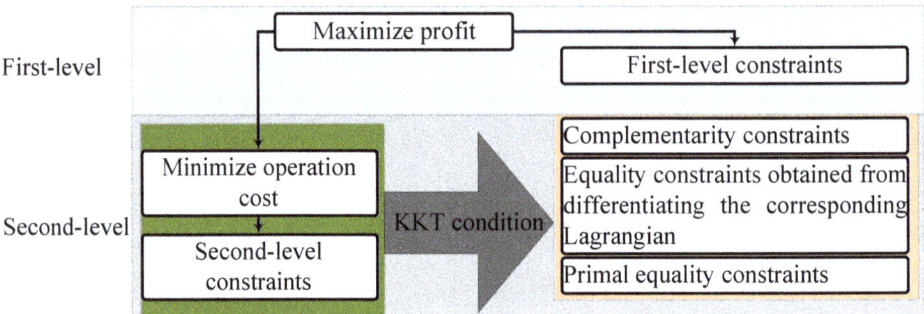

Figure 6. Converting the bi-level problem into one-level problem.

4. Mathematical Formulation

In the following sub-sections, the mathematical formulation of this paper is presented. In order to introduce the model, we define the following sets, parameters and decision variables.

4.1. The Bi-Level Model

The first-level represents the investment problem of the conventional producers who are seeking to maximize the present value of the total profit of investment (whether base or peak) and of operation. Due to the dynamic nature of the planning problem, dynamic dependency constraints exist in the first-level. The second-level problem represents the market-clearing. The clearing of the market for any given operating condition is represented as an optimization problem that identifies the operating decisions that maximize social welfare. In this model, maximizing the social welfare is equivalent to minimizing the generation cost. The market clearing problem is constrained by the DC power flow equations, transmission network limitations and units' capacity limits. The output of the second-level problem is nodal prices (dual variables associated with the power balance constraints), which are fed back to the first-level.

The multi-period stochastic investment problem is formulated using the following bi-level model, which comprises the first-level problem, i.e., Equations (4)–(12), and a collection of second-level problems, i.e., Equations (14)–(20).

First level: The objective function (i.e., Equation (4)) is the present value of the projected profit (expected revenue offset by investment cost) in the planning horizon, which comprises three terms. The first and second terms of profit function (i.e., Equation (4)) are associated with the investment cost of new units, including peak and base technologies and wind units, respectively. The third term of the profit function (i.e., Equation (4)) is the expected profit obtained by selling energy in the spot market. Equation (5) states that investment options for new base or peak units are only available in discrete blocks, just as that of wind investment modelled through Equation (6). These equations impose the constraint that only one technology is binding and determines the new technology to be installed at each bus of the system. It should be noted that, based on Equations (5) and (6), the producer can either open exactly one new plant each year, or choose one option for installing it. These constraints define the maximum capacity of wind units along a multi-stage horizon that can be constructed in each location with wind power facilities. The dynamic dependency constraints on the base or peak investment decision variables are represented in Equations (7) and (8), while Equation (9) is related to the dynamic dependency constraints applicable to the wind investment. The new units are operated in the next years as existing units in the network (Equations (7) and (9)). Equation (8) is related to the marginal cost of a new base or peak units in subsequent years. Equation (10) enforces that wind generation at each bus scenario is limited to the installed wind power at the corresponding bus multiplied by a factor that models the wind intensity at that bus and scenario. Equation (11) as the dynamic dependency constraint shows wind production constraints in the years after the establishment of new units. The available budget limitation is represented in Equation (12) for investment in wind and new thermal units.

Second level: The market clearing problems are represented by the negative social welfare expressed Equations (13)–(20). Equation (14) represents the energy balance at each bus, this being the associated dual variables' LMPs or nodal prices. Equations (15)–(19) impose power bounds for blocks of generation constraints, power flow and angle bounds. Equation (20) fixes the voltage angle at the reference bus. Dual variables are indicated at the relevant constraints following a colon:

First level objective function:

$$\min \ \sum_y \left(\frac{1}{1+y}\right)^y \sum X_s^{\text{TH}} k_s^{\text{TH}}$$

$$+ \sum_y \left(\frac{1}{1+f}\right)^y \sum_t (1-S) c_n^{\text{inv}} X_n^W$$

$$- \sum_y \left(\frac{1}{1+f}\right)^y \sum_t N_t^h \sum_w \gamma_{tw}$$

$$\sum_n P_{n,tw}^W \lambda_{n,tw}$$

$$+ \sum_n \sum_{y'} P_{yy'n,tw}^{\text{WY}} \lambda_{n,tw} \tag{4}$$

$$+ \sum_s P_{ystw}^{\text{TH}} \lambda_{n,tw} - \sum_s P_{ys,tw}^{\text{TH}} \text{CS}^{\text{TH}}$$

$$+ \sum_s \sum_{y'} P_{yy'stw}^{\text{THY}} \lambda_{n,tw}$$

$$- \sum_s \sum_{y'} P_{yy'stw}^{\text{THY}} \text{CS}_{Syy'}^{\text{THY}}$$

$$+ \sum_i P_{yitw}^{\text{ES}} - \sum_i P_{yitw}^{\text{ES}} \text{CS}_i^{\text{ES}}.$$

First level subjected to:
Subject to:

$$X_{ys}^{\text{TH}} = \sum_h u_{ysh}^{\text{TH}} X_{ysh}^{\text{TH}}, \ \sum_h u_{ysh}^{\text{TH}} = 1 \ u_{ysh}^{\text{TH}} \in \{0,1\} : \ \forall y, \forall s, \forall h, \tag{5}$$

$$X_{yn}^W = \sum_L u_{ynl}^W X_{nl}^W, \ \sum_L u_{ynl}^W = 1 \ u_{ynl}^W \in \{0,1\} : \ \forall y, \forall l, \forall n, \tag{6}$$

$$X_{yy's}^{\text{THY}} = X_{ys}^{\text{TH}}, \ \forall y, y' \subset \{X_{yy'}^{\text{TH}} > 0, y > y'\}, \forall s, \tag{7}$$

$$\text{CS}_{syy'}^{\text{THY}} = \text{CS}_s^{\text{TH}}, \ \forall y, y' \subset \{X_{yy'}^{\text{TH}} > 0, y > y'\}, \forall s, \tag{8}$$

$$X_{yy'n}^{\text{WY}} = X_{yn}^W, \ \forall y, y' \subset \{X_{y'n}^W > 0, y > y'\}, \forall n, \tag{9}$$

$$P_{yn,tw}^W \leq K_{n,tw} X_{yn}^W : \ \forall y, \forall n, \forall t, \forall w, \tag{10}$$

$$0 \leq P_{yy'n,tw}^{\text{WY}} \leq X_{yy'n}^{\text{WY}}, \ \forall y, y' \subset \{X_{y'n}^W > 0, y > y'\}, \forall t, \forall n, \forall w, \tag{11}$$

$$\sum_y \left(\frac{1}{1+f}\right)^y \left(\sum_n (1-S) C_n^{\text{inv}} X_n^W + \sum k_s^{\text{TH}} X_{ys}^{\text{TH}} \leq k^{\max}\right). \tag{12}$$

Second level objective function:

$$\min \ \sum_i \sum P_{yitw}^{\text{ES}} \text{CS}_i^{\text{ES}} + \sum_s \sum P_{yStw}^{\text{TH}} \text{CS}_S^{\text{TH}}$$

$$+ \sum_s \sum P_{yy'stw}^{\text{THY}} \text{CS}_{Sy'}^{\text{THY}} \qquad , \ \forall y, \forall t, \forall w. \tag{13}$$

Second level subjected to:

$$\sum_i \sum P_{\text{yitw}}^{\text{ES}} + \sum_s \sum P_{\text{ystw}}^{\text{TH}} + \sum P_{yy'n,tw}^{\text{WY}}$$

$$+ \sum P_{yy'stw}^{\text{THY}} + P_{\text{yntw}}^{W} - \sum_m B_{nm}^{\text{P.U}} S_b (\theta_{\text{ytnw}} - \theta_{\text{ytmw}}) \tag{14}$$

$$= \sum_j d_{\text{yitw}} : \lambda_{\text{ntw}} \quad \forall n, \forall t, \forall y, \forall w,$$

$$0 \le P_{\text{yitw}}^{\text{ES}} \le P_{\text{yi}}^{\text{ESmax}} : \phi_{\text{yitw}}^{\min}, \phi_{\text{yitw}}^{\max} \quad \forall y, \forall i, \forall t, \forall w, \tag{15}$$

$$0 \le P_{\text{ystw}}^{\text{TH}} \le X_{\text{ys}}^{\text{TH}} : \mu_{\text{ystw}}^{\min}, \mu_{\text{ystw}}^{\max} : \quad \forall y, \forall s, \forall t, \forall w, \tag{16}$$

$$0 \le P_{yy'stw}^{\text{THY}} \le X_{yy's}^{\text{THY}} : \mu_{yy'stw}^{\text{THYmin}}, \mu_{yy'stw}^{\text{THYmax}}$$

$$\forall y, y' \subset \{X_{y's}^{\text{TH}} > 0, y > y'\}, \forall t, \forall s, \forall w, \tag{17}$$

$$-F_{nm}^{\max} \le B_{nm}^{\text{P.U}} s_b (\theta_{\text{ytnw}} - \theta_{\text{ytmw}}) \le F_{nm}^{\max} :$$

$$v_{\text{ytnmw}}^{\min}, v_{\text{ytnmw}}^{\max} \quad \forall n, \forall m, \forall t, \forall y, \forall w, \tag{18}$$

$$-\pi \le \theta_{\text{ytnw}} \le \pi : \zeta_{\text{ytnw}}^{\min}, \zeta_{\text{ytnw}}^{\max}, \quad \forall n, \forall t, \forall y, \forall w, \tag{19}$$

$$\theta_{\text{ytnw}} = 0 : \zeta_{\text{ytw}}^{1} \quad n = 1, \forall t, \forall y, \forall w. \tag{20}$$

MPEC

The bi-level problem (i.e., Equations (4)–(20)) can be converted to a single level problem (MPEC) by enforcing KKT conditions to the second-level problems [14,20,27]. These are represented by Equations (21)–(35). This transformation is possible because the lower-level problems are continuous and linear (and thus convex). The resulting problem is an MPEC, which includes nonlinearities in the objective function and in the complementarity conditions. These nonlinearities could be transformed into equivalent linear terms. Using this linearization, the investment problem can be finally expressed by the following MILP problem, as presented in the algorithm in Figure 4. To find a linear expression for $\sum_{n \in \Omega^N} P_{n,tw}^{W} \lambda_{n,tw} + \sum_s P_{\text{ystw}}^{\text{TH}} \lambda_{n,tw} + \sum_i P_{\text{yitw}}^{\text{ES}} \lambda_{n,tw} + \sum_s \sum_{y'} P_{yy'stw}^{\text{THY}} \lambda_{n,tw}$ in Equation (4), we use the strong duality theorem and some of the KKT equalities [31]. The strong duality theorem says that, if a problem is convex, the objective functions of the primal and dual problems have the same value at the optimum.

Objective function of MPEC:

$$\min \sum_y (\tfrac{1}{1+f})^y \sum k_s^{\text{TH}} X_{\text{ys}}^{\text{TH}} + \sum_y (\tfrac{1}{1+f})^y \sum_t (1 - S) C_n^{\text{inv}} X_{\text{yn}}^{W}$$

$$- \sum_y (\tfrac{1}{1+f})^y \sum_t N_t^h \sum_w \gamma_{\text{tw}}$$

$$\left(\begin{array}{l} \sum \lambda_{\text{yn,tw}} \sum d_{\text{yj,tw}} - \sum_{i \in \Omega^N} \sum \phi_{\text{yitw}}^{\max} \\[4pt] - \sum_n V_{\text{ytnw}}^{\min} F_{nm}^{\max} - \sum_n V_{\text{ytnw}}^{\max} F_{nm}^{\max} - \sum_n (\zeta_{\text{ytnw}}^{\min} + \zeta_{\text{ytnw}}^{\max}) \pi \\[4pt] - \sum \sum_i P_{\text{yitw}}^{\text{Es}} \text{CS}_i^{\text{ES}} - \sum \sum_s P_{\text{ystw}}^{\text{TH}} \text{CS}_s^{\text{TH}} - \sum_s \sum_{y'} P_{yy'stw}^{\text{THY}} \text{CS}_{\text{Syy'}}^{\text{THY}} \end{array} \right). \tag{21}$$

MPEC subjected to:

- First-level constraints: Equations (5)–(12).
- Primal equality constraints equal to second-level: Equations (14)–(20).
- Equality constraints obtained from differentiating the corresponding Lagrangian:

$$\mathrm{CS}^{\mathrm{ES}}_i - \lambda_{\mathrm{yntw}} + \phi^{\max}_{\mathrm{yitw}} - \phi^{\min}_{\mathrm{yitw}} = 0 : \ \forall y, \forall i, \tag{22}$$

$$\mathrm{CS}^{\mathrm{TH}}_s - \lambda_{\mathrm{yntw}} + \mu^{\max}_{\mathrm{ystw}} - \mu^{\min}_{\mathrm{ystw}} = 0 : \ \forall y, \forall s, \tag{23}$$

$$\mathrm{CS}^{\mathrm{THY}}_{syy'} - \lambda_{\mathrm{yntw}} + \mu^{\mathrm{THY}\max}_{yy'\mathrm{stw}} - \mu^{\mathrm{THY}\min}_{yy'\mathrm{stw}} = 0 : \ \forall y, \forall s, \tag{24}$$

$$\sum_m B^{\mathrm{P.U}}_{nm} S_b (\lambda_{\mathrm{ytnw}} - \lambda_{\mathrm{ytmw}}) + \sum_m B^{\mathrm{P.U}}_{nm} S_b (V^{\max}_{\mathrm{ytnmw}} - V^{\max}_{\mathrm{ytmnw}})$$
$$+ \sum_m B^{\mathrm{P.U}}_{nm} S_b (V^{\min}_{\mathrm{ytnmw}} - V^{\min}_{\mathrm{ytmnw}}) + \zeta^{\max}_{\mathrm{ytnw}} - \zeta^{\min}_{\mathrm{ytnw}}. \tag{25}$$

- Complimentary constraints by KKT:

$$0 \leq P^{\mathrm{TH}}_{\mathrm{ystw}} \perp \mu^{\min}_{\mathrm{ystw}} \geq 0 \ \ \forall y, \forall t, \forall i, \forall w, \tag{26}$$

$$0 \leq P^{\mathrm{EX}}_{\mathrm{yitw}} \perp \phi^{\min}_{\mathrm{itw}} \geq 0 \ \ \forall y, \forall i, \forall t, \forall w, \tag{27}$$

$$0 \leq (X^{\mathrm{TH}}_{\mathrm{ys}} - P^{\mathrm{TH}}_{\mathrm{ystw}}) \perp \mu^{\max}_{\mathrm{ystw}} \geq 0 \ \ \forall y, \forall s, \forall t, \forall w, \tag{28}$$

$$0 \leq P^{\mathrm{THY}}_{yy'\mathrm{stw}} \perp \mu^{\mathrm{THY}\min}_{yy'\mathrm{stw}} \ \ \forall y, \forall s, \forall t, \forall w, \tag{29}$$

$$0 \leq (X^{\mathrm{THY}}_{yy's} - P^{\mathrm{THY}}_{yy'\mathrm{stw}}) \perp \mu^{\mathrm{THY}\max}_{yy'\mathrm{stw}} \ \ \forall y, \forall s, \forall t, \forall w, \tag{30}$$

$$0 \leq (P^{\mathrm{EX}\max}_i - P^{\mathrm{EX}}_{\mathrm{yitw}}) \perp \phi^{\max}_{\mathrm{itw}} \geq 0 \ \ \forall y, \forall i, \forall t, \forall w, \tag{31}$$

$$0 \leq [F^{\max}_{nm} - B^{\mathrm{P.U}}_{nm} + S_b(\theta_{\mathrm{ytnw}} - \theta_{\mathrm{ytmw}})] \perp V^{\max}_{\mathrm{ytnmw}} \geq 0$$
$$\forall y, \forall t, \forall n, \forall m, \forall w, \tag{32}$$

$$0 \leq [F^{\max}_{nm} + B^{\mathrm{P.U}}_{nm} + S_b(\theta_{\mathrm{ytnw}} - \theta_{\mathrm{ytmw}})] \perp V^{\min}_{\mathrm{ytnmw}} \geq 0$$
$$\forall y, \forall t, \forall n, \forall m, \forall w, \tag{33}$$

$$0 \leq (\pi - \theta_{\mathrm{ytnw}}) \perp \zeta^{\max}_{\mathrm{ytnw}} \geq 0 \ \ \forall y, \forall t, \forall n, \forall w, \tag{34}$$

$$0 \leq (\pi + \theta_{\mathrm{ytnw}}) \perp \zeta^{\min}_{\mathrm{ytnw}} \geq 0 \ \ \forall y, \forall t, \forall n, \forall w. \tag{35}$$

Equations (26)–(35) are linearized as follows [14,27]:
The complementarity condition

$$0 \leq a \perp b \geq 0 \tag{36}$$

can be replaced by

$$a \geq 0, b \geq 0, a \leq \tau M, b \leq (1 - \tau)M, \tau \in \{0,1\}, \tag{37}$$

where M is a large enough constant.

5. Case Studies

In this section, the efficiency of the proposed framework is examined through a real case study. The case study is the MAZANDARAN regional electric company (MREC) transmission network as a section of an IRAN interconnected power system.

Case Study: MREC Network

A single-line diagram representing the MREC transmission network is shown in Figure 7 [14,46]. The candidate buses for construction of the new base and peak units are assumed to be AMOL,

ALIABAD and NEKA, all of which operate at 230 kV. In addition, buses ROYAN and DARYASAR are considered as candidate sites for construction of the new wind units. The maximum wind capacity that can be installed at these buses is equal to 300 MW for each bus.

The cases considered for the MREC transmission network are characterized in Table 1. The second column of this table gives the type of planning (static or dynamic) and columns 3 and 4 identify the subsidy percentage and the available budget, respectively. In Cases 1–4, the type of planning is static, while Cases 5–8 refer to dynamic cases. In addition, in Cases 2, 4, 6 and 8, consideration has been made of the impact of subsidy on the wind investment. In these cases, the subsidy percentage is assumed to be 20%. The proposed model is solved using Solver XPRESS software GAMS (IBM ILOG CPLEX Solver, 11.0.1, Armonk, NY, USA) [47].

Table 1. The cases considered for the MREC transmission network.

Cases	Planning	Subsidy (%)	Budget (M$)
#1	Static	0	15.0
#2	Static	20	15.0
#3	Static	0	150
#4	Static	20	150
#5	Dynamic	0	150
#6	Dynamic	20	150
#7	Dynamic	0	1500
#8	Dynamic	20	1500

Case #1 total thermal investment is equal to 700 MW in the peak technology due to budget limitations and offset by their investment cost with respect to base technology. In this case, the total capacity added by the producer in the planning horizon has been established at 900 MW, resulting in the producer investing 200 MW in the wind technology and 700 MW in the peak technology. In this case, total thermal investment is attributable to peak technology due to budget limitation and less their investment cost with respect to base technology. The existing units supply 9.78 MMWh of the energy consumed by the network and thereby play an important role in the provision of energy to the MREC network;

Case #2 Investment in wind units in this case is increased by 50 MW compared with Case #1 due to a 20% subsidy. In addition, the investment cost of wind units is the same as Case #1. In this case, the production of wind units has increased by 25.58%. Therefore, the net profit in Case #2 has increased by 18.44% with respect to Case #1;

Case #3 The total capacity added by the producer in Case #3 has increased by 450 MW compared with Case #1 by increasing the budget from 15 M$ to 150 M$. Investment in wind units in this case has increased by 400 MW compared with the Case #1 due to an increase in the budget. In addition, the total thermal investment is base technology so that 750 MW is added to the MREC network. In Case #3, the production of wind and new thermal units have been increased by 201.62% and 98.15% compared with Case #1, respectively. In addition, the net profit in the planning horizon has been obtained equal to 108 M$ that it has been increased by 389.35% with respect to Case #1;

Case #4 In Case #4, investment in wind and thermal units is the same as Case #3. In this case, subsidy has no effect on wind investment because the total capacity of wind units added to network is the same as in Case #3. In addition, the production of different units is the same as Case #3. However, the investment cost of wind units has decreased by 20% due to the 20% subsidy. As a result, the investor's net profit has increased by 5.56% compared with Case #3;

Case #5 In this case, the total capacity added by the producer in the planning duration has been determined to be 1200 MW, resulting in the producer investing 500 MW in the base technology and 700 MW in the peak technology, while no wind unit was constructed in the MREC network. This is the result of the desire to invest in units that have a lower investment cost due to the budget limitations. In addition, distribution of the investment are as follows: 200 MW on peak technology in the first year, 500 MW on base technology in the fifth year, 250 MW on peak technology in the eighth year and 250 MW on peak technology in the ninth and tenth years. Due to lack of generation in the western region, the total capacity has been constructed in AMOL located in this region.

Case #6 All output of Case #6 is the same as Case #5; consequently, the consideration of a 20% subsidy has no influence on the desire to invest in wind units. It can be observed that the 20% subsidy is not enough to encourage investment in wind units and therefore the capacity of wind units is zero in this network.

Case #7 In Case #3, the total capacity added by the investor has increased by 1400 MW compared with Case #5 and Investment in wind units in this case have been increased by 600 MW compared with the Case #1. In addition, total thermal investment is base technology so that 2000 MW base technologies are added to the MREC network. In Case #7, the production of new thermal units has been increased by 75.89% compared with Case #5, while the production of existing units decreased by 81.04%. In addition, the production of wind units has been increased from 0 MMWh to 12.97 MMWh with respect to Case #5. Thus, the total net profit in the planning duration is predicted to be 973.28 M$, an increase of 188.86% with respect to Case #5.

Case #8 In this case, investment in wind and thermal units, and the production of different units, is the same as Case #7. However, investment cost of wind units has been decreased by 20% due to the 20% subsidy. As a result, the investor's net profit has been increased by 9.62% compared with Case #3.

Figure 8 shows the result of GEP for MREC network for: (a) the produced power, (b) the produced energy, and (c) investment. Figure 9 depicts the percentages related to the budget, investment cost and total net profit for each case in the planning period. For example, the percentage related to the budget, investment cost and total net profit in the Case #1 are shown to be equal 29%, 29% and 42%, respectively, corresponding to 15, 14.99 and 22.07 when expressed in M$. The net profit in Case #5 and Case #6 in the dynamic approach has been increased by 26.19% and 12.77% compared to Case #1 and Case #2 in the static approach, respectively. Furthermore, the investment cost has been decreased by 17.24% and 11.11%, respectively. In addition, the net profit in Case #3 and Case #4 in the static approach has been increased by 14.28% and 9.68% compared to Case #7 and Case #8 in the dynamic approach, respectively. Furthermore, the investment cost has been decreased by 14.28% and 12%, respectively. Therefore, it can be seen that the percentage profit as a result of increasing the available budget for both static and dynamic planning. For lower budgets, the percentage profit yielded by the dynamic approach exceeds that of the static approach so that the investment cost in the former is less than in the latter. However, if a larger budget is available, the percentage profit yielded by the static approach exceeds that of the dynamic approach, whilst the former requires a lower investment. The use of a subsidy can increase investor profit if it stimulates investment in wind technologies. In addition, the investment cost decreases in these cases. It can be seen that the degree of investment increases when the dynamic approach is applied, when compared to the static approach. Moreover, consideration of the dynamic versus static trade-offs when the planning of generation capacity leads to accurate and realistic results in the expansion planning.

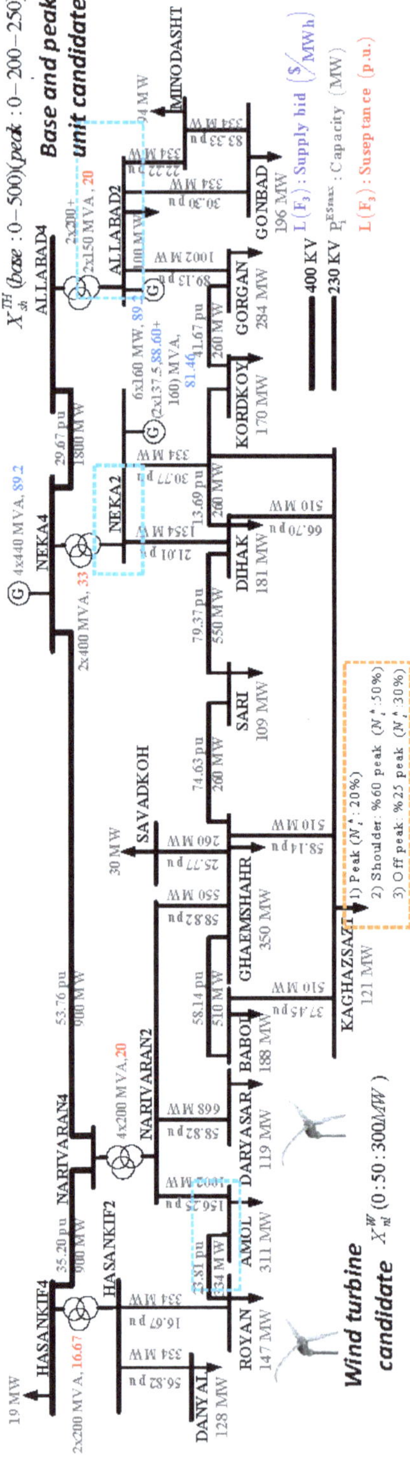

Figure 7. Single-line diagram of the MAZANDARAN Regional Electric Company (MREC) transmission network.

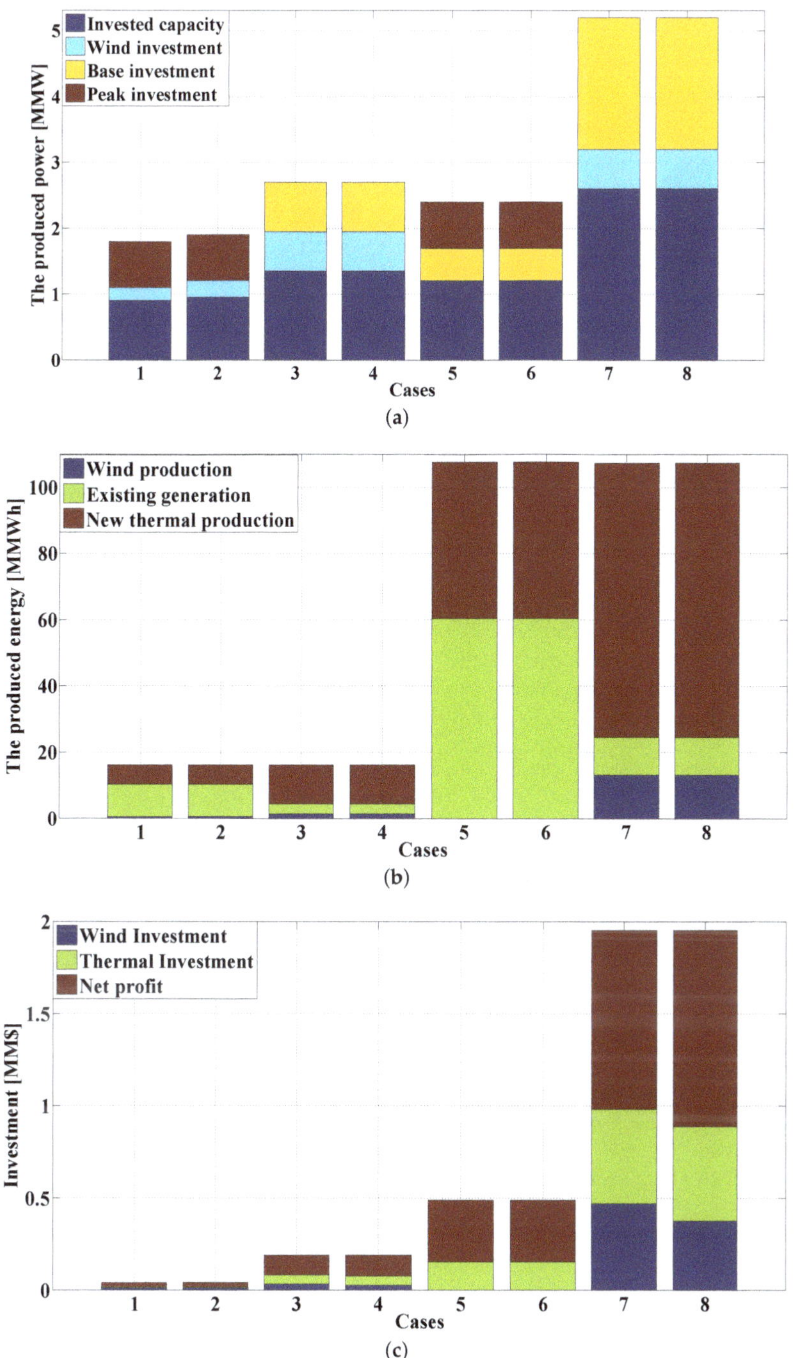

Figure 8. The result of GEP for MREC network. (**a**) the produced power; (**b**) the produced energy; (**c**) investment.

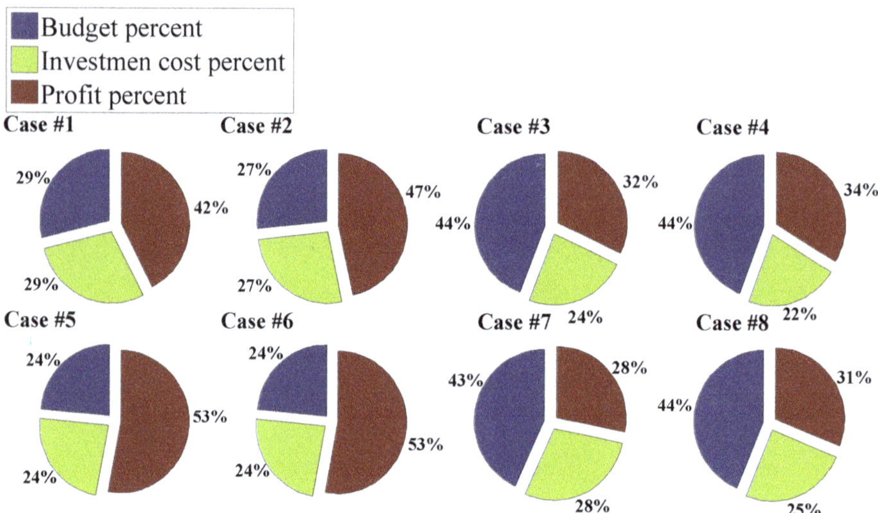

Figure 9. The percent related to the budget, investment cost and total net profit for each case.

6. Conclusions

This paper has presented a model for the expansion planning of wind resources in power systems at a multi-stage horizon. In this paper, the power system consists of a combination of fossil fuel technologies and wind resources for investment. Real case studies were considered and analyzed. The features of the proposed model and simulation results led to following conclusions:

- It can be seen that the percentage profit decreases as a result of increasing the available budget for both static and dynamic planning. At a lower budgetary level, the percentage profit yielded by the dynamic approach is more than that of the static approach so that the investment cost of the dynamic approach is less than the static. However, when the budgetary level is higher, the percentage profit in the static approach exceeds that of the dynamic approach while the static approach has a lower investment cost. The effect of a subsidy is to increase investor profit if the subsidy encourages investment on the wind technologies. In addition, the investment cost decreases in these cases. It can be seen that the total investor contribution has been increased in the dynamic approach with respect to the static one. Moreover, using the dynamic versus static approach in the planning of generation capacity leads to accurate and realistic results in the expansion planning.

- Additional work is underway to represent the strategic behaviour of market participants other than wind and new thermal producers. The proposed model can be adapted to consider the impact of transmission expansion plans, availability of gas transmission networks, tax policy, DSM plans and uncertainty in the demand growth.

Author Contributions: All authors contributed equally to this work and all authors have read and approved the final manuscript.

Funding: This research received no external funding.

Nomenclature

Indices

w	index for scenario
y/y'	indexes for stage (year)
t	index for demand blocks
$s/i/j$	indexes for new base or peak units/existing generation unit and demand
h/l	investment capacity of new base or peak unit s/wind power at bus n (MW)
n/m	indexes for bus

Acronyms

N_t^h	weight of demand block t in year y
γ	weight of scenario w in demand block t
$k_s^{TH}/c_n^{invyear}$	annualized investment cost of base or peak units/wind power (€/MW)
c_n^{inv}	investment cost of wind power at bus n
X_n^{Wmax}	maximum wind capacity that can be installed at bus n
X_{sh}^{Th}/X_{nl}^{W}	option h/l for investment capacity of new base or peak units s/wind power at bus n (MW)
p_i^{ESmax}	capacity of existing generation unit i of strategic producer (MW)
d_{yjtw}	load of demand j in block t and year y (MW)
CS_s^{TH}/CS_i^{ES}	price offered by new base or peak units/existing unit producer (€/MWh)
B_{nm}	susceptance of line n-m (p.u.)
F_{nm}^{max}	transmission capacity of line n-m (MW)
f	Discount rate
S	Subsidy percent

Decision variables

X_{ys}^{TH}/X_{yn}^{W}	capacity investment of new unit s/wind power at bus n (MW)
u_{ysh}^{TH}/u_{ynl}^{W}	binary variable that is equal to 1 if the h/lth investment option determines the base or peak unit/wind power is selected in year y
$X_{yy's}^{THY}/X_{yy'n}^{WY}$	available capacity of new unit s/wind power at bus n in year y', in the years after the installation in year y (MW)
$P_{ystw}^{TH}/P_{yitw}^{ES}/P_{yn,tw}^{W}$	power produced by new base or peak units s/existing unit i/wind power at bus n, in year y, demand block t and scenario w (MW)
$P_{ystw}^{TH}/P_{yitw}^{ES}/P_{yn,tw}^{W}$	power produced by new base or peak units s/wind power at bus n, in year y', in the years after the installation in year y (MW)
$CS_{Syy'}^{THY}$	price offered by new base or peak units producer a, in year y', In the years after the installation in year y (€/MWh)
θ_{ytnw}	voltage angle of bus n, in year y, demand block t and scenario w

References

1. Bagheri, A.; Monsef, H.; Lesani, H. Renewable power generation employed in an integrated dynamic distribution network expansion planning. *Electr. Power Syst. Res.* **2015**, *127*, 280–296. [CrossRef]
2. Kahraman, C.; Onar, S.C.; Oztaysi, B. A Comparison of Wind Energy Investment Alternatives Using Interval-Valued Intuitionistic Fuzzy Benefit/Cost Analysis. *Sustainability* **2016**, *8*, 118. [CrossRef]
3. Lumbreras, S.; Ramos, A.; Banez-Chicharro, F. Optimal transmission network expansion planning in real-sized power systems with high renewable penetration. *Electr. Power Syst. Res.* **2017**, *149*, 76–88. [CrossRef]
4. Gan, L.; Li, G.; Zhou, M. Coordinated planning of large-scale wind farm integration system and regional transmission network considering static voltage stability constraints. *Electr. Power Syst. Res.* **2016**, *136*, 298–308. [CrossRef]
5. Kamalinia, S.; Shahidehpour, M.; Khodaei, A. Security-constrained expansion planning of fast-response units for wind integration. *Electr. Power Syst. Res.* **2011**, *81*, 107–116. [CrossRef]
6. You, S.; Hadley, S.W.; Shankar, M.; Liu, Y. Co-optimizing generation and transmission expansion with wind power in large-scale power grids—Implementation in the US Eastern Interconnection. *Electr. Power Syst. Res.* **2016**, *133*, 209–218. [CrossRef]

7. Syed, I.M.; Raahemifar, K. Predictive energy management, control and communication system for grid tied wind energy conversion systems. *Electr. Power Syst. Res.* **2017**, *142*, 298–309. [CrossRef]

8. Saboori, H.; Hemmati, R. Considering Carbon Capture and Storage in Electricity Generation Expansion Planning. *IEEE Trans. Sustain. Energy* **2016**, *7*, 1371–1378. [CrossRef]

9. Nosair, H.; Bouffard, F. Flexibility Envelopes for Power System Operational Planning. *IEEE Trans. Sustain. Energy* **2015**, *6*, 800–809. [CrossRef]

10. Valinejad, J.; Marzband, M.; Busawona, K.; Kyyrä, J.; Pouresmaeil, E. Investigating Wind Generation Investment Indices in Multi-Stage Planning. In Proceedings of the 5th International Symposium on Environment Friendly Energies and Applications (EFEA), Rome, Italy, 24–26 September 2018.

11. Marzband, M.; Azarinejadian, F.; Savaghebi, M.; Pouresmaeil, E.; M.Guerrero, J.; Lightbody, G. Smart transactive energy framework in grid-connected multiple home microgrids under independent and coalition operations. *Renew. Energy* **2018**, *126*, 95–106. [CrossRef]

12. Tavakoli, M.; Shokridehaki, F.; Akorede, M.F.; Marzband, M.; Vechiu, I.; Pouresmaeil, E. CVaR-based energy management scheme for optimal resilience and operational cost in commercial building microgrids. *Int. J. Electr. Power Energy Syst.* **2018**, *100*, 1–9. [CrossRef]

13. Momoh, J.; Mili, L. *Economic Market Design and Planning for Electric Power Systems*; Wiley-IEEE Press: New York, NY, USA, 2009.

14. Valinejad, J.; Marzband, M.; Akorede, M.F.; Barforoshi, T.; Jovanović, M. Generation expansion planning in electricity market considering uncertainty in load demand and presence of strategic GENCOs. *Electr. Power Syst. Res.* **2017**, *152*, 92–104. [CrossRef]

15. Marzband, M.; Fouladfar, M.H.; Akorede, M.F.; Lightbody, G.; Pouresmaeil, E. Framework for smart transactive energy in home-microgrids considering coalition formation and demand side management. *Sustain. Cities Soc.* **2018**, *40*, 136–154. [CrossRef]

16. Marzband, M.; Javadi, M.; Pourmousavi, S.A.; Lightbody, G. An advanced retail electricity market for active distribution systems and home microgrid interoperability based on game theory. *Electr. Power Syst. Res.* **2018**, *157*, 187–199. [CrossRef]

17. Olsen, D.; Byron, J.; DeShazo, G.; Shirmohammadi, D.; Wald, J. Collaborative Transmission Planning: California's Renewable Energy Transmission Initiative. *IEEE Trans. Sustain. Energy* **2012**, *3*, 837–844. [CrossRef]

18. Soroudi, A.; Rabiee, A.; Keane, A. Information gap decision theory approach to deal with wind power uncertainty in unit commitment. *Electr. Power Syst. Res.* **2017**, *145*, 137–148. [CrossRef]

19. Baringo, L.; Conejo, A. Wind Power Investment: A Benders Decomposition Approach. *IEEE Trans. Power Syst.* **2012**, *27*, 433–441. [CrossRef]

20. Xiong, P.; Singh, C. Optimal Planning of Storage in Power Systems Integrated With Wind Power Generation. *IEEE Trans. Sustain. Energy* **2016**, *7*, 232–240. [CrossRef]

21. Sun, C.; Bie, Z.; Xie, M.; Jiang, J. Assessing wind curtailment under different wind capacity considering the possibilistic uncertainty of wind resources. *Electr. Power Syst. Res.* **2016**, *132*, 39–46. [CrossRef]

22. Li, S.; Coit, D.W.; Felder, F. Stochastic optimization for electric power generation expansion planning with discrete climate change scenarios. *Electr. Power Syst. Res.* **2016**, *140*, 401–412. [CrossRef]

23. Tavakoli, M.; Shokridehaki, F.; Mousa Marzband, R.G.; Pouresmaeil, E. A two stage hierarchical control approach for the optimal energy management in commercial building microgrids based on local wind power and PEVs. *Sustain. Cities Soc.* **2018**, *41*, 332–340. [CrossRef]

24. Hinojosa, V.H.; Velásquez, J. Improving the mathematical formulation of security-constrained generation capacity expansion planning using power transmission distribution factors and line outage distribution factors. *Electr. Power Syst. Res.* **2016**, *140*, 391–400. [CrossRef]

25. Valenzuela, J.; Wang, J. A probabilistic model for assessing the long-term economics of wind energy. *Electr. Power Syst. Res.* **2011**, *81*, 853–861. [CrossRef]

26. Zhang, T.; Baldick, R.; Deetjen, T. Optimized generation capacity expansion using a further improved screening curve method. *Electr. Power Syst. Res.* **2015**, *124*, 47–54. [CrossRef]

27. Pineda, S.; Morales, J.; Ding, Y.; Østergaard, J. Impact of equipment failures and wind correlation on generation expansion planning. *Electr. Power Syst. Res.* **2014**, *116*, 451–458. [CrossRef]

28. Brandi, R.B.S.; Ramos, T.P.; David, P.A.M.S.; Dias, B.H.; Marcato, A.L.M. Maximizing hydro share in peak demand of power systems long-term operation planning. *Electr. Power Syst. Res.* **2016**, *141*, 264–271. [CrossRef]

29. Jabr, R. Robust Transmission Network Expansion Planning With Uncertain Renewable Generation and Loads. *IEEE Trans. Power Syst.* **2013**, *28*, 4558–4567. [CrossRef]

30. Munoz, F.D.; Mills, A.D. Endogenous Assessment of the Capacity Value of Solar PV in Generation Investment Planning Studies. *IEEE Trans. Sustain. Energy* **2015**, *6*, 1574–1585. [CrossRef]

31. Valinejad, J.; Barforoushi, T. Generation expansion planning in electricity markets: A novel framework based on dynamic stochastic MPEC. *Int. J. Electr. Power Energy Syst.* **2015**, *70*, 108–117. [CrossRef]

32. Baringo, L.; Conejo, A. Strategic Wind Power Investment. *IEEE Trans. Power Syst.* **2014**, *29*, 1250–1260. [CrossRef]

33. Valinejad, J.; Marzband, M.; Barforoushi, T.; Kyyrä, J.; Pouresmaeil, E. Dynamic stochastic EPEC model for Competition of Dominant Producers in Generation Expansion Planning. In Proceedings of the 5th International Symposium on Environment Friendly Energies and Applications (EFEA), Rome, Italy, 24–26 September 2018.

34. Kim, J.-Y.; Kim, K.S. Integrated Model of Economic Generation System Expansion Plan for the Stable Operation of a Power Plant and the Response of Future Electricity Power Demand. *Sustainability* **2018**, *10*, 2417. [CrossRef]

35. Khaligh, V.; Buygi, M.O.; Anvari-Moghaddam, A.; Guerrero, J. A Multi-Attribute Expansion Planning Model for Integrated Gas–Electricity System. *Energies* **2018**, *11*, 2573. [CrossRef]

36. Ko, W.; Park, J.K.; Kim, M.K.; Heo, J.H. A Multi-Energy System Expansion Planning Method Using a Linearized Load-Energy Curve: A Case Study in South Korea. *Energies* **2017**, *10*, 1663. [CrossRef]

37. Hong, S.; Cheng, H.; Zeng, P. An N-k Analytic Method of Composite Generation and Transmission with Interval Load. *Energies* **2017**, *10*, 168. [CrossRef]

38. Zhou, X.; Guo, C.; Wang, Y.; Li, W. Optimal Expansion Co-Planning of Reconfigurable Electricity and Natural Gas Distribution Systems Incorporating Energy Hubs. *Energies* **2017**, *10*, 124. [CrossRef]

39. Li, R.; Ma, H.; Wang, F.; Wang, Y.; Liu, Y.; Li, Z. Game Optimization Theory and Application in Distribution System Expansion Planning, Including Distributed Generation. *Energies* **2013**, *6*, 1101–1124. [CrossRef]

40. Meng, K.; Yang, H.; Dong, Z.Y.; Guo, W.; Wen, F.; Xu, Z. Flexible Operational Planning Framework Considering Multiple Wind Energy Forecasting Service Providers. *IEEE Trans. Sustain. Energy* **2016**, *7*, 708–717. [CrossRef]

41. Flores-Quiroz, A.; Palma-Behnke, R.; Zakeri, G.; Moreno, R. A column generation approach for solving generation expansion planning problems with high renewable energy penetration. *Electr. Power Syst. Res.* **2016**, *136*, 232–241. [CrossRef]

42. Duong, M.Q.; Grimaccia, F.; Leva, S.; Mussetta, M.; Le, K.H. Improving transient stability in a grid-connected squirrel-cage induction generator wind turbine system using a fuzzy logic controller. *Energies* **2015**, *8*, 6328–6349. [CrossRef]

43. Conejo, A.J.; Carrión, M.; Morales, J.M. *Decision Making under Uncertainty in Electricity Markets*; Springer: Berlin/Heidelberg, Germany, 2010.

44. Kazempour, S.; Conejo, A.; Ruiz, C. Strategic Generation Investment Using a Complementarity Approach. *IEEE Trans. Power Syst.* **2011**, *26*, 940–948. [CrossRef]

45. Drud, A. CONOPT. Available online: https://ampl.com/products/solvers/solvers-we-sell/conopt/ (accessed on 31 July 2012).

46. Valinejad, J.; Oladi, Z.; Barforoshi, T.; Parvania, M. Stochastic Unit Commitment in the Presence of Demand Response Program Under Uncertainties. *IJE Trans. B Appl.* **2017**, *30*, 1134–1143.

47. Brooke, D.; Kendrick, A.; Meeraus, R.; Raman, R. *GAMS: A User's Guide*; GAMS Development Corp.: Washington, DC, USA, 1998.

 *sustainability*

Article

Valuing Improved Power Supply Reliability for Manufacturing Firms in South Korea: Results from a Choice Experiment Survey

Doo-Chun Kim, Hyo-Jin Kim and Seung-Hoon Yoo *

Department of Energy Policy, Graduate School of Energy & Environment, Seoul National University of Science & Technology, 232 Gongreung-Ro, Nowon-Gu, Seoul 01811, Korea; smartenergy@seoultech.ac.kr (D.-C.K.); hjinkim@seoultech.ac.kr (H.-J.K.)
* Correspondence: shyoo@seoultech.ac.kr

Received: 27 September 2018; Accepted: 26 November 2018; Published: 30 November 2018

Abstract: An outage of electricity may cause considerable economic damage to industrial sectors. Thus, South Korea electricity authorities demand information about the value of improved power supply reliability for the manufacturing sector to implement them in planning electricity supply. This article aims to measure the value using a specific case of South Korean manufacturing firms. The choice experiment (CE) approach is adopted for this purpose. A nationwide CE survey of 1148 manufacturing firms was undertaken. The firms revealed statistically significant willingness to pay for a decrease in the duration of interruption, avoiding interruption during daytime (9 a.m. to 6 p.m.) rather than off-daytime (6 p.m. to 9 a.m.), and preventing interruption during weekdays rather than weekend. For example, they accepted a 0.02% increase in the electricity bill for reducing one minute of interruption during electricity outage, a 2.98% increase in the electricity bill to avoid interruption during the daytime rather than off-daytime, and a 1.60% increase in electricity bill for preventing interruption during weekdays rather than weekends. However, they put no importance on the season of interruption. These results can be useful for policy-making and decision-making regarding improving electricity supply reliability.

Keywords: power supply reliability; electricity; manufacturing industry; choice experiment; willingness to pay

---

1. Introduction

In microeconomics, labor and capital are usually assumed as basic production factors. However, electricity is another important production factor in modern times. Even if labor and capital are plentiful, without electricity commodities cannot be produced because factories and various production facilities are operated using electricity. That is, electricity is an essential input to industrial production. In particular, the industrial sector may use more electricity as the industrial structure improves. For example, artificial intelligence, self-driving, and international data centers need a lot of power consumption. Thus, a stable power supply contributes to industrial production and further to economic development by increasing economic activities [1].

This is the case for South Korea [2–5]. As of 2017, 56.3% of the total power consumption was for industrial purposes. This percentage is the highest among OECD countries except Iceland. Steel, shipbuilding, semiconductor, automotive, and petrochemical industries, which mainly support the export-led South Korean economy, account for a significant portion of industrial electricity use. For example, a steel company consumes all of the electricity produced by a nuclear power plant with a capacity of 1 GW. If electricity is not supplied properly to these sectors, massive damage will shake the foundations of the national economy [6].

Even if a power outage occurs for just one minute, it can cause significant damage to the manufacturing firms without uninterrupted power supply (UPS). For example, in the case of a food factory, if the mechanical equipment is stopped for a short period of time, all products on the production line must be disposed of. In addition, a power outage in a semiconductor plant that requires ultraprecision microprocessing compared to other manufacturing operations can cause tremendous damage regardless of whether an UPS is installed or not.

South Korea experienced a nationwide rolling blackout in September 2011, with a sudden increase in power demand due to high temperatures and a decrease in power supply due to power plant maintenance. During the blackout, the industrial sector suffered great damage. Therefore, there is a consensus among the people that such a blackout should not occur again [7]. The Korea Electric Power Corporation (KEPCO), the only power distribution company in South Korea, and the South Korean government, which oversees KEPCO, are responsible for supplying electricity without any power outage. The government and KEPCO has made every effort to reliably supply electricity to the industrial sectors, making huge investments in power plants, transmission facilities, distribution facilities, and electricity storage systems. Not only the government and KEPCO but also the Korea Power Exchange (KPX) are responsible for reliably supplying electricity because KPX operates Korea power system.

The government is pushing for an energy transition policy to reduce the share of coal and nuclear power generation and increase the share of renewable energy generation from 2.2% in 2016 to 20.0% by 2030. Although public consensus has been formed on the promotion of the energy transition policy, there are also concerns about securing power supply stability due to the expansion of renewable energy. This is because electric power generation from renewable energy such as wind power and photovoltaic power has a nature of intermittency and uncertainty. Thus, a stable supply of electricity to the industrial sector will be the most important issue for the power authority, as renewable energy will be dramatically expanded in accordance with the government's energy transition policy. In particular, this is needed to secure additional backup power sources, such as gas-fired plants and pumping-up power plants, expand the installation of electricity storage devices, and drastically strengthen the power system. These require a large amount of investment. To justify the investment, the benefits of the investment must outweigh its costs.

Determining the optimal level of power supply reliability requires a function of cost needed to improve power supply reliability and a function of damage costs reduced by improving power supply reliability. The optimal level of power supply reliability is determined at a level that minimizes the sum of the two cost functions. In particular, the function of damage costs reduced due to improved power supply reliability is the same as the function of economic value resulting from improved power supply reliability. Therefore, it is necessary to develop a function that represents the economic value of improving the reliability of power supply.

The costs of increasing power supply reliability can be measured without particular difficulties. However, estimating the benefits or economic value arising from the investment for improving power supply reliability is a very difficult task. This is because the outcome of the investment is improved supply reliability of electricity, and power supply reliability is not a commodity traded in the market. It is necessary to apply techniques to create a hypothetical market for trading power supply reliability so that the reliability of power supply can be assessed by the consumers. Moreover, power supply reliability has several attributes, each of which should be valued. There are various types and periods of power outages, such as when the outages occur, how long they last, when they happen during the week or on weekend, and in which season they take place. In other words, power supply reliability is a multi-attribute good [8]. For example, power supply reliability has multiple attributes such as Information/Notice Provided, Continuous/Uninterrupted Supply, Frequency, Duration, Number, and Time of week.

There are two kinds of techniques to evaluate the power supply reliability. The first technique is to utilize actually revealed data. For example, one may directly investigate the economic damage incurred when a real power outage occurs and view this value as the economic value of improving the power supply reliability. Alternatively, a replacement cost approach that uses the cost information needed to install and operate an emergent backup generator that can reliably supply electricity in the event of a power outage could be applied. The second technique is to ask consumers directly or indirectly about the value of power supply reliability and analyze the responses using economic and econometric theories. In doing so, the application of specially designed economic method is required to value a multi-attribute good.

A typical way to do this is choice experiment (CE). CE is the most prevalent methodology for a multi-attribute good and has almost always been applied in some previous economic studies that have dealt with the valuation of improved power supply reliability [9–13]. In this study, the CE is applied to valuing improved power supply reliability for manufacturing firms in South Korea. The four attributes of power supply reliability considered in this study are duration of power outage, the season of power outage, the time of day when power outage occurs, and the day of the week when power outage occurs. These were identified as factors of interest to the power authority as well as consumers in managing power supply reliability.

As will be explained in more detail below, this study randomly selected 1148 manufacturing firms from all over the country under the supervision of a professional survey company to gather data on value judgments about improved power supply reliability through a CE survey of them. The subsequent composition of this paper is as follows. Section 2 describes in detail the methodology and application procedures used in this study. Section 3 explains the economic and statistical models for analyzing data collected through the CE survey. Section 4 reports some implications after reporting the results. The final section is devoted to presenting conclusions.

2. Methodology

2.1. CE Approach

Two techniques that have been widely employed for nonmarket good valuation in the literature are CE and contingent valuation (CV) [14–17]. The CE method asks the respondents to evaluate value trade-offs among some attributes and indirectly derives their willingness to pay (WTP). Usually, the CV method is applied to a single-attribute good while the CE method is applied to a multi-attribute good. Therefore, the CE method is more suitable for valuing a multi-attribute good than the CV method. The CE approach is theoretically grounded in the random utility maximization model. The model implies that if an individual chooses one alternative among several alternatives, the utility arising from the alternative is always more than the utility arising from the other alternatives. Therefore, the application of the approach requires a survey of consumers. CE is a useful method for estimating the relative values for different attributes of an environmental and nonmarket good or new product.

In general, respondents are required to choose the most preferred alternative out of several alternatives, which include a current status alternative, presented to them in the CE survey. Each alternative comprises several attributes of concern, including the price attribute. CE is a useful method to estimate the relative importance of several attributes for a good or service. Marginal WTP (MWTP) for increasing or decreasing the level of each attribute can be obtained through analyzing the data on respondents' choices and then interpreting or utilizing the results.

2.2. Attributes

In designing a CE, the first important thing to do is to determine the appropriate attributes and define their levels. An extensive literature review and consultation with experts enabled us to identify a preliminary list of attributes of power supply reliability. Most previous studies reveal that duration of power supply interruption, season of power supply interruption, power supply interruption time of day, and power supply interruption day of the week [10,13,18–22] have important implications for the value of improved power supply reliability. We then reviewed and revised the preliminary list of attributes derived from extensive literature reviews through extensive interviews with policy analysts, researchers, and professors. As a result, the final set of attributes was chosen by discussing with experts such as policy-makers, stakeholders, and environmental activists.

As reported in Table 1, the finally determined attributes are Duration of interruption, Season of interruption, Time of day, Day of week, and Price. A focus group interview with 30 people in the manufacturing industry was implemented to check for whether the survey questionnaire is fully meaningful, understandable, and persuasive to the respondents. Their responses were affirmative. The descriptions and levels of them are also explained in Table 1. They are assumed to be orthogonal in terms of valuation function rather than production function. Furthermore, all other attributes of power supply reliability are assumed to be the same in the course of the value judgments required in the CE survey.

Table 1. Descriptions and levels of four chosen attributes and price attribute used in this study.

Attributes	Descriptions	Levels
Duration of interruption	Duration of industrial electricity supply interruption	Level 1: 5 h [#] Level 2: 1 h Level 3: 20 min
Season of interruption	Season when industrial electricity supply interruption takes place	Level 1: Summer or winter [#] Level 2: Spring or fall
Time of day	Time when industrial electricity supply interruption occurs	Level 1: Day time [#] (9 a.m. to 6 p.m.) Level 2: Off-daytime (6 p.m. to 9 a.m.)
Day of week	Day when industrial electricity supply interruption happens	Level 1: Weekday [#] Level 2: Weekend
Price	Percentage of an additional payment for industrial electricity use (unit: %)	Level 1: 0 [#] Level 2: 1% Level 3: 5% Level 4: 10% Level 5: 20%

Note: [#] indicates the status quo of each attribute. Status quo is a Latin phrase meaning the current state.

The power supply reliability decreases when industrial electricity supply interruption occurs in summer or winter, daytime, and weekdays. Moreover, the longer the duration of the industrial electricity supply interruption the lower the power supply reliability. The status quo of Duration of interruption, Season of interruption, Time of day, and Day of week means the level with the most negative situation. In other words, the power supply interruption lasts for 5 h during the summer or winter, weekday, and daytime is assumed to be the current state, and the state that is improved in the current state is set to the level of each attribute. The level of the attribute for Price is explained as the percentage of an additional payment for industrial electricity use. Although there is no actual payment, we explained to the respondents that the power supply reliability for manufacturing firms could be improved by increasing the industrial electricity bill. The status quo of this attribute means that there is no additional payment in the most negative situation.

Since a number of alternatives can be created from Table 1, several alternatives from possible combinations of attributes should be derived. To this end, the orthogonal main effects design was employed and 16 alternatives were obtained. The orthogonal main effects design is effective in terms of isolating the effects of individual attributes on the choice. The ability to 'design in' this orthogonality is an important advantage over the revealed preference random utility models, in which attributes are often found to be highly correlated with one another [23]. The orthogonal main effect design was implemented using the SPSS 12.0 package. From these, eight choice sets were generated. Each choice set is made up of two alternatives and the current status alternative. Each interviewee was presented with eight choice sets and reported eight responses to the provided questions that indicated which alternatives were the most preferred among the three alternatives in each choice set.

2.3. Survey Instrument and Method

There are three parts to the survey instrument. Several questions about the power outage make up the first part to check respondents' perceptions before the CE survey on power supply reliability begins in earnest. To facilitate the respondents' understanding, a description of the features and effects of power supply reliability is provided, along with color photographs, shown in this section. Such work not only relieves respondents of the burden of a fully-fledged survey but also provides significant statistical data in itself. Explanations about the attributes and questions concerning the value trade-off work, which are conventionally required in a CE survey, are presented in the second part. The third part contains questions about the manufacturing firms' information. The main part of the survey questionnaire in this study is given in Appendix A. Figure A1 is an example of choice card we present to the respondents.

As in Makeen et al.'s [24] paper, supply reliability can be defined from an engineering perspective. However, since it may be difficult for interviewees affiliated with each firm to fully understand this in the CE survey, the authors attempted to define and explain the supply reliability in easier terms in the CE questionnaire. The power supply reliability was defined as the extent that the power system can reliably supply electric power to the consumer maintaining adequate voltage and frequency without interruption. In addition, it was targeted not only for certainty of quantitative supply of electricity but also for certainty of quality supply. These points were sufficiently conveyed to the respondents in the CE survey.

The number of observations to be analyzed in this study is 1148. Apparently, the sample size is small. However, accordingly to Statistics Korea [25], the population size of manufacturing industry of South Korea in 2016 was 64,885. Therefore, the sampling rate is approximately 1.77%. This figure is judged to be not small. This is because the Korea Ministry of Strategy and Finance and Korea Development Institute [26] recommended 1000 as a suitable sample size the total number of households in South Korea is about twenty million. Moreover, Arrow et al. (1993) recommended 1000 as an appropriate sample size of the United States households, although the United States has a much larger number of households than South Korea. Of course, budget constraints also affected the determination of the sample size. The cost of obtaining an observation through the survey was approximately USD 50. Therefore, more than $50,000 was invested to carry out the survey.

We have focused on sampling to ensure the representativeness of our sample in two aspects. First, a random sampling observed from an economy census by Statistics Korea, the Korean National Statistical Office. Our sample was classified by region and industry sector. Manufacturing can be divided into 25 sectors by Korean Standard Industrial Classification [27] and each section is denoted in Appendix A. Q1. Second, a field CE survey was done by interview experts from a professional polling firm during June and July. Trained interviewers visited the sampled manufacturing firms and, carried out total 1148 face-to-face interviews which require significantly higher costs than mail or telephone interviews. The interviews were conducted with randomly selected manufacturing firms to maximize the scope for detailed questions and answers.

3. Model

3.1. Utility Function

We assume that the utility function has a linear functional form. Let the levels of Duration of interruption, Season of interruption, Time of day, Day of week, and Price be X_t, where $t = 1, 2, 3, 4, p$. In addition, an alternative-specific constant (ASC) is introduced to capture the effect of any other factors not contained in the model. ASC represents a dummy for the respondent choosing the status quo option among three alternatives. ASC is one if the respondent chooses the third alternative (current status), zero otherwise. Let V_{jl} be the utility for interviewee j who chooses alternative l. The utility function is formulated as

$$
\begin{aligned}
V_{jl} &= W_{jl}(X_{jl}, T_j) + \varepsilon_{jl} \\
&= ASC_j + \beta_1 X_{1,jl} + \beta_2 X_{2,jl} + \beta_3 X_{3,jl} + \beta_4 X_{4,jl} + \beta_p X_{p,jl} + \varepsilon_{jl}
\end{aligned}
\tag{1}
$$

where W_{jl} and ε_{jl} are the deterministic and stochastic parts of the utility function, respectively. X_{jl} is a vector containing the levels of the attributes for alternative l given to respondent j. T_j is respondent j's characteristics, such as ASC. The β's are the coefficients that correspond to each attribute.

Omitting jl for simplicity, we can apply Roy's identity to Equation (1) and derive the $MWTP$ estimate, $MWTP_{X_t}$, as

$$
MWTP_{X_t} = (\partial W / \partial X_t) / (\partial W / \partial X_p) = \beta_t / \beta_p \text{ for } t = 1, 2, 3, 4
\tag{2}
$$

The $MWTP$s of each attribute represent the marginal rate of substitution between the price and each attribute.

3.2. How to Obtain the Utility Function

CEs share a common theoretical framework with other valuation approaches. Thus, in this study, the random utility model is used to explain individual choices by specifying functions for the utility that is derived from the available alternatives. Estimating the utility function implies estimating β's. This function can be estimated using the multinomial logit (MNL) model developed by McFadden [28]. Usually, MNL model has been most widely applied to obtain β in the literature. However, the MNL model inevitably assumes independence from irrelevant alternatives. Although the assumption seems to be somewhat restrictive, it has the advantage of enabling us to specify the log-likelihood function as a closed form. Thus, if the assumption is met, we can easily tackle the CE data. Let J be the number of interviewees and I_{jl} be a dummy variable that is defined as one if interviewee j selects alternative l; otherwise, I_{jl} is zero. The log-likelihood function for our MNL model is

$$
\ln L = \sum_{j=1}^{J} \ln \left[\frac{\prod_{k=1}^{3} \left(\exp(W_{jk}) \right)^{I_{jk}}}{\sum_{n=1}^{3} \exp(W_{jn})} \right]
\tag{3}
$$

4. Results and Discussion

4.1. Data

A nationwide CE survey of 1148 randomly chosen manufacturing firms was conducted by a professional polling firm through person-to-person interviews during June and July 2017. Each manufacturing company gave us eight observations. In other words, each interviewee was presented with eight choice sets and reported eight responses to the provided questions. Thus, we would get a data set size of 9184 (1148 respondents $\times$ 8 choice sets). Table 2 reports the definitions and sample statistics for some characteristics of the manufacturing firms. We selected three variables: Region, Size, and UPS. Since South Korea is classified into five mega-city regions (Seoul-Incheon-Gyunggi, Daejeon-Chungbuk-Chungnam, Gwangju-Jeonbuk-Jeonnam, and Pusan-Ulasn, Daegu-Gyungbuk-Gyungnam) and two special mega-city regions (Gangwon and Jeju), the variable Region identifies whether or not to locate in the five mega-city regions is introduced. Most manufacturing firms are located in the five mega-city regions. Size and UPS can also affect the outcomes of the CE. Figure 1 is a graph for samples classified by region.

Table 2. Definitions and sample statistics for some characteristics of the manufacturing firms.

Variables	Definitions	Mean	Standard Deviation
Region	Dummy for the interviewee's living in the five mega-city regions (0 = no; 1 = yes)	0.94	0.25
Size	The size of the firms (0 = large enterprise; 1= small and medium-sized enterprise)	0.94	0.22
UPS	Dummy for installing uninterrupted power supply (UPS) (0 = no; 1 = yes)	0.10	0.30

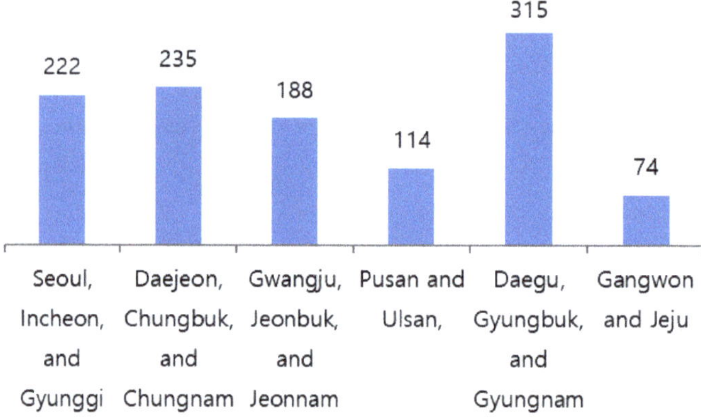

Figure 1. Samples classified by region.

4.2. Estimation Results

Table 3 reports the results of estimating the MNL model. All the coefficient estimates except for Season of interruption are statistically distinguishable from zero at the 1% level. The expected signs for coefficient estimates for the five attributes are all negative.

Table 3. Estimation results of the multinomial logit model.

Variables [a]	Multinomial Logit Coefficient Estimates [c]	
ASC [b]	0.6012 [#]	(8.27)
Duration of interruption	−0.0021 [#]	(−9.89)
Season of interruption	−0.0286	(−0.80)
Time of day	−0.3190 [#]	(−9.23)
Day of week	−0.1710 [#]	(−4.78)
Price	−0.1071 [#]	(−25.59)
Number of observations	9184	
Wald-statistic (*p*-value) [d]	1892.58 (0.000)	
Log-likelihood	−8812.59	

Notes: [a] The variables are defined in Table 1; [b] *ASC* refers to alternative-specific constants that represent dummies for the respondents choosing the status quo alternative; [c,#] indicates statistical significance at the 1% level, and *t*-values are reported in parentheses beside the estimates. *t*-value is the ratio of the departure of the estimated value of a parameter from its hypothesized value to its standard error; [d] The null hypothesis is that all the parameters are zero and the corresponding *p*-value is reported in parentheses beside the statistic.

The coefficient estimates for Duration of interruption, Time of day, and Day of week have negative signs. Thus, a one unit decrease in the level of Duration of interruption attribute increases the manufacturing firms' utility. Avoiding the status quo of Time of day and Day of week attributes also increases the utility. The coefficient for Price also has a negative sign. This implies that, as the price goes up the utility decreases. This result is quite reasonable, given that the price negatively contributes to the utility. The signs of all the estimated coefficients except for Season of interruption are consistent with our prior expectations.

4.3. MWTP Estimates for Each Attribute

Finally, the *MWTP* estimates for a decrease in the level of each attribute can be derived employing Equation (2). The results of estimating the *MWTP* values are provided in Table 4. The *MWTP* estimates for a one minute decrease in duration of interruption, avoiding interruption during daytime rather than off-daytime, and preventing interruption during weekdays rather than weekends are obtained as 0.02%, 2.98%, and 1.60%, respectively, of the electricity bill. These values are interpreted as the value of improved industrial electricity supply reliability in South Korea. Table 4 also presents the 95% confidence intervals for the *MWTP* estimates, which are computed using the procedures given in Krinsky and Robb [29].

Table 4. Estimation results of marginal willingness to pay (*MWTP*) values.

	MWTP per Manufacturing Firm per Month		
	Estimates	*t*-**Values**	**95% Confidence Intervals**
Duration of interruption	0.02% [#]	8.19	0.02% to 0.03%
Time of day	2.98% [#]	8.54	2.34% to 3.68%
Day of week	1.60% [#]	4.87	0.95% to 2.23%

Notes: [#] indicates statistical significance at the 1% level. The confidence intervals are computed using the procedures given in Krinsky and Robb [29].

4.4. Discussion of the Results

As explained above, this study applied the MNL model. However, other models such as nested logit [30] model and mixed logit model [31] are also applicable. The estimation results of the two models are given in Table 5. Although specific estimation results vary from model to model, there is no difference in signs of the estimated coefficients. Since the MNL model is most widely used in empirical CE studies, this study tries to use the estimation results from the MNL model and derive the MWTP estimates.

Table 5. Estimation results of the nested logit and mixed logit models.

Variables [a]	Nested Logit Coefficient Estimates [c]		Mixed Logit Coefficient Estimates [c]		
	Estimates	*t*-**Values**	**Assumed Distribution**	**Mean of the Coefficient Estimate**	**Variance of the Coefficient Estimate**
ASC [b]	−0.6548 [#]	9.63	Normal	−8.5119 [#]	−1.2917 [#]
Duration of interruption	−0.0020 [#]	−10.17	Normal	−0.0524 [#]	−2.9011 [#]
Season of interruption	0.0078	0.29	Normal	1.8101 [#]	0.6878 [#]
Time of day	−0.2266 [#]	−7.39	Normal	−18.2729 [#]	−5.1934 [#]
Day of week	−0.1792 [#]	−6.67	Normal	−2.8403 [#]	−1.1643 [#]
Price	−0.0785 [#]	−12.34	Normal	−1.3888 [#]	−5.7305 [#]
Inclusive value	0.6898 [#]	13.86			
Number of observations			9184		
Log-likelihood	−8796.99			−3751.88	

Notes: [a] The variables are defined in Table 1. [b] ASC refer to alternative-specific constants that represent dummies for the respondent's choosing status quo alternative. [c,#] indicates statistical significance at the 1% level.

Using the results presented in Table 4, we can estimate the value of improved power supply reliability, which is a combination of these attributes using the *MWTP* estimates for a decrease in the attributes. In other words, multiplying the figures reported in Table 4 by the levels of attributes gives us the value of the alternative for the hypothetical state of the industrial electricity supply interruption. As an illustration, the results of calculating the value of improved power supply reliability at which manufacturing firms assess several alternatives for the hypothetical state of the industrial electricity supply interruption are shown in Table 6. For example, the value of the third alternative, which means the hypothetical state in off-daytime and weekend, with a 280 min decrease in duration of interruption from the status quo, is computed as 10.18% of the electricity bill.

Table 6. The value of alternatives for hypothetical state of the industrial electricity supply interruption.

	Scenario A	Scenario B	Scenario C
Duration of interruption	5 h	1 h	20 min
Season of interruption	Summer or winter	Summer or winter	Summer or winter
Time of day	Off-daytime	Daytime	Off-daytime
Day of week	Weekend	Weekend	Weekend
The value of improved power supply reliability	4.58%	6.40%	10.18%

The analysis results presented in Table 6 have a variety of potential uses. First, by using these findings, one can identify which attributes manufacturing firms value. According to the estimated utility function, the absolute value of the coefficient estimate for Time of day among the four attributes was the greatest. On the other hand, the absolute value of the coefficient estimate for Duration of interruption was the smallest. Therefore, if the cost of improving the power supply reliability is the same, it would be better to concentrate on avoiding interruption during daytime rather than off-daytime to reduce interruption during electricity outage. Of course, the cost of improving system reliability may vary with the power system conditions. More specifically, operating reserve costs are highly dependent on the generation scheduling and demand status of facilities in the power system (i.e., outage of transmission lines or generators). Usually, the price for reserve might be higher during daytime than off-daytime. Using Tables 4 and 6 we cannot only calculate the value of improved power supply reliability for a variety of alternatives, but also make alternatives that result in specific value of power supply reliability. Alternatives may be proposed that satisfy the levels of acceptable value of power supply reliability within the scope of total costs not exceeding the total benefits.

The estimation results of improved power supply reliability may vary depending on the times in which a value judgment is made because the economic technique used in this study is a stated preference approach, which analyzes the data that is asked about the entity's preference. For example, when the economy is booming, the value of improved power supply reliability can be measured as being higher, whereas when the economy is in a recession, the value can be estimated as lower. In addition, the value of improved power supply reliability could be lowered if electricity-intensive firms are reduced as the industrial structure changes and more companies are using relatively little electricity. On the other hand, an increase in electricity-intensive companies could increase the value. In other words, the value of improved power supply reliability depends on the economic situation, industrial structure, social atmosphere, and power supply situation, so it is difficult to maintain the specific value. Therefore, it is necessary to conduct a study at regular intervals on the value of power supply reliability. It is possible to grasp the trajectory whose value changes with the lapse of time, and it is also possible to predict the future value with the results.

Additionally, this article seems to contribute to the literature from a research perspective. First, the article utilized a CE technique to look into the value of the attributes of improved power supply reliability and found that the application was successful, because the estimation results were statistically meaningful and the respondents actively participated in the CE survey. Improved power supply reliability is not just a problem for South Korea but an important issue for the worldwide, especially for developing countries [32–34]. Thus, comparison of the results from our work with those from future works that will be applied in other countries will yield new implications.

Our finding can be compared with a finding of London Economics [8]. The value of lost load for electricity in Great Britain (Final report for OFGEM and DECC) of London Economics [8] is a representative example for estimating the economic value of electricity use mainly by applying CE and CV. As a result of evaluating the economic value of industrial power using CE, the WTP estimates for the various choice scenarios for small and medium sized businesses range from around 2% to 4%

of the annual electricity bill. Under the same scenarios as the London Economics [8], the value of the industrial electricity supply interruption ranges from approximately 1.2% to 5.78% of the annual electricity bill in our study. Interestingly, the range of the results of both studies is similar.

However, the sample size used in this study is small compared to the population size. Thus, it is necessary to collect and analyze a larger number of observations through securing sufficient budget for the survey in the future. This work would also enable the derivation of industry-specific implications. Analysis of data classified by industry will enable us to derive industry-specific implications.

Author Contributions: All the authors participated in writing this paper. D.-C.K. collected and compiled the necessary background data for the study and prepared the CE questionnaire, which was essential for the data collection; H.-J.K. wrote a large part of the paper and educated the interviewers; and S.-H.Y. performed statistical analysis and supervised the entire process of writing the paper.

Acknowledgments: This work was supported by the Korea Institute of Energy Technology Evaluation and Planning (KETEP) and the Ministry of Trade, Industry & Energy (MOTIE) of the Republic of Korea (No. 20184030202230). In addition, the authors acknowledge that this research was financially supported by Korea Power Exchange (Grant Number: 2017-02).

Conflicts of Interest: The authors declare no conflict of interest.

Appendix A Main Part of the Survey Questionnaire

A.1. Questions About the Characteristics of the Manufacturing Firms

The interviewee was asked to respond to the characteristics of the company they are working with: Sale revenue, the number of employees, average salary, operation cost, and inventory value. Questions were all open-ended questions, and the question about the type of manufacturing firms was as follows:

Q1. Please check $\sqrt{}$ the type of manufacturing firms of your company.

① Manufacture of food products
② Manufacture of beverages
③ Manufacture of tobacco products
④ Manufacture of textiles, except apparel
⑤ Manufacture of wearing apparel, clothing accessories and fur articles
⑥ Manufacture of leather, luggage and footwear
⑦ Manufacture of wood and of products of wood and cork; except furniture
⑧ Manufacture of pulp, paper and paper products, printing and reproduction of recorded media
⑨ Printing and reproduction of recorded media
⑩ Manufacture of coke, briquettes and refined petroleum products
⑪ Manufacture of chemicals and chemical products; except pharmaceuticals and medicinal chemicals
⑫ Manufacture of pharmaceuticals, medicinal chemical and botanical products
⑬ Manufacture of rubber and plastics products
⑭ Manufacture of other non-metallic mineral products
⑮ Manufacture of basic metals
⑯ Manufacture of fabricated metal products, except machinery and furniture
⑰ Manufacture of electronic components, computer; visual, sounding and communication equipment
⑱ Manufacture of medical, precision and optical instruments, watches and clocks
⑲ Manufacture of electrical equipment
⑳ Manufacture of other machinery and equipment
㉑ Manufacture of motor vehicles, trailers and semitrailers
㉒ Manufacture of other transport equipment
㉓ Manufacture of furniture

㉔   Other manufacturing

㉕   Maintenance and repair services of industrial machinery and equipment

A.2. Questions About Marginal Willingness to Pay

Type A. Q1. Check with $\sqrt{}$ the only available alternative that you prefer among Alternative A, B, or the status quo.

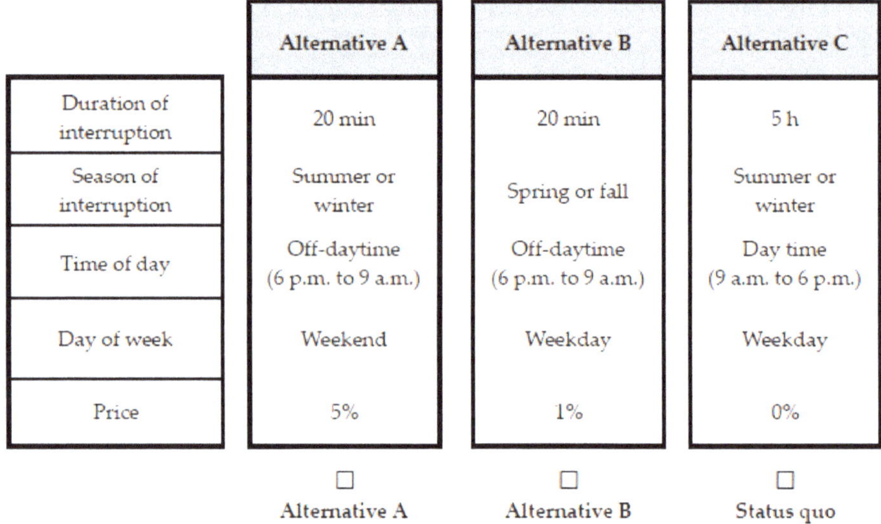

Figure A1. A sample choice set used in this study.

References

1. Ferguson, R.; Wilkinson, W.; Hill, R. Electricity use and economic development. *Energy Policy* **2000**, *28*, 923–934. [CrossRef]
2. Yoo, S.H. Electricity consumption and economic growth: Evidence from Korea. *Energy Policy* **2005**, *33*, 1627–1632. [CrossRef]
3. Yoo, S.H.; Lee, J.S. Electricity consumption and economic growth: A cross-country analysis. *Energy Policy* **2010**, *38*, 622–625. [CrossRef]
4. Ju, H.C.; Yoo, S.H.; Kwak, S.J. The electricity shortage cost in Korea: An input-output analysis. *Energy Sources Part B Econ. Plan. Policy* **2016**, *11*, 58–64. [CrossRef]
5. Lim, K.M.; Yoo, S.H. Economic value of electricity in the Korean manufacturing industry. *Energy Sources Part B Econ. Plan. Policy* **2016**, *11*, 542–546. [CrossRef]
6. Ghosh, R.; Goyal, Y.; Rommel, J.; Sagebiel, J. Are small firms willing to pay for improved power supply? Evidence from a contingent valuation study in India. *Energy Policy* **2017**, *109*, 659–665. [CrossRef]
7. Kim, K.; Cho, Y. Estimation of power outage costs in the industrial sector of South Korea. *Energy Policy* **2017**, *101*, 236–245. [CrossRef]
8. London Economics. The Value of Lost Load for Electricity in Great Britain. Final Report for OFGEM and DECC. 2013. Available online: https://www.ofgem.gov.uk/ofgem-publications/82293/london-economics-value-lost-load-electricity-gbpdf (accessed on 19 April 2018).
9. Hensher, D.A.; Greene, W.H. The mixed logit model: The state of practice. *Transportation* **2003**, *30*, 133–176. [CrossRef]
10. Carlsson, F.; Martinsson, P. Does it matter when a power outage occurs? A choice experiment study on the willingness to pay to avoid power outages. *Energy Econ.* **2008**, *30*, 1232–1245. [CrossRef]

11. Abdullah, S.; Mariel, P. Choice experiment study on the willingness to pay to improve electricity services. *Energy Policy* **2010**, *38*, 4570–4581. [CrossRef]

12. Hensher, D.A.; Shore, N.; Train, K. Willingness to pay for residential electricity supply quality and reliability. *Appl. Energy* **2014**, *115*, 280–292. [CrossRef]

13. Ozbafli, A.; Jenkins, G.P. Estimating the willingness to pay for reliable electricity supply: A choice experiment study. *Energy Econ.* **2016**, *56*, 443–452. [CrossRef]

14. Min, S.H.; Lim, S.Y.; Yoo, S.H. Consumer's willingness to pay a premium for eco-labeled LED TVs in Korea: A contingent valuation study. *Sustainability* **2017**, *9*, 814.

15. Lim, S.Y.; Kim, H.Y.; Yoo, S.H. Public willingness to pay for transforming Jogyesa Buddhist temple in Seoul, Korea into a cultural tourism resource. *Sustainability* **2016**, *8*, 900. [CrossRef]

16. Wang, J.; Ge, J.; Ma, Y. Urban Chinese consumers' willingness to pay for pork with certified labels: A discrete choice experiment. *Sustainability* **2018**, *10*, 603. [CrossRef]

17. Vanstockem, J.; Vranken, L.; Bleys, B.; Somers, B.; Hermy, M. Do looks matter? A case study on extensive green roofs using discrete choice experiments. *Sustainability* **2018**, *10*, 309. [CrossRef]

18. Moeltner, K.; Layton, D.F. A censored random coefficients model for pooled survey data with application to the estimation of power outage costs. *Rev. Econ. Stat.* **2002**, *84*, 552–561. [CrossRef]

19. Hensher, D.A.; Rose, J.; Greene, W.H. The implications on willingness to pay of respondents ignoring specific attributes. *Transportation* **2005**, *32*, 203–222. [CrossRef]

20. Morrison, M.; Nalder, C. Willingness to pay for improved quality of electricity supply across business type and location. *Energy J.* **2009**, *30*, 117–133. [CrossRef]

21. Meles, T.H. *Power outages, Increasing Block Tariffs and Billing Knowledge*; University of Gothenburg: Gothenburg, Sweden, 2017.

22. Oseni, M.O. Self-generation and households' willingness to pay for reliable electricity service in Nigeria. *Energy J.* **2017**, *38*, 165–194. [CrossRef]

23. Hanley, N.; Wright, R.E.; Adamowicz, V. Using choice experiments to value the environment. *Environ. Resour. Econ.* **1998**, *11*, 413–428. [CrossRef]

24. Makeen, P.; Swief, R.A.; Abdel-Salam, T.S.; El-Amary, N.H. Smart hybrid micro-grid integration for optimal power sharing-based water cycle optimization technique. *Energies* **2018**, *11*, 1083. [CrossRef]

25. Korea Ministry of Strategy and Finance and Korea Development Institute. *Overview of Preliminary Feasibility Evaluation System and Methodology*; Korea Ministry of Strategy and Finance and Korea Development Institute: Sejong, Korea, 2018. (In Korean)

26. Arrow, K.; Solow, R.; Portney, P.R.; Leamer, E.E.; Radner, R.; Schuman, H. Report of the NOAA panel on contingent valuation. *Fed. Regist.* **1993**, *58*, 4601–4614.

27. Korean Standard Statistical Classification. Korean Standard Industrial Classification. 2017. Available online: http://kssc.kostat.go.kr (accessed on 20 August 2018).

28. McFadden, D. Conditional logit analysis of qualitative choice behaviour in frontiers in econometrics. In *Frontiers in Econometrics*; Zarembka, P., Ed.; Academic Press: New York, NY, USA, 1973; pp. 105–142.

29. Krinsky, I.; Robb, A.L. On approximating the statistical properties of elasticities. *Rev. Econ. Stat.* **1986**, *68*, 715–719. [CrossRef]

30. Lim, S.Y.; Lim, K.M.; Yoo, S.H. External benefits of waste-to-energy in Korea: A choice experiment study. *Renew. Sustain. Energy Rev.* **2014**, *34*, 588–595. [CrossRef]

31. Lim, S.Y.; Kim, H.J.; Yoo, S.H. Assessing the external benefits of contaminated soil remediation in Korea: A choice experiment study. *Environ. Sci. Pollut. Res.* **2018**, *25*, 17216–17222. [CrossRef] [PubMed]

32. Sagebiel, J. Preference heterogeneity in energy discrete choice experiments: A review on methods for model selection. *Renew. Sustain. Energy Rev.* **2017**, *69*, 804–811. [CrossRef]

33. Reinders, A. Perceived and Reported Reliability of the Electricity Supply at Three Urban Locations in Indonesia. *Energies* **2018**, *11*, 140.

34. Jimenez, R.; Serebrisky, T.; Mercado, J. What does "better" mean? Perceptions of electricity and water services in Santo Domingo. *Utilities Policy* **2016**, *41*, 15–21. [CrossRef]

 *sustainability*

Article

Energy Life-Cycle Assessment of Fruit Products—Case Study of Beira Interior's Peach (Portugal)

João Pires Gaspar [1], Pedro Dinis Gaspar [1,2,*], Pedro Dinho da Silva [1,2], Maria Paula Simões [3,4] and Christophe Espírito Santo [5]

[1] Department of Electromechanical Engineering, University of Beira Interior, Rua Marquês d'Ávila e Bolama, 6201-001 Covilhã, Portugal; jfgaspar90@gmail.com (J.P.G.); dinho@ubi.pt (P.D.d.S.)

[2] C-MAST—Centre for Mechanical and Aerospace Science and Technologies, 6201-001 Covilhã, Portugal

[3] Agriculture School, Polytechnic Institute of Castelo Branco, 6001-909 Castelo Branco, Portugal; mpaulasimoes@ipcb.pt

[4] Centro de Recursos Naturais, Ambiente e Sociedade (CERNAS), Escola Superior Agrária de Coimbra, Bencanta, 3045-601 Coimbra, Portugal

[5] Agrofood Technological Center, Castelo Branco, Portugal; cespiritosanto@cataa.pt

* Correspondence: dinis@ubi.pt; Tel.: +351-275-329-759

Received: 8 August 2018; Accepted: 26 September 2018; Published: 1 October 2018

Abstract: Currently, there is a growing demand for cleaner and sustainable technologies due to environmental issues. In this sense, there is a necessity to manage the assessment of production processes and the rationalization of energy consumption. In this study, an Energy Life-Cycle Assessment (ELCA) was carried out through energy efficiency indicators, directed to the characterization and renewability of the peach production system life-cycle in the Portuguese region of *Beira Interior*. The study intends to investigate the non-renewable energy inputs from fossil fuels, as well as the emissions resulting from machinery. In addition, warehouse energy inputs are analyzed, mainly cooling systems of refrigerated chambers where fruits are preserved. This analysis aims to find opportunities for technological, environmental and best practices improvements. Test scenarios were analyzed and revealing soil groundcover maintenance is the operation with the largest impact in the energy consumption of the production process (3176 MJ·ha^{-1}). In the post-harvest processes, the energy consumption largest impact is given by the warehouse's operations (35,700 MJ·ha^{-1}), followed by transportation (6180 MJ·ha^{-1}). Concerning the emissions resulting from the fuels consumption, the largest impact is due to the plantation machinery and the transportation from warehouse to retailers.

Keywords: peach; Energy Life-Cycle Assessment; post-harvest

1. Introduction

Agriculture is considered both as a supplier and energy consumer [1,2]. In recent years, the agricultural sector in Europe has adopted intensive farming practices to increase plantations productivity and ultimately, meet population demands. However, this led to depletion of natural resources and climate change, among other consequences [3,4].

Thomassen and Boer [5] refer to the existence of a variety of tools and methods that can be used as comparison of environmental impacts and economic costs of agricultural production, namely the Life-Cycle Assessment (LCA).

The LCA is a standardized methodology [6] to assess sustainability of all industrial processes. However, it can lead to different results, even for very similar products. Following a standardized method helps to build the basics of an LCA (all inventory inputs, functional unit, etc.) however structuring an LCA needs to take into consideration the social impacts and economic viability [7–9]. Resulting sustainable solutions can be very efficient in the environmental aspect but impossible to accomplish due to social issues or economically inviable. Sustainability is an anthropocentric idea, before meeting the environmental criteria, it needs to meet the Human population requirements [7,10]. Through this methodology, an Energy Life Cycle Assessment (ELCA) is possible to be carried out, limiting only to energy issues [11]. Commercial and non-commercial production, where fruit production is included, can be converted and standardized in forms of power units. The ultimate goal for energy efficiency improvement is to contribute for a better economy, profitability and competitiveness, and, consequently, a more sustainable agriculture [1,2].

Therefore, having in mind the growing peach production in the Portuguese *Beira Interior* region, a study covering the ELCA will provide further insights to improve energy consumption in production and post-harvest processes, contributing for an environmental impact reduction.

In Europe, Spain is the major producer with more than 1,500,000 tons [12]. Portugal presents a production of 41,000 tons. The *Beira Interior* region is the leading Portuguese peach producing region with 1600 ha, 45.2% of the Portuguese area, and a production of 20,000 tons, 49.2% of the Portuguese production [13].

2. LCA Applied to Fruit Products: State of the Art

One of the first LCA projects applied to agriculture was developed by Weidema et al. [14], being an analysis of the environmental impact of a wheat culture in three distinct production systems: intensive, biological and integrated. Kramer et al. [15] developed a study in a Dutch agricultural production to calculate emissions of carbon dioxide (CO_2), methane (CH_4) and nitrous oxide (N_2O). Margini et al. [16] proposed a methodology based on LCA with the purpose of assessing the environmental impacts of the pesticides' application in agriculture.

Milà i Canals et al. [17] conducted a study based on apple orchards in New Zealand during 1999 and 2000, where they evaluated alternatives that could reduce the environmental impact associated with production. The use of agricultural machinery was the main energy consumer (30% to 50%). In addition, the harvest process was one of the operations that consumed most energy, due to be carried out using hydra-ladders, followed by irrigation, the phytosanitary treatments application, grass groundcover maintenance and pruning.

Alaphillipe et al. [18] analyzed the environmental impact through the LCA methodology in nine apple orchards located in the south of France. This analysis was based on data collected between 2006 and 2009 in three distinct production systems, namely, low-input orchards, organic and conventional. A mass-based functional unit (FU) was used with a value of 32 tonnes·ha^{-1} of apple. Energy expenditure was higher in operations involving plant chemical treatments and fertilization.

A study led by Royan et al. [1] aimed to analyze the energy consumption, by quantifying energy inputs, as well as the peach production efficiency in an Iran province. In addition, non-renewable energy and renewable energy inputs were considered, as well as the direct and indirect forms of energy. The fuel consumption determined for this application was 175.4 L·ha^{-1}, equivalent to 9879 MJ·ha^{-1}.

Hemmati et al. [19] held an energy analysis in olive production in flat and slope lands, accounting for the direct and indirect energy inputs, as well as renewable and non-renewable energies. This analysis considered energy inputs related to human labor, machinery, diesel consumption, chemical fertilizers, manure, pesticides, water for irrigation and electricity. In this study, diesel consumption in leveled land production was 957 MJ·ha^{-1}, while, in slope field was 935 MJ·ha^{-1}, thus higher fuel consumption was observed for flat land. However, total energy input was 15.9 GJ·ha^{-1} for flat systems and 23.3 GJ·ha^{-1} for slope field, showing a 46.5% higher energy input consumption for slope land.

Due to the increase of farming in Thailand, Soni et al. [20] developed a study to evaluate energy consumption, as well as the emissions associated with the agricultural practice in 46 rainfed places. This study considered the total energy consumption, including direct or indirect, renewable or non-renewable energy sources. This study concluded that for different cultures, energy inputs are also distinct. In all cultures tested, the fossil fuel (diesel) was the most significant energy input.

In a study by Ingrao et al. [21], LCA was conducted for peach production to identify and quantify aspects that can be improved and increase the environmental and economic sustainability. This analysis was performed between 2002 and 2011 and divided into two distinct phases. In a first phase, the LCA considered the whole cycle, including the production process with a land-based FU. Whereas the second phase was carried out with an assessment of the possible environmental improvement through a sensitivity analysis. In that study, an average annual production of 31.5 ton·ha^{-1} was considered. The consumption of fossil fuels was 8.42 kg·ha^{-1}, considering that transportation was only assumed as energy input associated with the fertilizers application and packaging in post-harvest process. The irrigation revealed the largest environmental impact due to the high amount of water consumption and large electrical and fossil energy consumption required to pump water.

Keyes et al. [22] conducted a study on apple production in the Nova Scotia province, in Canada. The study quantified and evaluated resources, such as the energy required for production, storage and transportation. The largest contributors for environmental impacts were associated with harvest operations, diesel consumption and the application of phytosanitary treatments. The energy consumption associated with apple production resulted from the use of fuels was 398 MJ·tonne^{-1}, from the orchard to the warehouse was 57.6 MJ·tonne^{-1}, from the warehouse was of 2200 MJ·tonne^{-1} and finally, transport to the consumer (about 103 km) was 283 MJ·tonne^{-1}.

Vinyes et al. [23] conducted an LCA of peach plantations from the start (tree planting) until the end (extinction, about 15 years later), in order to provide the whole perspective on peach production environmental impacts. Data was obtained in a plantation located in the northeast of Spain. The FU was 1 kg of peaches with four distinct scenarios specified according to the different periods of production. Namely, low production, high production, growing phase and average over the years. The study revealed a variation from 7% to 69% between scenarios, depending on the indicator. In addition, the mass-based study during a year of production might show alterations, especially in a low production scenario. The fuel consumption was (on average) 6.28 kg·ha^{-1} for the four scenarios. Authors referred that the operation with the largest energy consumption was the phytosanitary treatment application.

Longo et al. [24] developed a study applying the LCA for organic apples production in Northern Italy. The study analyzed the inputs of raw materials and energy sources, as well as the production and post-harvest processes and consumer distribution. Most of the energy consumption was due to fertilizers and pesticides application. Post-harvest operations had a negligible energy consumption when compared to the production process.

Also, Ghatrehsamani et al. [25] developed an energy analysis of peach production in Bakhitari province, in Iran, in order to determine the amount of incoming and outgoing system energy. The direct and indirect sources of energy inputs and the forms of non-renewable and renewable energy were considered. Data collected through questionnaires and interviews to 100 different producers in the region, revealed that diesel consumption was 74.7 L·ha^{-1}, corresponding to 4206 MJ·ha^{-1} of energy consumption.

Due to the growing importance of peach production in Portugal, in particular in the *Beira Interior* region, this study aims to carry out an energy analysis of the peach life-cycle to determine energy inputs importance in production and post-harvest processes. Considering the major findings of the above-mentioned studies, alternative scenarios were established for result comparison, giving a broader perspective on new energy efficiency opportunities and rationalization improvements.

3. Case Study: ELCA of Beira Interior' Peach

3.1. Functional Unit

The functional unit (FU) is a measure of performance and can be classified as mass-based or land-based [26]. The use of a mass-based FU can be complemented with the FU land-based for fruit production, this will provide two distinct results. Both FU are dependent on the study method. The former is used in land-based while the latter is more used for agricultural production guidance [27]. According to Cerutti et al. [26], comparing different functional units to the same fruit production system, can provide different situations, especially in cultures with a higher yield, in which there is a significantly better environmental performance using a mass-based FU. On the other hand, to use a land-based FU is more advantageous in fruit crops with a lower yield. To develop the peach ELCA study in the Portuguese *Beira Interior* region, the utilized FU was land-based during the production process, since the energy consumption is quite similar in high or low production, due to the fact that the operations carried out in the plantation are not production dependent. In the harvest and post-harvest period, mass-based FU was essential to use since the energy consumption is dependent on the processed fruit amount.

3.2. System Area

The *Beira Interior* region peach production was only carried out as an energy analysis, being considered the energy and fossil fuel consumption in the defined borders, in a regional farm, *Quinta dos Lamaçais*. Therefore, the experimental data was obtained for each operation carried out in the plantation, the time taken by the tractor to cover each area, the amount of machinery used in each operation, as well as the number of operations carried out in a year vegetative cycle. The dimension of the studied area was the plantation area (3 m × 5 m), the number of trees per row and the number of rows where operations occurred related to the production process. It must be highlighted that this study only covered the impacts associated with one peach production year. A complete study will require monitoring and collect data throughout the peach trees life-cycle, i.e., since the plantation until the tree removal, which would require approximately 15 years for data collection [21].

3.3. Inventory Analysis

The energy consumption considered operations performed during cultivation and post-harvest periods. Different phases were considered during the period of cultivation, namely, pruning, soil maintenance, spraying, herbicide application, irrigation and harvest. Post-harvest operations include refrigerated storing, processing, packaging and transportation as represented in Figure 1.

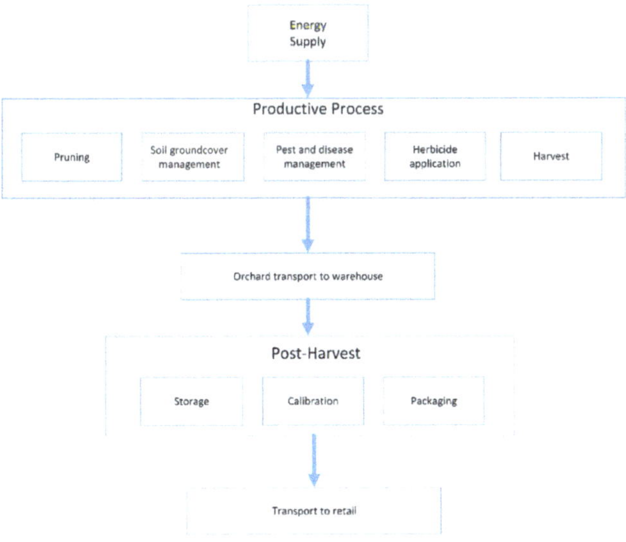

Figure 1. System limits associated with the cultivation of peach.

The inventory data collected was carried out during the years 2015 and 2016, which included field collection of data covering field operations and data gathering from previous years, related to the total number of performed plantation operations and the average production per hectare. The energy consumption was determined by calculating the fuel consumed amount per hectare for each operation performed. According to Ghatrehsamani et al. [25], the equivalent energy factor in Liters of fuel is expressed in Mega Joules (MJ), as shown by Equation (1).

$$E = 56.31 \times C \tag{1}$$

where,

E = Energy consumed in each production operation (MJ·ha^{-1})

C = Fuel consumed in each operation (L)

A tractor with 60 horse power (hp) and a density of 0.844 L·kg^{-1} (ADP, 2009) was considered in the calculation of fuel consumption. According to Freitas [28], the linear relationship between diesel consumption (expressed L·h^{-1}) and machine power (expressed in hp) is given by Equation (2).

$$C_{ijk} = 0.1 \times P_{ik} \tag{2}$$

where,

C_{ijk} = Fuel consumption j during operation i execution using machine k (L·h^{-1})

P = Power of the machine k used in operation i (hp)

3.4. Post-Harvest Processes

After the harvesting process, the peaches are transported in vans to the warehouse, where they are stored, processed and packaged. Subsequently, these are placed in cooling chambers, staying there until being transported to the retailer. During storage and packaging processes, peaches may suffer some changes if there is a high number of big plastic boxes waiting to be stored and packed. In this way, the peaches are placed in the cooling chambers so there is no quality loss, because the ambient temperature during the harvest season is far superior compared to the storage temperature.

In the farm, peach cooling is done only by refrigerated chambers, not suffering any kind of pre-cooling prior to its entry in these chambers. Once packed, peaches are transported by a forklift to the cooling chamber, where these are stored at a temperature of approximately 2 °C up to a maximum of 15 days, counting from the harvesting date.

The farm under study achieved an annual production of 2553 tons and a power consumption of 509.2 MWh, where the most energy is consumed using the refrigeration chambers [29–31]. Therefore, the energy consumption was obtained with the conversion factor of Ghatrehsamani et al. [25] as shown in Equation (3).

$$E = 11.93 \times CE \tag{3}$$

where,

E = Energy consumed (MJ·tonne^{-1})

CE = Electricity consumed (kWh)

Considering a FU of 1 ton, the annual consumption at the warehouse is 199.5 kWh·tonne^{-1}, that corresponds to 2380 MJ·tonne^{-1}.

The annual warehouse energy consumption was determined by energy audits [32–34]. The values shown in Table 1 include the power consumption of all machines used during processing, storage and packaging of peaches, i.e., all energy consumed by the transporter mats, brushing, stackers and cooling chambers.

Table 1. Diesel energy consumption during production (FU = 1 ha of plantation; Quinta dos Lamaçais).

Flows	Diesel Consumption (kg·ha^{-1})	Energy (MJ·ha^{-1})
Diesel	175.24	11,693.3

3.5. Market Transport

Similar to the study of Pereira et al. [35,36], the calculation of transport emissions can be calculated based on *Tier* 2 and *Tier* 3 methods. In the case of peaches transport, this calculation is divided into two sections, one directed to the Light Duty Vehicles, LDV < 3.5 tons and other for the Heavy-Duty Vehicles, HDV > 3.5 tons. Emission factors were taken from tables considering the year of the vehicle, being 1990 to 1995 for LDV and HDV, respectively, and the fuel type used, more precisely diesel [37]. Additionally, the calculation of fuel consumption was performed according to EMEP/CORINAIR [37], which takes into consideration the type of consumed fuel and an average speed of 50 km·h^{-1} for the LDV and 90 km·h^{-1} for the HDV.

The transport summary of the transportation during post-harvest is shown in Table 2, which demonstrates the type of vehicle used in each path, the fuel consumption for each vehicle, according to EMEP/CORINAIR [37], as well as the total distance travelled between the warehouse and orchard with the LDV and between warehouse and retailer for the HDV.

Table 2. Transport fuel consumption and distance traveled (Quinta dos Lamaçais).

Parameter	LDV < 3.5 tonne	HDV > 3.5 tonne
Consumption [g·km^{-1}]	69.235	183.939
Distance travelled [km]	4	500

The fuel consumption is calculated according to the distance travelled from the plantation to the warehouse. Since this distance is the same for each operation, fuel consumption will also be identical.

3.6. Test Case Scenarios

3.6.1. Scenarios #1 and #2—Low and High Production Scenarios

Peach production amount obtained per hectare depends of several factors, in particular, climatic conditions and pests and diseases incidence that cause significant production losses [38]. Thus, annual peach production is important to evaluate in two distinct scenarios, 23 tonne·ha^{-1} for high, and 8 tonne·ha^{-1} for low production.

The production process is based on the FU of 1 hectare of land. Yet, for the comparison of two scenarios, ELCA uses a FU based on mass (ton). Regarding energy consumption, this remains constant in both production scenarios since all operations are still required.

During post-harvest process, the transportation is considered from plantation to warehouse and from warehouse to retailer. Electricity consumed in the warehouse must be considered. The ELCA outputs considered, include fuel consumption, from plantation processes and transportation.

3.6.2. Scenario 3—Production Cycle with Irrigation Pump

During peach production process, irrigation is a determining factor for tree and fruit development, contributing to a high production quality and quantity. According to Simões et al. [38], irrigation process depends from meteorological factors, culture, phenological state, root development and soil properties.

In the *Beira Interior* region, according to the Directorate General for Agriculture and Rural Development of Portugal (DGADR), the annual water needs for peach trees of *Cova da Beira* region are around 4700 m^3·ha^{-1}, where in July most water gets consumed (1600 m^3·ha^{-1}), although the water requirement is cultivar dependent. Simões et al. [38], for *Beira Interior* region the water needs is

3080 m^3·ha^{-1} for early cultivars and 3340 m^3·ha^{-1} for late cultivars. Since the farm under this study, use the Cova da Beira's irrigation infrastructure system that consists irrigation by gravity, consequently no irrigation pumps are used, this case study allows assessing the energy costs for peach plantation not covered by the State irrigation system.

Regarding the energy consumption per hectare, this is calculated based on Vinyes et al. [23] method. A specific energy consumption of 1300 kWh·ha^{-1} of water pumps for the plantation irrigation and a 7600 m^3·ha^{-1} water consumption was determined. According to Vinyes et al. [23], a necessity of 4700 m^3 of water per year has a specific energy consumption of 804 kWh·ha^{-1}.

3.6.3. Scenario 4—Pruning Operation with Electric Shears

Pruning operation of fruit trees requires several tools, including small and large shears, pneumatic shears, knives, saws, axes, scythes, stairs, among others [20]. Commonly, pruning is performed using hydra-ladders connected to an air compressor. A working tractor is constantly required to power the air compressor for using hydra-ladders and pneumatic shears, this requirement generates high costs due to fuel consumption. However, most frequent peach tree pruning technique adopted in the region is the vase (open center) systems, this leads to small trees. Thus, pruning can be performed without the use of hydra-ladders and pneumatic shears. In order to reduce the energy consumption during the pruning operation, electric shears linked to a portable battery carried in a backpack on the operator could replace the current method. Makita shears [39], have a performance of 10,000 cuts load cycle, the equivalent to 7 work hours. This operation reaches an energy consumption of approximately 0.826 kW·ha^{-1}, the equivalent to 9.85 MJ, considering 10 shear and a pruning operation time (hours) per hectare. Therefore, energy consumption during peach orchard pruning operation will be much lower by using electric shears compared to pneumatic shear.

4. Results and Discussion

Field energy consumption from different operations resulted in 175.24 kg·ha^{-1} of Diesel Consumption, corresponding to 11,693 MJ·ha^{-1} (see Table 1). Warehouse energy consumption is 2380 MJ·tonne^{-1} (Table 3). Transportation energy consumption to the market depends on the type of Light or Heavy-Duty Vehicles (see Table 4).

Table 3. Warehouse energy consumption (FU = 1 peach ton; Quinta dos Lamaçais).

Energy Consumption [kWh·tonne^{-1}]	Energy Consumption [MJ·tonne^{-1}]
199.5	2380.0

Table 4. Transport fuel consumption and distance traveled (Quinta dos Lamaçais).

	LDV < 3.5 ton	HDV > 3.5 ton
Distance travelled [km]	4	500
Truck consumption [g·km^{-1}]	69.235	183.939

Figure 2 resumes all energy consumption per hectare values for real and test scenarios. Operations conducted in the field has a constant energy consumption per hectare regardless the production level during the production period (low or high production, scenario 1 and 2, respectively). The soil groundcover maintenance is the operation that shows largest energy consumption, since the tractor needs to cut 3 times the grass in each row. The tree line spacing is 5 m, if the flail mower has 2.5 m, the cutting number is reduced to 2 times, consequently this will incur in a third of the energy consumption. For scenario 3, with a pumping irrigation system (instead of gravity irrigation), shows an increase of energy consumption of 9723 MJ·ha^{-1}. In Scenario 4 considers replacing pneumatic shears by electric shears, show an energy saving of approximately 99.6% in this operation.

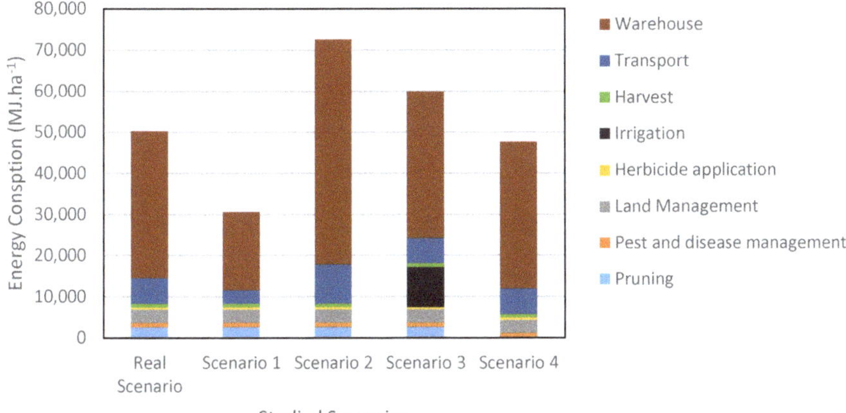

Figure 2. Energy consumption per hectare (MJ·ha^{-1}). according to field technique as pruning, pesticide treatments, soil groundcover management, herbicide application, watering and harvesting, and in the post-harvest process (transport and storage), for the real and test case scenarios (productive processes): Scenario 1: low production; Scenario 2: high production; Scenario 3: using water pump; and Scenario 4: Pruning with electric shears.

Through the analysis of Table 5, the use of electric shears in the pruning operation are beneficial, resulting in a reduction of the energy consumption in 2614 MJ·ha^{-1}. In the post-harvest processes, transport and warehouse energy consumptions and warehouse are altered in different scenarios, because energy consumptions are related to annual production per hectare (tonne·ha^{-1}) variations. Comparing the real scenario with the low and high production scenarios, there is an energy difference of 19,544 MJ·ha^{-1} and 22,336 MJ·ha^{-1}, respectively. Warehouse energy consumption corresponds to the largest share. This fact is largely due to the refrigeration system to extend peach useful life, ensuring the product freshness, quality and food safety. Transportation energy consumption is relevant, although not as significant as the warehouse energy consumption.

The energy requirements according to field technique are shown in Figure 3, using a defined plantation area the values reflect the energy consumed to produce 1 peach ton (MJ·tonne^{-1}). In order to analyze the obtained production yield, a based-mass functional unit reveals representative values, highlighting the energy consumption variation between low and high production scenarios. Having a lower production, energy consumption is higher, while in high production scenario, power consumption is lower (Figure 3). In accordance with Hemmati et al. [13], energy ratio can vary, either because the harvest yield increases, or energy input consumption reduces.

To evaluate energy consumption inherent to *Beira Interior* region, peach cultivation, *Quinta dos Lamaçais* real production was considered, which averages 15 tonne·ha^{-1} (for a real scenario), a low production scenario is 8 tonne·ha^{-1} and a high production scenario is 23 tonne·ha^{-1} (Table 5).

Regarding energy consumption, for the real scenario is 50.21 GJ·ha^{-1}, while for low production scenario the value is 30.67 GJ·ha^{-1}. The high production scenario shows an energy consumption of 72.55 GJ·ha^{-1}, which provides greater profitability, because the production per hectare is superior.

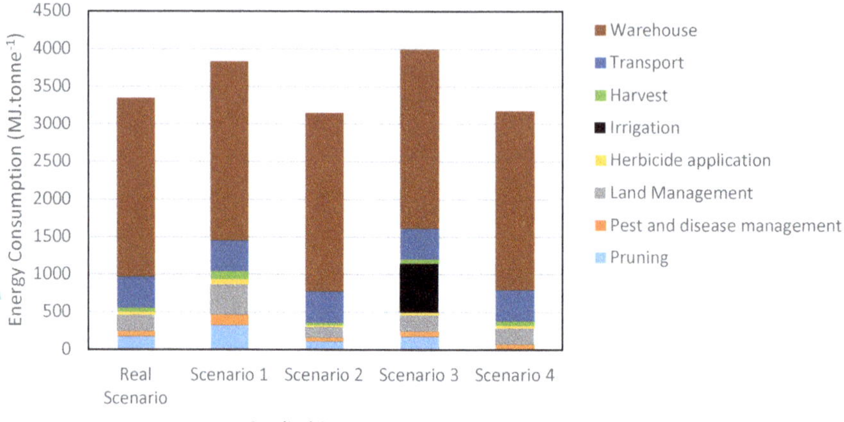

Figure 3. Energy required to obtain one ton of peaches (energy consumption per ton (MJ·tonne^{-1})) according to field technique as pruning, pesticide treatments, soil groundcover management, herbicide application, watering and harvesting, and post-harvest processes (transport and storage), for each of the different scenarios (productive and post-harvest processes): Scenario 1: low production; Scenario 2: high production; Scenario 3: using water pump; and Scenario 4: Pruning with electric shears.

Table 5. Power consumption (inputs) to the three production scenarios.

Scenarios (Inputs)	Units	Real Production	Low Production	High Production
Production	Tonne·ha^{-1}	15	8	23
Production Process				
	kg	175.24	175.24	175.24
Diesel (tractors)	L	147.90	147.90	147.90
	GJ	8.33	8.33	8.33
Post-harvest				
Electricity	kWh	2992.50	1596.00	4589.00
	GJ	35.70	19.00	54.74
	kg	130.04	69.36	199.40
Diesel (transport)	l	109.76	58.54	168.29
	GJ	6.18	3.30	9.48
Total	GJ	50.21	30.67	72.55

As diesel consumption is the main responsible for the machinery energy consumption [18], Figure 4 shows the fuel consumption comparison between several studies and data obtained from *Quinta dos Lamaçais* during production. Royan et al. [1] performed an energy analysis for peach production in Iran, in which was obtained a fuel consumption of 175 L·ha^{-1}, corresponding to 9.88 GJ·ha^{-1}. In turn, Ghatrehsamani et al. [25] determined a fuel consumption of 74.7 L·ha^{-1}, equivalent to 4.21 GJ·ha^{-1} in the province of Bakhitari, in Iran. Thus, the variations of energy consumption do not only depend on the age of the orchard, but also on other variables, in particular, fruit type, geographical location, climatic conditions, irrigation system, phytosanitary treatments application, as well as, pest and disease incidence and control optimization [23]. As provided by Keyes et al. [22], fuel consumption is higher in the apple cultivation, especially during pruning and harvest operations, because hydra-ladders are required. Fuel consumption for post-harvest period is not compared, as it depends on several factors with high variability, including speed and distance traveled.

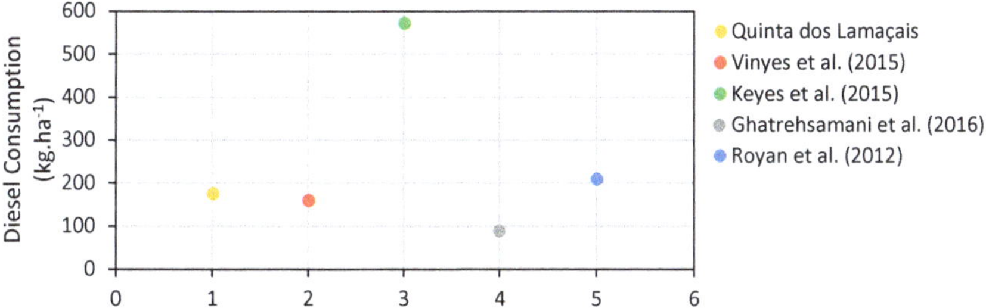

Figure 4. Comparison of fuel consumption (kg·ha^{-1}) in the production process between several studies and the data obtained in Quinta dos Lamaçais.

Concerning peach cultivation, fuel consumption, 175.24 kg·ha^{-1}, does not vary significantly across all studies observed, consequently values obtained at *Quinta dos Lamaçais* were expected regarding the fuel consumption per hectare.

5. Conclusions

Here, *Beira Interior* (Portugal) region peach cultivation, production and post-harvest processes, main impacts were identified. The ELCA results are presented as energy efficiency, in order to be used as a tool for improvement.

Based on energy evaluation, the energy consumption per hectare, during the production processes, is the same regardless the production level. Considering the operations carried out at the orchard, the grass groundcover maintenance is the highest energy consumption procedure, followed by pruning, pesticide treatment, and finally, herbicide application. For the post-harvest period, the warehouse had the highest energy consumption, mainly due to cooling fruits in refrigeration chambers for conservation. Transportation can also be quantified for its energy consumption, because of the high distance traveled between warehouse and retailer.

In order to carry out a sensitivity analysis of the main system, alternative scenarios were outlined. Considering the results obtained, it is also possible to verify that when comparing the real scenario with the low and high production scenarios, an energy difference of 19.54 GJ·ha^{-1} and 22.34 GJ·ha^{-1} was verified, respectively. In addition to the low and high production scenarios, a test case scenario was also included in which there was used a pumping irrigation system in the orchard to evaluate the increase of energy consumption. In view of this situation, it was noted that the energy consumption with the usage of irrigation pumps had an increase of 9.72 GJ·ha^{-1}. Finally, another test case scenario was considered aiming the energy consumption reduction during pruning. This operation is performed with pneumatic shears, which entails the constant operation of the tractor. One way to reduce energy consumption is to replace the current method by electric shears connected with a portable battery. This replacement achieved a reduction of 2.61 GJ·ha^{-1}.

Author Contributions: Conceptualization, P.D.G. and M.P.S.; methodology, P.D.G. and M.P.S.; validation, P.D.G., M.P.S., P.D.d.S. and C.E.S.; formal analysis, , P.D.G., M.P.S., P.D.d.S.; investigation, J.P.G.; resources, P.D.G.; data curation, P.D.G., M.P.S. and P.D.d.S.; writing—original draft preparation, J.P.G., P.D.G., M.P.S. and P.D.d.S.; writing—review and editing, P.D.G., M.P.S., P.D.d.S., C.E.S.; visualization, J.P.G.; supervision, P.D.G. and M.P.S.; project administration P.D.G.

Conflicts of Interest: The authors declare no conflict of interest.

References

1. Royan, M.; Khojastehpour, M.; Emadi, B.; Mobtajer, H. Investigation of energy inputs for peach production using sensitivity analysis in Iran. *Energy Convers. Manag.* **2012**, *64*, 441–446. [CrossRef]
2. Marchi, B.; Zanoni, S. Supply Chain Management for Improved Energy Efficiency: Review and Opportunities. *Energies* **2017**, *10*, 1618. [CrossRef]

3. Martínez-Blanco, J.M.; Antón, A.; Rieradevall, J.; Castellari, M.; Muñoz, P. Comparing nutritional value and yield as functional units in the environmental assessment of horticultural production with organic or mineral fertilization. *Int. J. Life Cycle Assess.* **2011**, *16*, 12–26. [CrossRef]

4. Van der Werf, H.; Garnett, T.; Corson, M.; Hayashi, K.; Huisingh, D.; Cederberg, C. Towards eco-efficient agriculture and food systems: Theory, praxis and future challenges. *J. Clean. Prod.* **2014**, *73*, 1–9. [CrossRef]

5. Thomassen, M.A.; de Boer, I.J. Evaluation of indicator to assess the environmental impact of dairy production systems. *Agric. Ecosyst. Environ.* **2005**, *111*, 185–199. [CrossRef]

6. ISO. *Environmental Management-Life Cycle Assessment-Principles and Framework (ISO 14040: 2006)*; International Organization for Standardization: Geneva, Switzerland, 2006.

7. Schaubroeck, T.; Rugani, B. A revision of what life cycle sustainability assessment should entail: Towards modeling the Net Impact on Human Well-Being. *J. Ind. Ecol.* **2017**, *21*, 1464–1477. [CrossRef]

8. Curran, M.A. Life cycle assessment: A review of the methodology and its application to sustainability. *Curr. Opin. Chem. Eng.* **2013**, *2*, 273–277. [CrossRef]

9. Weidema, B. The integration of economic and social aspects in life cycle impact assessment. *Int. J. Life Cycle Assess.* **2006**, *11*, 89–96. [CrossRef]

10. Falcone, P.M.; Imbert, E. Social Life Cycle Approach as a Tool for Promoting the Market Uptake of Bio-Based Products from a Consumer Perspective. *Sustainability* **2018**, *10*, 1031. [CrossRef]

11. Malça, J.; Freire, F. Life cycle energy analysis for bioethanol: Allocation methods and implications for energy efficiency and renewability. In *Energy-Efficient, Cost-Effective and Environmentally Sustainable Systems and Processes*; Rivero, R., Monroy, L., Pulido, R., Tratsaroins, G., Eds.; Instituto Mexicano del Petrol: Gustavo A. Madero, Mexico, 2004.

12. Iglesias. Exposición comentada de variedades de melocotón, nectarine, melocotón plano y nectarina plana. In Proceedings of the XXII Jornada Frutícola, Mollerussa, Spain, Country, 25–26 October 2017; pp. 109–114.

13. INE. Produção e Superfície das Principais Culturas Agrícolas por Localização Geográfica, Especie e Ano. Instituto Nacional de Estatística (INE), 2015. Available online: https://www.ine.pt/ngt_server/attachfileu.jsp?look_parentBoui=271434441&att_display=n&att_download=y (accessed on 15 November 2015).

14. Weidema, B.P.; Wesnæs, M.S. Data quality management for life cycle inventories—An example of using data quality indicators. *J. Clean. Prod.* **1996**, *4*, 167–174. [CrossRef]

15. Kramer, K.; Moll, H.; Nonhebel, S. Total greenhouse gas emissions related to the Dutch crop production system. *Agric. Ecosyst. Environ.* **1999**, *72*, 9–16. [CrossRef]

16. Margini, M.; Rossier, D.; Crehaz, P.; Jolliet, O. Life Cycle impact assessment of pesticides on human health and ecosystems. *Agric. Ecosyst. Environ.* **2002**, *93*, 379–392. [CrossRef]

17. Milà i Canals, L.; Burnip, G.M.; Cowell, S.J. Evaluation of the environmental impacts of apple production using Life Cycle Assessment (LCA): Case study in New Zealand. *Agric. Ecosyst. Environ.* **2006**, *114*, 226–238. [CrossRef]

18. Alaphillippe, A.; Simon, S.; Brun, L.; Hayer, F.; Gaillard, G. Life cycle analysis reveals higher agroecological benefits of organic and low-input apple production. *Agron. Sustain. Dev.* **2013**, *33*, 581. [CrossRef]

19. Hemmati, A.; Tabatabaeefer, A.; Rajabipour, A. Comparison of energy flow and economic performance between flat land and sloping land olive orchards. *Energy* **2013**, *61*, 472–478. [CrossRef]

20. Soni, P.; Taewichit, C.; Salokhe, V.M. Energy consumption and CO_2 emissions in rainfed agricultural production systems of Northeast Thailand. *Agric. Syst.* **2013**, *116*, 25–36. [CrossRef]

21. Ingrao, C.; Matarazzo, A.; Tricase, C.; Clasadonte, M.; Huisingh, D. Life Cycle Assessment for highlighting environmental hotspots in Sicilian peach production systems. *J. Clean. Prod.* **2015**, *92*, 109–120. [CrossRef]

22. Keyes, S.; Tyedmers, P.; Beazley, K. Evaluating the environmental impacts of conventional and organic apple production in Nova Scotia, Canada, through life cycle assessment. *J. Clean. Prod.* **2015**, *104*, 40–51. [CrossRef]

23. Vinyes, E.; Gasol, C.; Asin, L.; Alegre, S.; Munõz, P. Life Cycle Assessment of multiyear peach production. *J. Clean. Prod.* **2015**, *104*, 68–79. [CrossRef]

24. Longo, S.; Mistretta, M.; Guariano, F.; Cellura, M. Life Cycle Assessment of organic and conventional apple supply chains in the North of Italy. *J. Clean. Prod.* **2017**, *140*, 654–663. [CrossRef]

25. Ghatrehsamani, S.; Ebrahimi, R.; Kazi, S.; Badry, A.; Sadeghinezhad, E. Optimization model of peach production relevant to input energies—Yield function in Chaharmahal va Bakhtiari province, Iran. *Energy* **2016**, *99*, 315–321. [CrossRef]

26. Cerutti, A.; Beccaro, G.; Bruun, S.; Bosco, S.; Donno, D.; Notarnicola, B.; Bounous, G. Life-cycle assessment application in the fruit sector: State of the art and recommendations for environmental declarations of fruit products. *J. Clean. Prod.* **2014**, *73*, 125–135. [CrossRef]
27. Hayashi, K. Practical recommendations for supporting agricultural decisions through life cycle assessment based on two alternative views of crop production: The example of organic conversion. *Int. J. Life Cycle Assess.* **2013**, *18*, 331–339. [CrossRef]
28. Freitas, C. *Análise dos Encargos Horários com a Utilização das Máquinas Pesadas na Agricultura e Floresta*; Gabinete de Gestão do Parque de Máquinas, Direção de Serviços de Projetos e Obras, Instituto de Desenvolvimento Rural e Hidráulica: Lisboa, Portugal, 2004. (In Portuguese)
29. Nunes, J.; Neves, D.; Gaspar, P.D.; Silva, P.D.; Andrade, L.P. Predictive tool of energy performance of cold storage in agrifood industries: The portuguese case study. *Energy Convers. Manag.* **2014**, *88*, 758–767. [CrossRef]
30. Silva, P.D.; Gaspar, P.D.; Nunes, J.; Andrade, L.P. Specific electrical energy consumption and CO_2 emissions assessment of agrifood industries in the central region of Portugal. *Appl. Mech. Mater.* **2014**, *675–677*, 1880–1886. [CrossRef]
31. Gaspar, P.D.; Silva, P.D.; Nunes, J.; Andrade, L.P. Characterization of the specific electrical energy consumption of agrifood industries in the central region of Portugal. *Appl. Mech. Mater.* **2014**, *590*, 878–882. [CrossRef]
32. Neves, D.; Gaspar, P.D.; Silva, P.D.; Nunes, J.; Andrade, L.P. Computational tool for the energy efficiency assessment of horticultural industries—Case study of inner region of Portugal. In *Computational Science and Its Applications—ICCSA 2014*; Lecture Notes in Computer Science 8584 (LNCS); Murgante, B., Misra, S., Rocha, A.M.A.C., Torre, C.M., Rocha, J.G., Falcão, M.I., Taniar, D., Apduhan, B.O., Gervasi, O., Eds.; Springer: Cham, Switzerland, 2014; Part VI; pp. 87–101.
33. Nunes, J.; Silva, P.D.; Andrade, L.P.; Gaspar, P.D.; Domingues, L.C. Energetic evaluation of refrigeration systems of horticultural industries in Portugal. In Proceedings of the 3rd IIR International Conference on Sustainability and Cold Chain (ICCC 2014), London, UK, 23–25 June 2014; pp. 397–404.
34. Gaspar, P.D.; Silva, P.D.; Andrade, L.P.; Nunes, J.; Domingues, C. Best practices in refrigeration applications to promote energy efficiency—The Portuguese case study. In *Food Industry: Assessment, Trends and Current Issues*; Series Food and Beverage Consumption and Health; Cunningham, D., Ed.; Nova Publishers: Hauppauge, NY, USA, 2016; Chapter 3.
35. Pereira, T.; Seabra, T.; Pina, A.; Freitas, L.; Amaro, A. *Portuguese Informative Inventory Report 1990–2012*; Portuguese Environmental Agency: Amadora, Portugal, 2014.
36. Pereira, T.; Seabra, T.; Pina, A.; Freitas, L.; Amaro, A. *Portuguese Informative Inventory Report 1990–2013*; Portuguese Environmental Agency: Amadora, Portugal, 2015.
37. EMEP/EEA. *Emission Inventory Guidebook*. 2012. Available online: https://www.eea.europa.eu/publications/emep-eea-guidebook-2016 (accessed on 10 September 2016).
38. Simões, M.P.; Barateiro, A.; Duarte, A.C.; Dias, C.; Ramos, C.; Alberto, D.; Ferreira, D.; Calouro, F.; Vieira, F.; Silvino, P.; et al. *+Pêssego. Guia prático da Produção*; Centro Operativo e Tecnológico Hortofrutícola Nacional (COTHN), Empresa Diário do Porto, Lda: Porto, Portugal, 2016; Volume I. (In Portuguese)
39. Makita®. Tesoura de Poda 24V. Available online: http://www.makportugal.com/tesouras/18-tesoura-de-poda-24v-makita-4604dw.html (accessed on 16 August 2016).

 *sustainability*

Review

Future Perspectives of Biomass Torrefaction: Review of the Current State-Of-The-Art and Research Development

Jorge Miguel Carneiro Ribeiro [1], Radu Godina [2], João Carlos de Oliveira Matias [1,3] and Leonel Jorge Ribeiro Nunes [1,3,*]

[1]   DEGEIT—Department of Economics, Management, Industrial Engineering and Tourism, University of Aveiro, 3810-193 Aveiro, Portugal; radugodina@gmail.com (J.M.C.R.); jmatias@ua.pt (J.C.d.O.M.)
[2]   C-MAST—Centre for Aerospace Science and Technologies—Department of Electromechanical Engineering, University of Beira Interior, 6201-001 Covilhã, Portugal; rd@ubi.pt
[3]   GOVCOPP—Research Unit on Governance, Competitiveness and Public Policies, University of Aveiro, 3810-193 Aveiro, Portugal
*   Correspondence: leonelnunes@ua.pt

Received: 31 May 2018; Accepted: 4 July 2018; Published: 5 July 2018

Abstract: The growing search for alternative energy sources is not only due to the present shortage of non-renewable energy sources, but also due to their negative environmental impacts. Therefore, a lot of attention is drawn to the use of biomass as a renewable energy source. However, using biomass in its natural state has not proven to be an efficient technique, giving rise to a wide range of processing treatments that enhance the properties of biomass as an energy source. Torrefaction is a thermal process that enhances the properties of biomass through its thermal decomposition at temperatures between 200 and 300 °C. The torrefaction process is defined by several parameters, which also have impacts on the final quality of the torrefied biomass. The final quality is measured by considering parameters, such as humidity, heating value (HV), and grindability. Studies have focused on maximizing the torrefied biomass' quality using the best possible combination for the different parameters. The main objective of this article is to present new information regarding the conventional torrefaction process, as well as study the innovative techniques that have been in development for the improvement of the torrefied biomass qualities. With this study, conclusions were made regarding the importance of torrefaction in the energy field, after considering the economic status of this renewable resource. The importance of the torrefaction parameters on the final properties of torrefied biomass was also highly considered, as well as the importance of the reactor scales for the definition of ideal protocols.

Keywords: renewable energy; biomass; torrefaction; grindability; rotary reactor

---

1. Introduction

Currently, fossil fuels, such as oil, natural gas, and coal, represent the primary energy sources existent on the planet. However, these resources are limited and their shortage is predicted for the next 50 years [1–5]. Other than their potential shortage, fossil fuels contribute considerably to negative environmental impacts. Therefore, reductions in carbon dioxide (CO_2) emissions, one of the main gases responsible for greenhouse effects (GHG), through the use of renewable energy sources, is a main target, with goals to reduce GHGs emissions from 1990 to 2030 by 40% and to reduce GHGs emissions by 80–95% by 2050 [5–9].

With the advent of the concept of sustainability, the idea of using natural resources to meet the needs of the present without compromising on the satisfaction of future needs has become a first priority [10–12]. One of the mechanisms developed for better conservation efficiency is the use of renewable energy sources that have little or no direct impact on the environment. The use of renewable energy is important because of the economic factor, where the use of cheaper resources for energy production favors the preservation of the environment, since most use natural, abundant, and reusable means for the production of electric energy [13–15]. The demand for ecologically acceptable substitutes for fossil fuels has been accelerated both by increasing the use of renewable energy sources and by predicting the declining supply of non-renewable energy. Biomass is considered the oldest energy source, and its energy (bioenergy) is becoming more and more promising due to a set of characteristics that allow the substitution of fossil fuels, thus, reducing GHG emissions [16–18].

The composition of biomass includes cellulose, hemicellulose, and lignin, thus, comprising of these main components and other organic and inorganic components, such as minerals, that are present in lower quantities [19]. The mass percentages of the three main components of biomass depend on its origin [20]. Due to their different structural compositions, cellulose and hemicellulose have different behaviors during thermal decomposition. In general, the thermal decomposition temperature (TDT) of hemicellulose is smaller, occurring in temperatures between 220 and 315 °C whereas cellulose decomposes between 315 and 400 °C. Lignin has a more gradual thermal decomposition, with its TDT fluctuating between 160 and 900 °C [20]. Biomass derived energy can be used directly, but can also be converted to a secondary energy resource through a chain of thermal and biochemical processes, such as gasification, torrefaction, liquefaction, pyrolysis, anaerobic digestion, fermentation, and transesterification [2,16,21].

Some of the problems associated with the energy production of biomass in its original form when in comparison with fossil fuels are its lower density, higher moisture content, and hydrophilic nature, which cause its HV to decrease, making it difficult to use biomass for large scale productions [3,22,23]. All these parameters influence the logistics of the energy production process using biomass, as well as its energetic efficiency [24].

Torrefaction is one of several processes of biomass improvement, thus, modifying it physical and chemical composition. This process consists of the slow heating of biomass through a range of temperatures between 200 and 300 °C in a controlled atmosphere without the presence of oxygen (O), [24]. Torrefaction enhances the performance of biomass during co-combustion and gasification [25–27].

Presently, there are several studies regarding the effects of different torrefaction parameters on the final physical and chemical composition of biomass for energy production [2,22,24,28].

There are also innumerous studies showing the importance of laboratory scale reactors for the development of ideal protocols to mimic when using other scales, such as pilot and commercial. Laboratory scale reactors are considered the most important due to their price, which is considerably lower than in other scales, allied with the fact that this scale is enough to define the right parameters to use even when applying them to different scales [29].

In this article, such studies are summarized and reviewed to study the concept of torrefaction and the current and innovative attempts to improve the outcomes of this process, specifically, the quality of torrefied biomass using different techniques and the different sources of raw biomass.

2. Torrefaction Process

As was previously mentioned, torrefaction (specifically, dry torrefaction) consists of a thermal treatment of biomass, where biomass is heated in a non-oxidizing atmosphere to improve the biomass' energy density through the increase of its HV and hydrophobicity [16,30–32].

Taking into account a more chemical approach, the principle of this process rests on the removal of oxygen (O) and hydrogen (H), with the production of a final solid product with lower oxygen-carbon (O/C) and hydrogen-carbon (H/C) ratios as shown in Figure 1 [2,33].

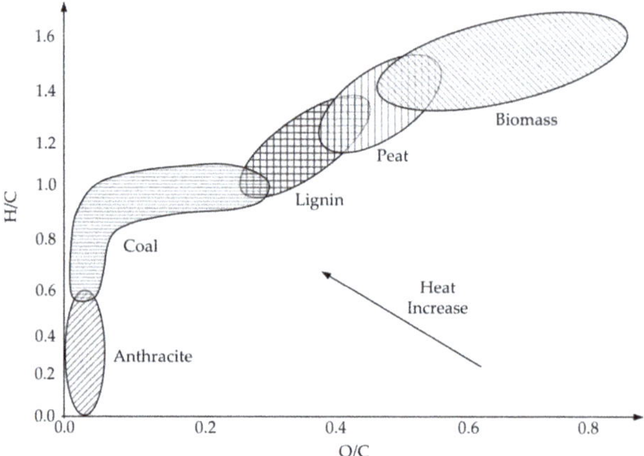

Figure 1. van Krevelen diagram (adapted from [33]).

Several studies have proven that after the torrefaction process, biomass properties are extensively modified and improved [2,34–37].

This type of thermal treatment not only destroys the fibrous nature of biomass and, consequently, its tenacity, but also increases its HV [2,16]. Torrefaction also increases the hydrophobicity of biomass, which means that biomass becomes more resistant to water adsorption, resulting in an improvement in the control of storage conditions due to the fact that torrefied biomass is more resistant to bacterial and fungi attacks, and, thus, more resistant to rotting [2]. During its torrefaction, biomass suffers mass loss, maintaining, however, its energy yield [19].

Other properties, such as O/C and grindability, allied with the fact that the characteristics of torrefied biomass are more uniformly distributed, make biomass more appealing when compared with non-torrefied biomass, as can be observed in Figure 2 [16].

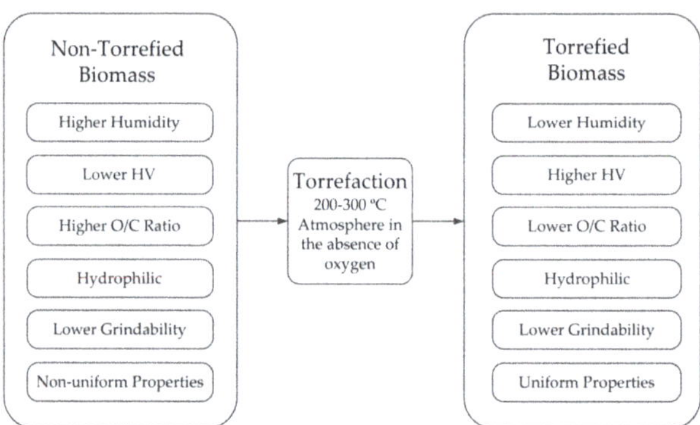

Figure 2. Schematic representation of biomass characteristics compared with torrefied biomass characteristics.

2.1. Torrefaction Process

The torrefaction process (dry torrefaction) can be divided into distinct phases: Heating, drying, torrefaction, and cooling [2]. According to Bergam et al. (2005), the drying process is subdivided into two phases, making torrefaction a process comprised of five different phases, as explained in Table 1 [38].

Table 1. Description of the different torrefaction phases (adapted from [38]).

	Phases	Description
1.	Heating	Biomass is heated until the drying temperature is obtained and the biomass' humidity starts to evaporate.
2.	Pre-drying	Occurs at 100 °C when the free water present on biomass evaporates at a stable temperature.
3.	Post-drying	The temperature is increased until it reaches 200 °C. The remaining water present on biomass chemical bonds is completely evaporated. This phase is responsible for mass loss due to the evaporation of several biomass components.
4.	Torrefaction	Main phase of the torrefaction process. It occurs at 200 °C and is responsible for the main mass lost. The torrefaction temperature (TT) is given by the maximum stable temperature used during the process.
5.	Cooling	The final product is cooled down to a temperature below 200 °C, which is the temperature of wood auto-ignition, before it contacts the air and until room temperature is reached.

Figure 3 represents the different stages of biomass heating during the torrefaction process, starting at room temperature until the TT is reached (T_{tor}), followed by cooling.

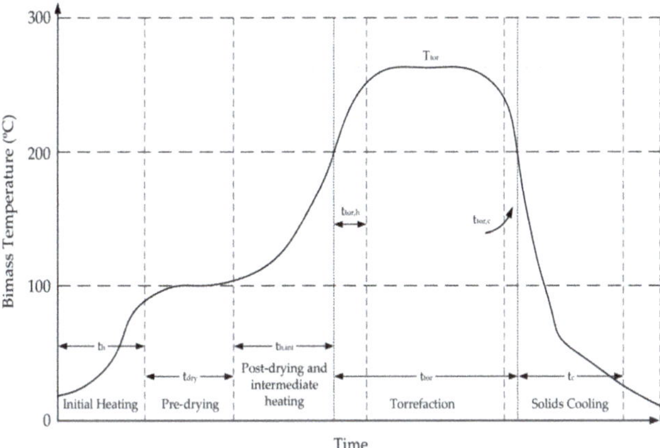

Figure 3. Schematic description of the different torrefaction stages (adapted from [38]). Where t_h represents the time for drying to start; t_d represents the time for drying; $t_{h,int}$ is the intermediate heating time from drying to torrefaction; t_{tor} is the reaction time at the desired torrefaction temperature, T_{tor}; $t_{tor,c}$ is the cooling time from the desired T_{tor} to 200 °C; and t_c is the cooling time to room temperature.

2.2. Parameters that Influence the Torrefaction Process

There are a wide range of parameters that influence not only the torrefaction process, but also the final product characteristics. These parameters include the temperature and residence time, heating rate, operating atmospheric composition, controlling of the torrefaction process instability, and the type of reactor.

Although there are not many studies regarding the optimization of such parameters, the ideal process would be determined by their combination, while trying to maximize the quality of the torrefied biomass production, depending on the type of biomass in use [38,39].

2.2.1. Temperature and Residence Time

The perception of biomass components, as well as it chemical composition, enables the study of biomass behavior during the heating processes [22]. The exposure of biomass to high temperatures leads to thermal degradation of it physical structure and, thus, to mass loss. The degradation of biomass depends on the duration of heating and the maximum temperature obtained [40]. The different components of biomass have distinct characteristics and, therefore, behave in various ways, depending on their origin, and also interact differently depending on the thermal process and its temperature [22].

Other variables that influence the torrefaction process also take part in the changes that occur to the composition and structure of biomass, such as particle size, heating rate, and pressure [41].

Residence time mainly affects the decomposition of hemicellulose, whereas cellulose loses mass depending on the reaction time [38].

The final product characteristics are more affected by temperature than residence time. Temperature defines the kinetics of the torrefaction reaction while residence time affects process characteristics, but only during some temperature ranges [42,43].

2.2.2. Heating Rate

The heating rate ($^\circ$C/min) used during the torrefaction process influences the secondary degradation reactions, which affect the final solid, liquid, and gas product distribution ([44], 2018).

Strezov et al. observed that the energy yield of liquid pyrolysis of *Pennisetum purpureum* is higher when the heating rate is also higher whereas coal does not suffer any changes. The main reason for the distribution of reaction products is the reduction of the number of secondary reactions when using higher heating rates [45].

Kumer et al. also suggested that by increasing the heating rate, there would be a reduction of the effects of heat and mass transfers between particles [46].

2.2.3. Operating Atmospheric Composition

The torrefaction process can be affected by the gas flow used during the process [47]. This occurs due to the secondary interactions between the gases of the torrefaction process, such as water vapor, air, and other atmospheric compounds [48].

According to several studies, carbon monoxide (CO) is the main gas released during the torrefaction process and it is formed during a secondary reaction between water vapor (H_2O (g)), CO_2 and solid torrefaction products when the temperature increases [42,43]. The minerals present in biomass can also serve as catalysts for this reaction [44], hence, the ratio between CO_2 and CO decreases with a higher residence time [42,43].

There are no substantial changes in biomass reactivity depending on the O_2 presence in the atmosphere, nor substantial changes in the solid reaction products [49].

2.2.4. Controlling Torrefaction Process Instability

Temperature is one of the most influential parameters of the torrefaction process and, therefore, the most substantial to control. There are some difficulties associated with controlling the torrefaction process' temperature, which influences the inertia of the process, making it faster or slower. To maintain the quality of the final torrefaction products, it is of essence to keep temperature conditions as stable as possible [44].

It is also important to consider that, during the torrefaction process, the emission of volatile compounds, both non-condensable and condensable, occurs. In case their removal does not occur, especially the removal of condensable products, the cooling process will promote the formation of tar and other hydrocarbon-based compounds that can interfere with the self-ignition of torrefied biomass due to their low ignition temperature [44].

Condensable compounds originate from biomass with higher contents of condensable materials. One solution for this problem would be the implementation of a pre-treatment protocol for biomass to reduce the compounds released during the torrefaction process [44].

2.2.5. Type of Reactor

Using green biomass for combustion processes has many disadvantages, such as its instability during this process due to the humidity and size of the reactor chamber [29].

The types of reactors used can have three distinct scales: Pilot, commercial, and laboratory [29].

The laboratory scale reactors are considered the most important reactors for the development of studies to test parameters of the torrefaction process for later applications on the pilot and commercial scales [50]. This type of torrefaction reactor can be subdivided into four types: (i) Fixed bed torrefaction reactor, (ii) microwave torrefaction reactor, (iii) rotary drum reactor, and (iv) fluidized bed torrefaction reactor [29].

(i)   Fixed bed torrefaction reactor

The fixed bed torrefaction reactor is considered the simplest reactor. The fixed amount of raw biomass is filled inside the reactor and is heated by heat conduction from the electrical heater around the surface of the reactor [29].

(ii)   Microwave reactor

The microwave torrefaction reactor uses high frequency electromagnetic waves, namely microwaves. These microwaves create a vibration of water molecules inside the biomass, resulting in an increase of its temperature. Wang et al. constructed and tested the microwave reactor in [51].

(iii) Rotary drum reactor

The rotary drum reactor is the most common type of reactor, which could be directly and indirectly heated and can be observed in Figure 4. The rotary drum reactor is a continuous reactor and process that can be considered a proven technology for a wide range of applications. This type of rotary drum is constructed in such a way that it can receive biomass near the inlet end and displays a discharge port near the outlet end. In the case of torrefaction, the biomass in the reactor can be heated directly or indirectly with superheated steam or exhaust gas that is produced by the combustion of volatile organic compounds (VOCs). When the process of torrefaction ends, the torrefied biomass is then discharged from one or more ports on the shell of the drum [52,53].

(iv)   Fluidized bed torrefaction reactor

The raw biomass is placed on a grate and the hot inert gas flows from the bottom through the raw biomass bed. At a suitable inert gas velocity, the raw biomass floats and behaves like a fluid. This results in a uniform temperature distribution throughout the raw biomass bed [54].

Although there are many types of reactors, further research regarding the ideal reactor design for minimum energy use are mainly concentrated on laboratory scale reactors due to the reproducibility of these reactors to a pilot and commercial scale [29].

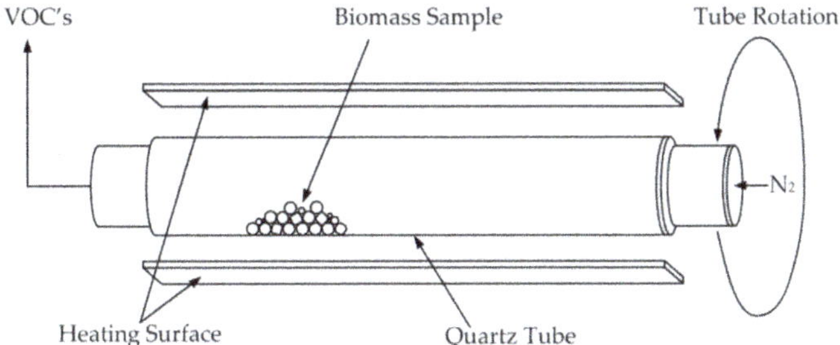

Figure 4. Diagram scheme of the laboratorial NABERTHERM rotary reactor [53].

3. Torrefied Biomass

3.1. Properties of Torrefied Biomass

As previously mentioned, torrefaction modifies certain properties of biomass, making it more appealing than non-torrefied biomass [16]. The most significant properties are the moisture content, bulk density and energy density, grindability, particle size distribution, sphericity and specific surface area, and the heating value [16,22].

3.1.1. Moisture Content

Normally, the moisture content of non-torrefied biomass ranges from 10–50%. However, since torrefaction is a process that occurs at high temperatures, including phases with specific biomass drying, the moisture content of the non-torrefied biomass is reduced to about 1–3% [22]. This is an important parameter to consider as high moisture content translates into energy loss when burning biomass [16]. Therefore, to increase energy efficiency and the quality of the final product, it is necessary to reduce its moisture content by reducing emissions during the thermo-chemical process of energy conversion [55].

Reducing the humidity of the biomass has many positive consequences in relation to its transport and storage, since, thanks to its low water content, it becomes lighter and less susceptible to rot [56].

During the torrefaction process, the hydroxyl groups present in wood are partially destroyed by dehydration, which prevents the formation of hydrogen bonds, causing the torrefied biomass to become hydrophobic [25].

3.1.2. Bulk Density and Energy Density

The loss of mass in the form of solids, liquids, and gases during torrefaction of the biomass causes an increase in the porosity of the same. This results in a significant decrease in the volumetric density of the biomass, usually ranging from 180 to 350 kg/m^3, depending on the initial density of the non-torrefied biomass [16]. Despite this decrease in bulk density, the energy density of the torrefied biomass increases after torrefaction ($\pm$30% increase) [25].

3.1.3. Grindability

Biomass is extremely fibrous and tenacious. During torrefaction, biomass loses its toughness, mainly due to the destruction of its hemicellulose matrix and cellulose depolymerization [16], which leads to a decrease in the length of their fibers. The length of each particle also decreases without changing its diameter, which results in better grinding characteristics, handling, and greater creep during processing and transport [22].

These changes in biomass microstructure increase the mass percentages (wt.%) of the fine particles, subject to the same grinding conditions [1,57].

There is also a 90% decrease in energy consumption for the grinded torrefied biomass chips compared to non-torrefied wood chips [58].

3.1.4. Particle Size Distribution, Sphericity, and Specific Surface Area

Parameters, such as particle size distribution curves, sphericity, and specific surface area (SSA), are crucial parameters in the perception of creep and behavior during the co-combustion of the torrefied biomass [16,28]. Recent studies show that torrefaction of biomass allows the formation of narrower particles, resulting in more uniform sizes, which is a characteristic similar to coal. Authors, such as Phanphanic et al. (2011), observed that the particle size distribution curve for torrefied biomass has its maximum distribution along smaller sizes as the torrefaction temperature increases [58].

As for sphericity and SSA, these are also two parameters significantly affected by torrefaction. According to Phanphanic et al., from 300 °C the sphericity and SSA values also increase [58].

3.1.5. Heating Value

The amount of H and O lost by the biomass during the torrefaction process is higher than the amount of C lost by it, which causes an increase in the heating value of the torrefied biomass [59]. The HV of the torrefied biomass is higher than that of the non-torrefied biomass, since there is an increase of the fixed carbon, in contrast to the output of the oxygen compounds, leaving more carbon available to be oxidized and thus to release energy [22].

4. Economic and Social Analysis

Due to the biomass properties which allow it to be used preferentially in relation to fossil fuels, since it is not only renewable energy but also its CO_2 emissions balance is neutral, it is considered to be the fourth primary source of energy used in the world [60].

As already mentioned, torrefaction is a process of improving biomass properties to form a more homogeneous product, which is densified through palletization, resulting in biomass with a higher HV and energy—Torrefied pellets (TOP's, according to ECN; or TBP's, according to [24]), which have properties very similar to those of coal [61]. In this way, torrefied biomass has a wide range of potential uses in industries where coal is typically used as an energy source [62].

The increasing interest in biomass and, consequently, its demand, causes an increase in the price of most biomass fuels for conversion into thermal energy [36].

Several studies suggest that biomass plantations in developing countries can contribute, in the long term, to increased biomass production. However, other authors suggest that this contribution of biomass is minimal [44].

In any case, these plantations are important and, therefore, it is necessary to consider factors, such as the availability of soil and yield of biomass production among other parameters, such as the climate and types of soil [63].

Despite the potential of biomass to replace fossil fuels, there is a risk that its planting will affect ecosystems by creating negative impacts on the quality of water and soil and affecting food chains [64]. Another potential consequence of biomass production is its overproduction, which leads to deforestation and a consequent decrease in biodiversity [65].

Radics et al. concluded that the profitability of making torrefied pellets essentially depends on sale prices and feedstock costs. However, the costs associated with a scale-up and long-term operations are still very uncertain [66]. This is one of the reasons why laboratory scale reactors pose such an important role in developing torrefaction protocols.

5. Analysis of the Research Literature

Torrefaction is a slow and low temperature pyrolysis process that is not very different from the one used in charcoal production cells, which were used as a reducing agent in the earlier stages of metallurgical processes at the beginning of the industrial revolution. However, the development of the torrefaction process only began with the production of coffee in the late nineteenth century, as documented in the first patents of Thiel (1897) [67] and Offrion (1900) [68]. Other patents in the area of torrefaction were patented in the following years and can be seen in Table 2.

Table 2. Other patents in the area of torrefaction and their respective period of patenting.

Period of Time	Number of Patents
1922–1925	3
1930–1932	3
1939–1952	10

Some research on torrefaction, still in the 1930s, was devoted to the production of gaseous fuels. During the first half and the middle of the twentieth century, only a few works sporadically appeared that were dedicated to the torrefaction of biomass for energy. However, more information and fundamental data on heat treatments of lignocellulosic materials can be found from this period, mainly on high temperature drying, dry distillation, thermal degradation, pyrolysis, thermal stabilization, and preservation of wood.

5.1. Research Focused on Torrefaction

The development of modern works in the area of torrefaction can be divided into the pioneering French publications documented by *ARMINES Assoc pour Recherche et Dev des Methodes et Process Ind* [69] and Bourgois et al. [70,71], from 1981 to 1989, and the recent and extensive efforts of a large number of groups initiated by the work of scientists and engineers at the Eindhoven University of Technology and the Dutch Center for Energy (ECN) [72]. In the late 1980s, initial research implemented in France resulted in a specialized unit in France, where torrefaction was utilized to create a reducing agent for the metallurgical industry. The unit was built by the Pechiney enterprise and operated for a few years until it was dismantled for economic reasons. It must be mentioned and acknowledged that other scientific research was carried out during this period, in parallel with the French and Dutch works [73].

The torrefaction of biomass has been attracting a considerable amount of attention in the research community in the last few years [27,74–78]. Thus, the authors proceeded to the quantification of this interest. By examining all publications that mention biomass torrefaction for this study the authors sought to gather the current research. The aim of the analysis made in this study was to highlight further advances in the use of torrefaction. In this review, journals, conference proceedings, and book chapters were all considered to make this study as broad as possible. In the case of this study, the most significant databases were utilized in search of related papers. The databases used for this search are: Elsevier, Springer, MDPI, IEEE Xplore, Taylor & Francis, Wiley, Emerald Insight, Nature, and Inderscience Online. These databases were chosen as they comprise almost all the publications on this subject.

After compiling and categorizing all found publications, more than a few conclusions can be reached. The first finding is that 2304 publications were found in the above mentioned databases. The second finding is that the majority of the publications concerning the torrefaction of biomass have been published relatively recently, as can be seen in Figure 5. Also, by observing this figure, it is possible to deduce that the number of publications has been increasing almost exponentially and academic interest in this topic has been steadily growing. In Figure 6, the distribution of publication by database can be seen. By carefully observing Figure 6 it can be concluded that the majority of biomass torrefaction publications can be found in the Elsevier database.

5.2. Future Perspectives and Research Developments

The increasing interest of the industry in the use of fuels also causes an increase in the studies that involve this subject to improve the production of biomass while reducing the costs of this process [25].

As previously mentioned, torrefaction parameters affect the final properties of biomass, thus, studies are required that investigate such parameters to find the best set to obtain an ideal torrefied biomass sample [16,22].

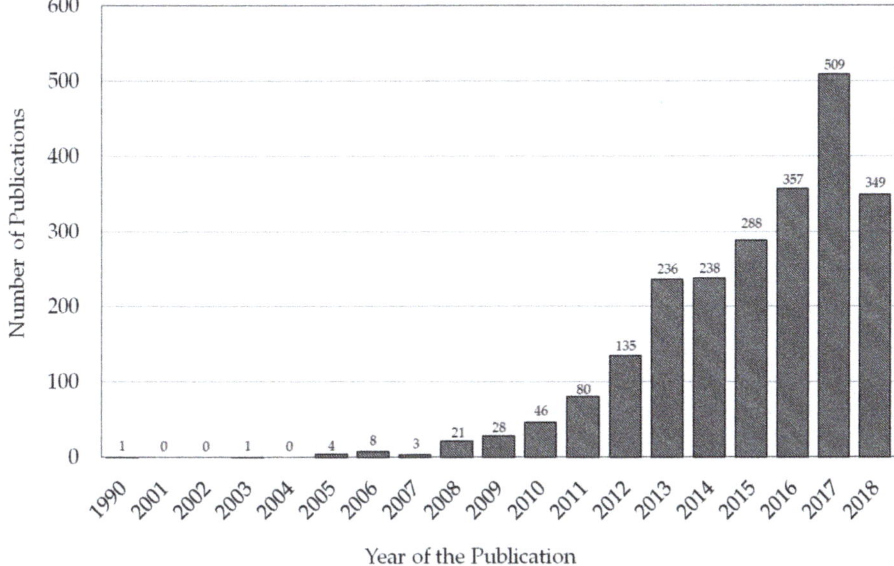

Figure 5. The year of publication of all the publications concerning the torrefaction of biomass.

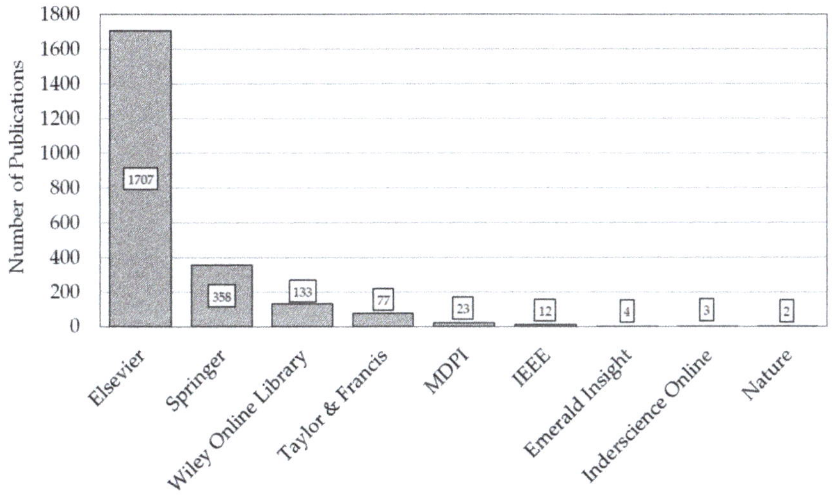

Figure 6. The number of biomass torrefaction publications per database.

To investigate the influence of temperature and residence time on the final biomass quality, Grigiante et al. analyzed the effect of different temperature pathways versus time to obtain mass yield rate values for samples of pine biomass. The results allowed the conclusion that, regardless of the route used, for the same temperature and time values, the TYR is the same. These parameters affected, to a large extent, the final results of the TYR, however, considering the low temperature range chosen, the temperature oscillation did not provoke oscillations in the final biomass energy parameters [79]. Chen et al. studied cellulose, lignin, and hemicellulose at a range of torrefaction temperatures based on the properties of their three-phase products, and a substantial difference in torrefaction characteristics was found due to their different molecular structures [75]. In [77], the characterization of biomass waste torrefaction under conventional and microwave heating was studied and the conclusions indicated that microwave torrefaction is more efficient for biomass upgrading and densification than conventional torrefaction.

Little attention has been directed to the definition of a numerical representation of the quality of biomass torrefaction. In addition to not having a complex database for biomass torrefaction, there is also an index that quantifies the torrefaction degree and shows the effect of the type of biomass used, as well as the effect of torrefaction parameters [80]. Almeida et al. noticed a linear relationship between mass loss and energy yield and carbon fixation of torrential biomass for a range of temperatures [81]. Li et al. observed a linear relationship between mass yield and energy yield, whereas Peng et al. used the mass loss as an indicator of the torrefaction state and developed a linear relationship between energy density and mass loss [54,82]. However, these parameters would have to be measured separate from the torrefaction process. Thus, Basu et al. attempted to develop a quantitative parameter to measure the degree of torrefaction by specifically targeting its goal for the energy industry. Three torrefaction regimes were then defined: Mild, medium, and severe, according to the temperatures used and the torrefaction rates obtained [80]. In [78], different kinetic, volatile release, and solid composition models were analyzed through numerical simulations and optimized different biomass.

Hence, torrefaction is a technique that exploits biomass characteristics to the maximum extent and it is important to consider efficient conversion techniques. With increasing knowledge about the torrefaction of plants, such as bamboo or sugar cane, it is necessary to make the most of the properties of these plants, since they have high growth rates, low ash formation, low alkaline index, and low heating rates. Therefore, Rousset et al. tested different properties of torrefied bamboo, comparing them with the properties of other solid fuels. The results allowed the conclusion that this type of biomass has much improved energy properties after its torrefaction [83]. The torrefaction of many different plant volatile organic compounds are starting to attract the attention of the research community. Poudel et al. conducted a comparative study of the torrefaction of empty fruit bunches and palm kernel shell performed in a horizontal tubular reactor at a temperature ranging from 150 to 600 °C [84]. They concluded that 290–320 °C is the required temperature range for optimum torrefaction of empty fruit bunches and 300–320 °C is the optimum range for palm kernel shell. In [85], the energy densification of sugarcane bagasse through torrefaction under minimized oxidative atmosphere was studied. The results indicated that torrefaction improved several fuel characteristics, making the sugarcane bagasse suitable for both domestic and industrial applications. Other possibilities of torrefaction have been applied and studied, such as for microalgae [86], Black Lilac (*Sambucus nigra* L.) [87], *Prosopis juliflora* [88], corncob [89], almond and walnut shells [90], bamboo sawdust [91], rice husk [92], and cotton stalk [93], among many others [94].

Brachi et al. have studied an innovative torrefaction process based on fluidized bed technology at various temperatures and with different residence times, using tomato peels as the raw material. This shows that it is possible to take advantage of low quality raw materials and, through torrefaction, improve them, thus, reducing the costs of obtaining raw material. The results of this study showed that it is possible to increase the energetic quality of the starting material by maintaining its TYR. This study also showed that this technology has numerous possibilities for its use in torrefaction when it comes to non-wood raw materials, allowing a uniform and consistent quality of the final torrefied products [95].

Álvarez et al. addressed the non-oxidative torrefaction of biomass to enhance its fuel properties in which both the hydrophobicity and the fixed carbon were increased [76]. Chen et al. researched the effect of torrefaction pretreatment on the pyrolysis of rubber wood sawdust through pyrolysis-gas chromatography/mass spectrometry and concluded that the contents of oxy-compounds, such as acids and aldehydes, decreased with rising torrefaction temperature [74]. Finally, in [86], the effects of torrefaction on physical properties, chemical composition, and reactivity of microalgae were studied. For the development of an ideal torrefaction protocol, Rodrigues et al. analyzed the effects of torrefaction undergoing normal conditions of temperature (265 °C) and residence time (15 min) in an N atmosphere and during a total 1 h 45 min heating period on a set of sixteen woody biomasses provenient from poplar short rotation coppice (SRC) and other Portuguese roundwoods [53].

5.3. The Current State of the Built Production Units

Of the more than 60 announced torrefaction initiatives and the 15 large-scale units scheduled to start by 2011, very few were built and hardly ever achieved a steady total industrial production and commercial status. The assumptions and expectations for the start-up were initially very high. Most equipment suppliers tend to overstate their capabilities and underestimate the time and effort required. Technological entrepreneurs with limited biomass experience have also encountered difficulties in the face of simple challenges, such as food, transport, storage, and quality of raw materials. Another issue is the relatively high total costs. "Drying and re-drying a little more", seems very simple, and attracted a huge group of serious entrepreneurs, but also the so-called "fortune hunters". However, torrefaction is a more complex process than initially anticipated.

Torrefaction must be conducted intelligently, with controlled costs, and is entirely directed towards the progress and success of the marketing. There are currently a number of challenges in systems and processes that require careful research and development (R&D) and intelligent solutions, such as: Processing and control of the atmosphere; gas production for inertization; heat transfer; control and moderation of exothermic reactions; cooling of products; behavior of the torrefaction gases, their deposition, and use; integration of systems and processes; energy optimization and exergetics; densification; and the optimization of the entire process supply chain. Perhaps, the most important part is the diagnosis and control of the process. Due to the close relationship between temperature and residence time with the product quality and standardization, careful control of these process variables is critical.

The material produced should be completely homogeneous, with respect to the degree of torrefaction, and preferably dark brown (not over-torrefied), to allow sufficient yield and to facilitate densification. There are few initiatives that are paving the way for the torrefaction industry. There are currently five industrial torrefaction units constructed and functioning in Europe [33] and at least eight units of torrefaction are planned and ready to begin functioning in the near future. These torrefaction industrial units are presented in Table 3.

Table 3. Torrefaction industrial units planned to begin operation in the near future.

Name of the Facility	Country	Capacity of Production in Tons Per Year
Biolake	The Netherlands	9000
Thermya	Spain	20,000
ECN/Andritz	The Netherlands	8–16,000
Fox Coal	The Netherlands	35,000
EBES/Andritz	Austria	8000
Bio Endev	Sweden	16,000
Rotawave	USA	100,000
Advanced Fuel Solutions	Portugal	96,000

6. Conclusions

The potential of biomass as a replacement to fossil fuels leads to a number of issues, such as the increase in their price and environmental consequences due to their excessive harvesting and plantation. This leads to an increase of studies that address this topic to improve the production of biomass while reducing its cost. The torrefaction of biomass has proved to be an ideal process for improving the biomass characteristics as this energy source has proved to be a good alternative to the use of fossil fuels.

Nowadays, the torrefaction of biomass is still an experimental technology, but due to the characteristics of the resulting products, it seems to be a promising technology that generates the curiosity and interest of the sector's investors. From this perspective, the potential development of biomass thermal conversion technologies, such as torrefaction and/or carbonization, is considered promising regarding the utilization of new forms of biomass, namely, the less environmentally friendly, more abundant, and faster growing forms, as is the case of shrub plants.

The main goal of this article was to present fresh information regarding the conventional torrefaction process, as well as to analyze the literature and study the innovative techniques that have been in development for the improvement of torrefied biomass qualities. The results show that the publications regarding this topic have been strongly increasing, which suggests a strong academic and industrial interest for this subject in the last few years. Several of these studies were analyzed and the future perspectives and research developments were consequently addressed.

However, some effects are less positive, since there is a risk that the widespread consumption of biomass could affect various ecosystems by creating negative impacts on the quality of water and soil and affecting food chains. Another potential consequence of biomass production is its overproduction, which leads to deforestation and a consequent decrease in biodiversity. These less positive aspects were also addressed.

Author Contributions: J.M.C.R. performed parts of the literature review. R.G. handled the writing and editing of the manuscript and contributed with parts of the literature review. J.C.d.O.M. and L.J.R.N. supervised, revised and corrected the manuscript.

Acknowledgments: The present work was supported by Fundação para a Ciência e Tecnologia (FCT), under the project UID/EMS/00151/2013 C-MAST, with reference POCI-01-0145-FEDER-007718. This research was also funded by the following project: Centro-01-0145-FEDER-000017-EMaDeS-Energy, Materials and Sustainable Development, co-financed by the Portugal 2020 Program (PT 2020), comprised by the Regional Operational Program of the Center (CENTRO 2020) and the European Union through the European Regional Development Fund (ERDF).

Conflicts of Interest: The authors declare no conflict of interest.

References

1. Chen, W.H.; Kuo, P.C. Torrefaction and co-torrefaction characterization of hemicellulose, cellulose and lignin as well as torrefaction of some basic constituents in biomass. *Energy* **2011**, *36*, 803–811. [CrossRef]
2. Van der Stelt, M.J.C.; Gerhauser, H.; Kiel, J.H.A.; Ptasinski, K.J. Biomass upgrading by torrefaction for the production of biofuels: A review. *Biomass Bioenergy* **2011**, *35*, 3748–3762. [CrossRef]
3. Vassilev, S.V.; Vassileva, C.G.; Vassilev, V.S. Advantages and disadvantages of composition and properties of biomass in comparison with coal: An overview. *Fuel* **2015**, *158*, 330–350. [CrossRef]
4. Elizondo, A.; Pérez-Cirera, V.; Strapasson, A.; Fernández, J.C.; Cruz-Cano, D. Mexico's low carbon futures: An integrated assessment for energy planning and climate change mitigation by 2050. *Futures* **2017**, *93*, 14–26. [CrossRef]
5. Meeus, L.; Azevedo, I.; Marcantonini, C.; Glachant, J.-M.; Hafner, M. EU 2050 Low-Carbon Energy Future: Visions and Strategies. *Electr. J.* **2012**, *25*, 57–63. [CrossRef]
6. Saidur, R.; Abdelaziz, E.A.; Demirbas, A.; Hossain, M.S.; Mekhilef, S. A review on biomass as a fuel for boilers. *Renew. Sustain. Energy Rev.* **2011**, *15*, 2262–2289. [CrossRef]
7. Fragkos, P.; Tasios, N.; Paroussos, L.; Capros, P.; Tsani, S. Energy system impacts and policy implications of the European Intended Nationally Determined Contribution and low-carbon pathway to 2050. *Energy Policy* **2017**, *100*, 216–226. [CrossRef]
8. Corradini, M.; Costantini, V.; Markandya, A.; Paglialunga, E.; Sforna, G. A dynamic assessment of instrument interaction and timing alternatives in the EU low-carbon policy mix design. *Energy Policy* **2018**, *120*, 73–84. [CrossRef]
9. Schanes, K.; Jäger, J.; Drummond, P. Three Scenario Narratives for a Resource-Efficient and Low-Carbon Europe in 2050. *Ecol. Econ.* **2018**. [CrossRef]
10. Schirone, L.; Pellitteri, F. Energy Policies and Sustainable Management of Energy Sources. *Sustainability* **2017**, *9*, 2321. [CrossRef]

11. Bollino, C.A.; Asdrubali, F.; Polinori, P.; Bigerna, S.; Micheli, S.; Guattari, C.; Rotili, A. A Note on Medium- and Long-Term Global Energy Prospects and Scenarios. *Sustainability* **2017**, *9*, 833. [CrossRef]

12. Sung, B.; Park, S.-D. Who Drives the Transition to a Renewable-Energy Economy? Multi-Actor Perspective on Social Innovation. *Sustainability* **2018**, *10*, 448. [CrossRef]

13. Ntanos, S.; Kyriakopoulos, G.; Chalikias, M.; Arabatzis, G.; Skordoulis, M. Public Perceptions and Willingness to Pay for Renewable Energy: A Case Study from Greece. *Sustainability* **2018**, *10*, 687. [CrossRef]

14. Jiang, Y.; van der Werf, E.; van Ierland, E.C.; Keesman, K.J. The potential role of waste biomass in the future urban electricity system. *Biomass Bioenergy* **2017**, *107*, 182–190. [CrossRef]

15. Muench, S. Greenhouse gas mitigation potential of electricity from biomass. *J. Clean. Prod.* **2015**, *103*, 483–490. [CrossRef]

16. Chen, W.; Peng, J.; Bi, X.T. A state-of-the-art review of biomass torrefaction, densification and applications. *Renew. Sustain. Energy Rev.* **2015**, *44*, 847–866. [CrossRef]

17. Andreoli Bonazzi, F.; Cividino, S.R.S.; Zambon, I.; Mosconi, E.M.; Poponi, S. Building Energy Opportunity with a Supply Chain Based on the Local Fuel-Producing Capacity. *Sustainability* **2018**, *10*, 2140. [CrossRef]

18. Dietrich, R.-U.; Albrecht, F.G.; Maier, S.; König, D.H.; Estelmann, S.; Adelung, S.; Bealu, Z.; Seitz, A. Cost calculations for three different approaches of biofuel production using biomass, electricity and CO_2. *Biomass Bioenergy* **2018**, *111*, 165–173. [CrossRef]

19. Proskurina, S.; Heinimö, J.; Schipfer, F.; Vakkilainen, E. Biomass for industrial applications: The role of torrefaction. *Renew. Energy* **2017**, *111*, 265–274. [CrossRef]

20. Lu, K.M.; Lee, W.J.; Chen, W.H.; Liu, S.H.; Lin, T.C. Torrefaction and low temperature carbonization of oil palm fiber and eucalyptus in nitrogen and air atmospheres. *Bioresour. Technol.* **2012**, *123*, 98–105. [CrossRef] [PubMed]

21. Joshi, Y.; De Vries, H.; Woudstra, T.; De Jong, W. Torrefaction: Unit operation modelling and process simulation. *Appl. Therm. Eng.* **2015**, *74*, 83–88. [CrossRef]

22. Shankar Tumuluru, J.; Sokhansanj, S.; Hess, J.R.; Wright, C.T.; Boardman, R.D. REVIEW: A review on biomass torrefaction process and product properties for energy applications. *Ind. Biotechnol.* **2011**, *7*, 384–401. [CrossRef]

23. Singh, R.; Krishna, B.B.; Kumar, J.; Bhaskar, T. Opportunities for utilization of non-conventional energy sources for biomass pretreatment. *Bioresour. Technol.* **2016**, *199*, 398–407. [CrossRef] [PubMed]

24. Nunes, L.J.R.; Matias, J.C.O.; Catalão, J.P.S. A review on torrefied biomass pellets as a sustainable alternative to coal in power generation. *Renew. Sustain. Energy Rev.* **2014**, *40*, 153–160. [CrossRef]

25. Bergman, P.C.A.; Kiel, J.H.A. Torrefaction for Biomass Upgrading. In Proceedings of the 14th European Biomass Conference, Paris, France, 17–21 October 2005; pp. 17–21.

26. Chen, Q.; Zhou, J.S.; Liu, B.J.; Mei, Q.F.; Luo, Z.Y. Influence of torrefaction pretreatment on biomass gasification technology. *Chin. Sci. Bull.* **2011**, *56*, 1449–1456. [CrossRef]

27. Xin, S.; Mi, T.; Liu, X.; Huang, F. Effect of torrefaction on the pyrolysis characteristics of high moisture herbaceous residues. *Energy* **2018**, *152*, 586–593. [CrossRef]

28. Acharya, B.; Sule, I.; Dutta, A. A review on advances of torrefaction technologies for biomass processing. *Biomass Convers. Biorefinery* **2012**, *2*, 349–369. [CrossRef]

29. Dhungana, A.; Basu, P.; Dutta, A. Effects of Reactor Design on the Torrefaction of Biomass. *J. Energy Resour. Technol.* **2012**, *134*, 41801. [CrossRef]

30. Tran, K.Q.; Luo, X.; Seisenbaeva, G.; Jirjis, R. Stump torrefaction for bioenergy application. *Appl. Energy* **2013**, *112*, 539–546. [CrossRef]

31. Basu, P. Chapter 4—Torrefaction. In *Biomass Gasification, Pyrolysis and Torrefaction*, 3rd ed.; Basu, P., Ed.; Academic Press: Cambridge, MA, USA, 2018; pp. 93–154, ISBN 978-0-12-812992-0.

32. Da Silva, C.M.S.; Carneiro, A.D.C.O.; Vital, B.R.; Figueiró, C.G.; de Freitas Fialho, L.; de Magalhães, M.A.; Carvalho, A.G.; Cândido, W.L. Biomass torrefaction for energy purposes—Definitions and an overview of challenges and opportunities in Brazil. *Renew. Sustain. Energy Rev.* **2018**, *82*, 2426–2432. [CrossRef]

33. Nunes, L.J.; Matias, J.C.O.; Catalão, J.P.S. *Torrefaction of Biomass for Energy Applications: From Fundamentals to Industrial Scale*, 1st ed.; Academic Press: Cambridge, MA, USA, 2017; ISBN 978-0-12-809462-4.

34. Chew, J.J.; Doshi, V. Recent advances in biomass pretreatment—Torrefaction fundamentals and technology. *Renew. Sustain. Energy Rev.* **2011**, *15*, 4212–4222. [CrossRef]

35. Singh, K.; Zondlo, J. Characterization of fuel properties for coal and torrefied biomass mixtures. *J. Energy Inst.* **2017**, *90*, 505–512. [CrossRef]

36. García, R.; Pizarro, C.; Lavín, A.G.; Bueno, J.L. Biomass sources for thermal conversion. Techno-economical overview. *Fuel* **2017**, *195*, 182–189. [CrossRef]

37. Acharya, B.; Dutta, A.; Minaret, J. Review on comparative study of dry and wet torrefaction. *Sustain. Energy Technol. Assess.* **2015**, *12*, 26–37. [CrossRef]

38. Bergman, P.C.A.; Boersma, A.R.; Zwart, R.W.R.; Kiel, J.H.A. *Torrefaction for Biomass Co-Firing in Existing Coal-Fired Power Stations*; Report No. ECNC05013; Energy Research Centre of The Netherlands (ECN): Petten, The Netherlands, 2005; p. 71.

39. Isa, K.M.; Daud, S.; Hamidin, N.; Ismail, K.; Saad, S.A.; Kasim, F.H. Thermogravimetric analysis and the optimisation of bio-oil yield from fixed-bed pyrolysis of rice husk using response surface methodology (RSM). *Ind. Crops Prod.* **2011**, *33*, 481–487. [CrossRef]

40. Esteves, B.; Marques, A.V.; Domingos, I.; Pereira, H. Influence of steam heating on the properties of pine (*Pinus pinaster*) and eucalypt (*Eucalyptus globulus*) wood. *Wood Sci. Technol.* **2007**, *41*, 193–207. [CrossRef]

41. Lipinsky, E.S.; Arcate, J.R.; Reed, T.B. Enhanced wood fuels via torrefaction. *ACS Div. Fuel Chem. Prepr.* **2002**, *47*, 408–409.

42. Prins, M.J.; Ptasinski, K.J.; Janssen, F.J.J.G. Torrefaction of wood. Part 1. Weight loss kinetics. *J. Anal. Appl. Pyrolysis* **2006**, *77*, 28–34. [CrossRef]

43. Prins, M.J.; Ptasinski, K.J.; Janssen, F.J.J.G. Torrefaction of wood. Part 2. Analysis of products. *J. Anal. Appl. Pyrolysis* **2006**, *77*, 35–40. [CrossRef]

44. Nunes, L.J.; Matias, J.C.; Catalão, J.P. Torrefaction of Biomass for Energy Applications, 1st Edition. Available online: https://www.elsevier.com/books/torrefaction-of-biomass-for-energy-applications/nunes/978-0-12-809462-4 (accessed on 4 July 2018).

45. Strezov, V.; Evans, T.J.; Hayman, C. Thermal conversion of elephant grass (*Pennisetum Purpureum Schum*) to bio-gas, bio-oil and charcoal. *Bioresour. Technol.* **2008**, *99*, 8394–8399. [CrossRef] [PubMed]

46. Kumar, G.; Panda, A.K.; Singh, R. Optimization of process for the production of bio-oil from eucalyptus wood. *J. Fuel Chem. Technol.* **2010**, *38*, 162–167. [CrossRef]

47. Medic, D.; Darr, M.; Shah, A.; Potter, B.; Zimmerman, J. Effects of torrefaction process parameters on biomass feedstock upgrading. *Fuel* **2012**, *91*, 147–154. [CrossRef]

48. Neves, D.; Thunman, H.; Matos, A.; Tarelho, L.; Gómez-Barea, A. Characterization and prediction of biomass pyrolysis products. *Prog. Energy Combust. Sci.* **2011**, *37*, 611–630. [CrossRef]

49. Rousset, P.; Aguiar, C.; Volle, G.; Anacleto, J.; De Souza, M. Torrefaction of Babassu: A potential utilization pathway. *BioResources* **2013**, *8*, 358–370. [CrossRef]

50. Pecha, B.; Garcia-Perez, M. *Bioenergy*; Elsevier: New York, NY, USA, 2015; ISBN 978-0-12-407909-0.

51. Wang, M.J.; Huang, Y.F.; Chiueh, P.T.; Kuan, W.H.; Lo, S.L. Microwave-induced torrefaction of rice husk and sugarcane residues. *Energy* **2012**, *37*, 177–184. [CrossRef]

52. Thorn, M.; Bennett, A.; Griend, S.V. Rotary Torrefaction Reactor. U.S. Patent 9150790B2, 3 November 2011.

53. Rodrigues, A.; Loureiro, L.; Nunes, L.J.R. Torrefaction of woody biomasses from poplar SRC and Portuguese roundwood: Properties of torrefied products. *Biomass Bioenergy* **2018**, *108*, 55–65. [CrossRef]

54. Li, H.; Liu, X.; Legros, R.; Bi, X.T.; Lim, C.J.; Sokhansanj, S. Torrefaction of sawdust in a fluidized bed reactor. *Bioresour. Technol.* **2012**, *103*, 453–458. [CrossRef] [PubMed]

55. Narayanasamy, L.; Murugesan, T. Degradation of Alizarin Yellow R using UV/H_2O_2 Advanced Oxidation Process. *Environ. Sci. Technol.* **2014**, *33*, 482–489.

56. Pang, S.; Mujumdar, A.S. Drying of woody biomass for bioenergy: Drying technologies and optimization for an integrated bioenergy plant. *Dry. Technol.* **2010**, *28*, 690–701. [CrossRef]

57. Arias, B.; Pevida, C.; Fermoso, J.; Plaza, M.G.; Rubiera, F.; Pis, J.J. Influence of torrefaction on the grindability and reactivity of woody biomass. *Fuel Process. Technol.* **2008**, *89*, 169–175. [CrossRef]

58. Phanphanich, M.; Mani, S. Impact of torrefaction on the grindability and fuel characteristics of forest biomass. *Bioresour. Technol.* **2011**, *102*, 1246–1253. [CrossRef] [PubMed]

59. Uslu, A.; Faaij, A.P.C.; Bergman, P.C.A. Pre-treatment technologies, and their effect on international bioenergy supply chain logistics. Techno-economic evaluation of torrefaction, fast pyrolysis and pelletisation. *Energy* **2008**, *33*, 1206–1223. [CrossRef]

60. Shen, D.K.; Gu, S.; Luo, K.H.; Bridgwater, A.V.; Fang, M.X. Kinetic study on thermal decomposition of woods in oxidative environment. *Fuel* **2009**, *88*, 1024–1030. [CrossRef]

61. Batidzirai, B.; Mignot, A.P.R.; Schakel, W.B.; Junginger, H.M.; Faaij, A.P.C. Biomass torrefaction technology: Techno-economic status and future prospects. *Energy* **2013**, *62*, 196–214. [CrossRef]

62. Agar, D.; Wihersaari, M. Bio-coal, torrefied lignocellulosic resources—Key properties for its use in co-firing with fossil coal—Their status. *Biomass Bioenergy* **2012**, *44*, 107–111. [CrossRef]

63. De Siqueira Ferreira, S.; Nishiyama, M.Y.; Paterson, A.H.; Souza, G.M. Biofuel and energy crops: High-yield Saccharinae take center stage in the post-genomics era. *Genome Biol.* **2013**, *14*. [CrossRef] [PubMed]

64. Mola-Yudego, B.; Dimitriou, I.; Gonzalez-Garcia, S.; Gritten, D.; Aronsson, P. A conceptual framework for the introduction of energy crops. *Renew. Energy* **2014**, *72*, 29–38. [CrossRef]

65. Milbrandt, A.R.; Heimiller, D.M.; Perry, A.D.; Field, C.B. Renewable energy potential on marginal lands in the United States. *Renew. Sustain. Energy Rev.* **2014**, *29*, 473–481. [CrossRef]

66. Radics, R.I.; Gonzalez, R.; Bilek, E.M.; Kelley, S.S. Systematic review of torrefied wood economics. *BioResources* **2017**, *12*, 6868–6884. [CrossRef]

67. Thiel, F.C. New or Improved Roaster or Torrefier for Coffee and other Vegetable Substances. 1898. Available online: https://patents.google.com/patent/GB189710658A/en?q=New&q=Improved&q=Roaster&q=Torrefier&q=Coffee&q=Vegetable&q=Substances&oq=New+or+Improved+Roaster+or+Torrefier+for+Coffee+and+other+Vegetable+Substances (accessed on 4 July 2018).

68. Offrion, V.F.O. Improvements in the Process of and Apparatus for Rationally and Continuously Treating or Torrefying Coffee. 1900. Available online: https://patents.google.com/patent/US20150068113A1/en?q=Improvements&q=Process&q=Apparatus&q=Rationally&q=Continuously&q=Treating&q=Torrefying&q=Coffee&oq=Improvements+in+the+Process+of+and+Apparatus+for+Rationally+and+Continuously+Treating+or+Torrefying+Coffee (accessed on 4 July 2018).

69. Schwob, Y. Fuel Pellets or Briquettes of High Heating Value mfd. from Wood—By Baking Dry, Grinding, opt. Adding oil, and Pressing. 1983. Available online: https://patents.google.com/patent/FR2525231A1/en?q=Fuel+Pellets&q=Briquettes&q=High&q=Heating&q=Value&q=mfd.&q=Wood&oq=Fuel+Pellets+or+Briquettes+of+High+Heating+Value+mfd.+from+Wood (accessed on 4 July 2018).

70. Bourgeois, J.P.; Doat, J. Torrefied Wood from Temperate and Tropical Species. Advantages and Prospects. In *Bioenergy 84. Proceedings of Conference 15–21 June 1984, Goteborg, Sweden. Volume III. Biomass Conversion*; Elsevier: New York, NY, USA, 1984; pp. 153–159.

71. Bourgois, J.; Guyonnet, R. Characterization and analysis of torrefied wood. *Wood Sci. Technol.* **1988**, *22*, 143–155. [CrossRef]

72. Energy Research Centre of the Netherlands (ECN). Available online: http://www.aebiom.org/jwdmembers/energieonderzoek-centrum-nederland-ecn/ (accessed on 4 July 2018).

73. Pentananunt, R.; Rahman, A.N.M.M.; Bhattacharya, S.C. Upgrading of biomass by means of torrefaction. *Energy* **1990**, *15*, 1175–1179. [CrossRef]

74. Chen, W.-H.; Wang, C.-W.; Kumar, G.; Rousset, P.; Hsieh, T.-H. Effect of torrefaction pretreatment on the pyrolysis of rubber wood sawdust analyzed by Py-GC/MS. *Bioresour. Technol.* **2018**, *259*, 469–473. [CrossRef] [PubMed]

75. Chen, D.; Gao, A.; Cen, K.; Zhang, J.; Cao, X.; Ma, Z. Investigation of biomass torrefaction based on three major components: Hemicellulose, cellulose, and lignin. *Energy Convers. Manag.* **2018**, *169*, 228–237. [CrossRef]

76. Álvarez, A.; Nogueiro, D.; Pizarro, C.; Matos, M.; Bueno, J.L. Non-oxidative torrefaction of biomass to enhance its fuel properties. *Energy* **2018**, *158*, 1–8. [CrossRef]

77. Ho, S.-H.; Zhang, C.; Chen, W.-H.; Shen, Y.; Chang, J.-S. Characterization of biomass waste torrefaction under conventional and microwave heating. *Bioresour. Technol.* **2018**, *264*, 7–16. [CrossRef] [PubMed]

78. Gul, S.; Ramzan, N.; Hanif, M.A.; Bano, S. Kinetic, volatile release modeling and optimization of torrefaction. *J. Anal. Appl. Pyrolysis* **2017**, *128*, 44–53. [CrossRef]

79. Grigiante, M.; Antolini, D. Experimental results of mass and energy yield referred to different torrefaction pathways. *Waste Biomass Valorization* **2014**, *5*, 11–17. [CrossRef]

80. Basu, P.; Kulshreshtha, A.; Acharya, B. An Index for Quantifying the Degree of Torrefaction. *BioResources* **2017**, *12*, 1749–1766. [CrossRef]

81. Almeida, G.; Brito, J.O.; Perré, P. Alterations in energy properties of eucalyptus wood and bark subjected to torrefaction: The potential of mass loss as a synthetic indicator. *Bioresour. Technol.* **2010**, *101*, 9778–9784. [CrossRef] [PubMed]
82. Peng, J.H.; Bi, X.T.; Sokhansanj, S.; Lim, C.J. Torrefaction and densification of different species of softwood residues. *Fuel* **2013**, *111*, 411–421. [CrossRef]
83. Rousset, P.; Aguiar, C.; Labbé, N.; Commandré, J.M. Enhancing the combustible properties of bamboo by torrefaction. *Bioresour. Technol.* **2011**, *102*, 8225–8231. [CrossRef] [PubMed]
84. Poudel, J.; Ohm, T.-I.; Gu, J.H.; Shin, M.C.; Oh, S.C. Comparative study of torrefaction of empty fruit bunches and palm kernel shell. *J. Mater. Cycles Waste Manag.* **2017**, *19*, 917–927. [CrossRef]
85. Conag, A.T.; Villahermosa, J.E.R.; Cabatingan, L.K.; Go, A.W. Energy densification of sugarcane bagasse through torrefaction under minimized oxidative atmosphere. *J. Environ. Chem. Eng.* **2017**, *5*, 5411–5419. [CrossRef]
86. Phusunti, N.; Phetwarotai, W.; Tekasakul, S. Effects of torrefaction on physical properties, chemical composition and reactivity of microalgae. *Korean J. Chem. Eng.* **2018**, *35*, 503–510. [CrossRef]
87. Butlewski, K.; Golimowski, W.; Gracz, W.; Marcinkowski, D.; Waliński, M.; Podleski, J. Torrefaction of the Black Lilac (*Sambucus nigra* L.) as an Example of Biocoal Production from Garden Maintenance Waste. In *Renewable Energy Sources: Engineering, Technology, Innovation*; Springer Proceedings in Energy; Springer: Cham, Switzerland, 2018; pp. 345–356, ISBN 978-3-31-972370-9.
88. Natarajan, P.; Suriapparao, D.V.; Vinu, R. Microwave torrefaction of *Prosopis juliflora*: Experimental and modeling study. *Fuel Process. Technol.* **2018**, *172*, 86–96. [CrossRef]
89. Li, S.-X.; Chen, C.-Z.; Li, M.-F.; Xiao, X. Torrefaction of corncob to produce charcoal under nitrogen and carbon dioxide atmospheres. *Bioresour. Technol.* **2018**, *249*, 348–353. [CrossRef] [PubMed]
90. Chiou, B.-S.; Cao, T.; Valenzuela-Medina, D.; Bilbao-Sainz, C.; Avena-Bustillos, R.J.; Milczarek, R.R.; Du, W.-X.; Glenn, G.M.; Orts, W.J. Torrefaction kinetics of almond and walnut shells. *J. Therm. Anal. Calorim.* **2018**, *131*, 3065–3075. [CrossRef]
91. Zhang, S.; Su, Y.; Xu, D.; Zhu, S.; Zhang, H.; Liu, X. Assessment of hydrothermal carbonization and coupling washing with torrefaction of bamboo sawdust for biofuels production. *Bioresour. Technol.* **2018**, *258*, 111–118. [CrossRef] [PubMed]
92. Zhang, S.; Su, Y.; Ding, K.; Zhu, S.; Zhang, H.; Liu, X.; Xiong, Y. Effect of inorganic species on torrefaction process and product properties of rice husk. *Bioresour. Technol.* **2018**, *265*, 450–455. [CrossRef] [PubMed]
93. Zeng, K.; Yang, Q.; Zhang, Y.; Mei, Y.; Wang, X.; Yang, H.; Shao, J.; Li, J.; Chen, H. Influence of torrefaction with Mg-based additives on the pyrolysis of cotton stalk. *Bioresour. Technol.* **2018**, *261*, 62–69. [CrossRef] [PubMed]
94. Christoforou, E.A.; Fokaides, P.A. Recent Advancements in Torrefaction of Solid Biomass. *Curr. Sustain. Energy Rep.* **2018**, *5*, 163–171. [CrossRef]
95. Brachi, P.; Miccio, F.; Miccio, M.; Ruoppolo, G. Torrefaction of Tomato Peel Residues in a Fluidized Bed of Inert Particles and a Fixed-Bed Reactor. *Energy Fuels* **2016**, *30*, 4858–4868. [CrossRef]

Article

Evaluation of the Physical, Chemical and Thermal Properties of Portuguese Maritime Pine Biomass

Helder Filipe dos Santos Viana [1,2,*], Abel Martins Rodrigues [3,4], Radu Godina [5], João Carlos de Oliveira Matias [6,7] and Leonel Jorge Ribeiro Nunes [6,7]

[1]   Centre for the Research and Technology of Agro-Environmental and Biological Sciences, CITAB, University of Trás-os-Montes and Alto Douro, UTAD, 5001-801 Vila Real, Portugaly

[2]   Agrarian Superior School, Polytechnic Institute of Viseu, 3504-510 Viseu, Portugal

[3]   Departamento de Tecnologia e Inovação, INIAV—Instituto Nacional de Investigação Agrícola e Veterinária, 2780-157 Oeiras, Portugal; abel.rodrigues@iniav.pt

[4]   Maretec Research Centre, Instituto Superior Técnico, Av. Rovisco Pais, 1049-001 Lisboa, Portugal

[5]   C-MAST—Centre for Aerospace Science and Technologies—Department of Electromechanical Engineering, University of Beira Interior, 6201-001 Covilhã, Portugal; rd@ubi.pt

[6]   DEGEIT—Department of Economics, Management, Industrial Engineering and Tourism, University of Aveiro, 3810-193 Aveiro, Portugal; jmatias@ua.pt (J.C.d.O.M.); leonelnunes@ua.pt (L.J.R.N.)

[7]   GOVCOPP—Research Unit on Governance, Competitiveness and Public Policies, University of Aveiro, 3810-193 Aveiro, Portugal

*   Correspondence: hviana@esav.ipv.pt; Tel.: +351-232-446-600

Received: 30 July 2018; Accepted: 9 August 2018; Published: 13 August 2018

Abstract: A characterisation of *Pinus pinaster Aiton.* (Maritime Pine) woody biomass and ashes is presented in this study. Physical, thermal and chemical analysis, including density, moisture content, calorific value, proximate and ultimate analysis, were carried out. The fuel Energy Density (E_d) and the Fuelwood Value Index (FVI) were assessed by ranking the fuelwood quality. Furthermore, the determination of the ash metal elementals was performed. The results from this study indicated, for *Pinus pinaster* biomass tree components, carbon content ranging from 46.5 to 49.3%, nitrogen content from 0.13 to 1.18%, sulphur content from 0.056 to 0.148% and hydrogen content around 6–7%. The ash content in the tree components ranged from 0.22 to 1.92%. The average higher heating value (HHV) was higher for pine needles (21.61 MJ·kg^{-1}). The E_d of 8.9 GJ·m^{-3} confirm the good potential of *Pinus pinaster* biomass tree components as fuel. The FVI ranked the wood stem (4658) and top (2861.8) as a better fuelwood and pine needles (394.2) as inferior quality. The chemical composition of the ashes revealed that the elemental contents are below the national and most European countries legislation guidelines for the employment of ash as a fertiliser.

Keywords: biomass; *Pinus pinaster*; fuel; heating value; fuelwood value index; energy density; ash recovery

---

1. Introduction

Maritime pine (*Pinus pinaster Aiton.*) is the second species in terms of occupied area in Portugal, accounting for around 23% of forest area, occurring mostly in the north and central regions of Portugal [1]. An estimation made in 2010 for NUTS II (Nomenclature of Territorial Units for Statistics) reported a yearly average of 579.9 thousand dry tons, ranging between 400.96 thousand dry tons per year and 673.54 thousand dry tons per year [2].

In 2017, forest fires were particularly severe, with a forest area of about 500,000 ha burned in the country and, presently, a large amount of biomass supply is available for conversion [3–5].

The energy acquired from biomass is distinguished and remarkable in that it can be obtained without difficulty and is also renewable [6]. It promotes the protection of the environment, since biomass is abundant, natural, and reusable [7]. The chemical composition and physical properties of biomass ashes and biomass components, wherein the latter is strictly linked to its anatomic polymeric structure, are determinant for the performance of thermochemical conversion of the feedstocks [8–12]. Biomass shows highly variable chemical composition and properties regarding moisture, structural and inorganic components, which is related to the myriad plant growth processes and growing conditions [13]. The main and less important elements of biomass, in diminishing order of quantity, are normally the following: *C, O, H, N, Ca, K, Si, Mg, Al, S, Fe, P, Cl* and *Na*, in addition to *Mn, Ti* and other trace elements. Typical proximate analysis of maritime pine components shows values ranging between 65% to 72% for volatiles, 12.5% to 21% for fixed carbon, and 0.2% to 2.6% for ash content. Typical values for ultimate analysis of *C, H, N*, and *O* range between 46% and 56%, 5% and 6%, 0.1% and 0.9%, and 31% and 37%, respectively. Sulphur amounts are negligible. Finally, the High Heating Value (HHV) ranges from 18 MJ·kg^{-1} to 17 MJ·kg^{-1} [14].

As a comparison, for coal, the carbon content is, on average, about 80% or higher, the oxygen about 8% or lower, and the *N* and *S* are about 1.8% and 0.8% or higher, respectively. Fixed carbon and volatile amounts for coal are of the order of 50% and 44%, respectively. HHV and LHV for coal are of 35 MJ·kg^{-1} and 33 MJ·kg^{-1}, respectively. This chemical profile fits the overall picture that woody biomasses posit advantages over fossil fuels such as coal related to carbon neutrality and reduced emissions of SO_2 and NO_x [13]. However, handicaps of woody biomass, such as lower energy density and calorific values, hygroscopic properties, and high moisture content leading to degradation and self-heating, make the handling and transportation costlier and more complex.

Torrefaction of the feedstock is an option for minimising these drawbacks and making the chemical profile of torrefied biomass closer to that of coal. Biomass torrefaction is a thermochemical pre-treatment, carried out in the absence of oxygen at temperature ranges of about 220 °C to 320 °C, delivering a product with lower O/C and H/C ratios, closer to typical coal ratios [15]. The torrefied products are more homogeneous and show higher energy density and grindability [16]. Also, the similarity of the torrefied pulverised product to coal powder makes possible its co-firing, at amounts as high as 40%, with coal in power plants [17]. Experimental evidence with torrefaction of maritime pine carried out at a lab scale made it possible to obtain a torrefied product with a LHV of 22.6 MJ·kg^{-1}, corresponding to a gain of about 20%, a C amount of 65% (15% gain), a fixed carbon amount of 30% (35% gain) and 29% for oxygen amount, corresponding to a decrease of 21% from raw biomass [18]. Thereby, the ratio of O/C decreased by 32% for torrefied maritime pine biomass. The ratio of torrefied pine biomass also decreased by 18%. The LHV in conifers is on average 2% higher than in hardwoods. This difference occurs due to a higher lignin content, and sometimes due to the higher resin, oil and wax content present in conifers [18]. Indeed, while LHV for cellulose ranges between 17.2 MJ·kg^{-1} and 17.5 MJ·kg^{-1} and is about 16 MJ·kg^{-1} for hemicelluloses, in lignin, LHV is higher, ranging between 26 MJ·kg^{-1} and 27 MJ·kg^{-1}. A slight variability in the calorific value can be witnessed and occurs due to certain variability in the elementary *H* content. It also occurs as a result of a much larger variability in ash contents [19].

Low heating value is a variable which allows evaluating the enthalpy released in forest fires through the combustion of fuels available in the field surface, thereby providing additional information about the easiness of fire propagation. The LHV of the different forest canopy components is thereby one valuable indicator of the energetic status of forest biomass which helps to optimise the management of the energetic forest resources in the field. In [20], data is presented for the seasonal variability of biomass flammability and the heating value of maritime pine woody biomass components in NW Spain.

The Fuelwood Value Index (FVI), expressed in MJ·m^{-3} and defined as the ratio between the products between the calorific power and density and between the ash and moisture contents, is another relevant variable for ranking the fuel aptitude of woody biomasses for different species [21].

Energy density (E$_a$), expressed in GJ·m^{-3}, defined as the product between LHV and bulk density, is another variable that can be assessed in order to identify the potential and to choose a fuel for use in small-scale heating plants and households. The energy density is relevant for woody fuels, because its storage and transport could be cheaper and more efficient at a higher energy density.

Another factor that could be taken into consideration is the evaluation of biomass ashes, concerning either its chemical composition as the mechanisms and quantities produced in the thermo-chemical conversions. Biomass ashes are made up of *Cl* and *S*, with major elements (*Al, Ca, Fe, K, Mg, Na, P, Ti, Si*) and minor or trace elements (*As, Ba, Cd, Co, Cr, Cu, Hg, Mn, Mo, Ni, Sb, Pb, Tl, V, Zn*). In comparison with coal, biomass usually shows higher amounts, in decreasing order, of *Mn, K, P, Cl, Ca, Mg, Na, O* and *H*, and lower amounts of ash, *Al, C, Fe, N, S, Si*, and *Ti*.

Elements included in biomass ashes can be divided in two classes: intrinsic or inherent chemicals bonded to a carbon structure, and entrained elements that can come with biomass as mineral soil particles that have been incorporated into biomass during plant growth, or taken away during harvesting and transport [22]. These elements are relevant for aerosol emissions, ash melting, deposit formation, fly ash and corrosion, and also for the utilisation/disposal of the ashes.

During combustion, some volatilisation of ash-forming compounds occurs, with the volatised fraction depending on factors such as the technology employed, chemical composition of the fuel, operative temperature, and surrounding gas atmosphere. Indeed, volatilisation of heavy metals such as *Pb, Cd* and *Zn* can happen under high operative temperatures and low oxidising atmospheres. Toxicity due to heavy metals (*Cu, Zn, Cr, Cd, Mn, Ni* or *Pb*) can cause serious consequences in soils located in sites corresponding to former agricultural areas with excessive use of fertilisers, fungicides and insecticides. Dangerous trace elements in the form of salts are highly mobile on the ashes, but issues related to their accessibility and availability still lack some fundamental knowledge. For example, these trace elements may be water soluble or bound into glass, which is formed from the fusion of inorganic material in biomass [14].

Some toxic elements such as *As, Cd, Cr, Pb, Cu, Mn, Fe, Ni*, and *Zn* are described in the literature with concentrations as high as 243 ppm, 657 ppm, 0.17%, 5%, 0.24%, 4.7%, 25 ppm, 0.5 ppm and 16.4% in fly ashes, respectively (e.g., [23]). Plant growth and biomass can thereby be drastically affected by release of these elements.

Typical ranges in ash element composition of biomass tree components (wood, bark, leaves, tops or branches) for *Fe, Al, As, Cd, Pb, Co, Cu, Cr, Mn, Zn*, and *Ni* are 3000–40,000, 4700–74,000, 3–60, 0–25, 15–650, <1–20, 15–400, 10–250, 1000–30,000, 15–4400 and 6–200 mg·kg^{-1}, respectively [14]. In industrial areas, nickel toxicity is particularly acute, at high soil concentrations above a threshold of 200 ppm with drastic physiological consequences. Excess *Cd* exerts similar effects in plants, and damage to root cells has been reported on grey poplar. Plant growth, biomass production, shoots and root lengths has been ascribed to arsenic contamination in soils as a result of the widespread use of arsenical pesticides. Thereby, heavy metals in biomass ash, e.g., *Zn, Cu, Cd, Cr* and *Pb* above certain limits, are surely toxic to plants, precluding their use as fertiliser, insofar as its application to soil can constitute a potential source of contamination for aquatic and terrestrial ecosystems [24,25].

Under this negative environmental context, there are strict regulations in some countries, e.g., in North and Central Europe, specifying thresholds on amounts of elements such as *Co, Cr, K, N, Cu, V, Zn, Ca, Cd, Cl, Ni, S* or *Pb* in biomass fuels and ashes, considering all possible forms of conversion [26,27].

The assessment of ash use in forestry and agriculture differs between European states, to some extent due to different conditions [28]. While in some countries, mainly the Nordic countries of Europe, the use of ashes is a key factor for the replenishment of nutrients to acidic and poor soils; in other countries, such problems are not so deep and are therefore less worrisome. On the other hand, the logging of forest residues for energy production is made in a superior extent, extracting a large amount of nutrients in biomass. For instance, in Sweden, the recycling of ash is deemed to be a significant measure for forestry sustainability, while in Finland, the ash is considered to be a fertiliser and is used to increase the growth of forests growing on peatlands. As another example, in Denmark, ash recycling is deemed to be a method to make up for the loss of phosphorous and potassium [29]. As a consequence, several producers of wood in Northern America and Europe are beginning to recycle ash on an operational scale. For instance, in Finland, over 10% of wood and bark ash created by forest industry returns to the forest [30], as the wood ash is dispersed at high rates in many forests (in Finland 3–5 t·ha^{-1} and in Sweden 1–3 t·ha^{-1}).

The present work, under all of the above context, aims to evaluate the chemical composition and physical properties of the maritime pine biomass, considering proximate, ultimate and ash analysis in the tree components stem, tops, branches and needles, sampled from 16 plots in the Northern Portuguese county of Viseu.

2. Materials and Methods

2.1. Stand Measurements and Biomass Samples for Fuel and Ash Analysis

The biomass samples for fuel and ash analysis were collected from 16 maritime pine stands, located in the Viseu country in Northern-Central Portugal, where maritime pine mainly occurs.

The forest stands were measured and the structural stand variables such as number of trees per hectare (N), dominant tree height class (h_{dom}), basal area (G), stand age (t), crown closure class (Cc) and site index (SI) were calculated, and the aboveground biomass quantified by destructive approach.

In each field plot, a biomass sample of each tree component (wood stem, top, branches and leaves) was collected and the sampled mixed biomass gave a typical tree sample per plot, totalling thereby sixteen typical trees, a concept useful for heat and mass balance evaluations in a burned forest area of maritime pine such as that. The collected samples were positioned inside hermetically closed containers with the intention of avoiding the loss of moisture, and were then sent to the laboratory.

The preparations and homogenisation of the samples, previous to the analytical measurements, were carried out by following the technical specifications of the CEN/TS 14780:2005 standard for sample preparations. The results on physical properties, proximate, and ultimate analysis were with respect to a typical tree per plot, according to the biomass sampling mentioned above.

2.2. Analytical Measurements

2.2.1. Physical Properties

The moisture content, as wet basis, was assessed according to the European Standard CEN EN 14774-1:2009, which consists of the drying of the sample in a drying oven at a temperature within the range of 105 ± 2 °C in atmospheric air until a constant mass is reached. The percentage of moisture is assessed from the mass loss of the sample. Density (D) and basic density on dry basis (D_b) of tree components was assessed by using the technique of water displacement and is presented as the weight per unit volume [31].

2.2.2. Proximate Analysis

The assessment of the ash content (in the case of the dry basis) was performed according to CEN EN 14775:2009 standard. The ash content (%) was calculated from the mass of the residue remaining after 1g of oven-dried sample inside a platinum crucible was heated in a muffle furnace at 550 °C ± 10 °C.

The volatile matter content (dry basis) was determined according to CEN EN 15148:2009, by burning 1 g of oven-dried sample in a fused silica crucible with lid on and heated in a muffle furnace at 900 °C ± 10 °C for 7 min. The overall percentage of the volatile matter was assessed from the mass loss of the test sample. The fixed carbon content (%) was assessed as the difference between the totality of volatile matter and ash contents from 100%. A minimum of 3 test runs were performed for every analysis, and the average of the two closest results (less than 5% variation) was taken as the measured value.

2.2.3. Ultimate Analysis

The elemental composition of the biomass with respect to carbon (C), hydrogen (H) and nitrogen (N) and sulphur (S) was measured by the standard methods of analysis. CEN/TS 15104:2005 and

CEN/TS 15289:2006 describe the instrumental method for determining CHN and S, respectively, in solid biofuels. The concurrent assessment of the CHN was performed with the aid of Leco TruSpec® Elemental Determinator [32].

A sample of wood fuel powder, of circa 0.1 g, is burned in an oxygen/carrier gas mixture by strictly following several conditions that guarantee a full combustion and that also guarantee the conversion of a few by-products to water vapour, carbon dioxide and nitrogen for gas analysis. The determination of sulphur was carried out in a Leco SC-144DR® machine by employing infrared detection and direct combustion. Duplicate calibration and determination techniques were made to guarantee a robust result consistency. The values of oxygen content were achieved by subtracting, as percentages, from 100% the total of *C*, *H*, *N*, *S* and ash contents.

2.2.4. Determination of the Low and High Heating Values

High Heating Value (HHV) and Low Heating Value (LHV) were assessed by using a Parr 6400® calorimeter, and strictly following the EN 15296 [33] and EN 14918 [34] standards and also by following the standard operating and calculating instructions of the abovementioned equipment.

2.2.5. Fuelwood Value Index (FVI) and Energy Density (E_d)

The Fuel Value Index (FVI) is a parameter for ranking fuelwood species. FVI is calculated as the product of calorific value (MJ·kg^{-1}) and density (g·cm^{-3}) of the biomass by the product of ash content (g·g^{-1}) and water content (g·g^{-1}) as presented in Equation (1) [35,36]:

$$FVI = \frac{\text{Calorific Value} \left(MJ \cdot Kg^{-1}\right) \times \text{Density} \left(g \cdot cm^{-3}\right)}{\text{Ash Content} \left(g \cdot g^{-1}\right) \times \text{Moisture Content} \left(g \cdot g^{-1}\right)} \tag{1}$$

Since the content of the moisture of the wood fluctuates with the proportions of the plant and also fluctuates with the season of the year and other variables, the water content is not treated as a factor of the intrinsic value of a species as a fuel [37]. Therefore, FVI was calculated using the modified formula (Equation (1)), ignoring the moisture content.

The density of the energy as received (E_d) was assessed by utilising the LHV as received and the bulk density of each tree component, according to (2):

$$E_d = \frac{1}{3600} \times q_{p,net,ar} \times BD_{ar} \tag{2}$$

where E_d represents the energy density of the biofuel as received (MWh·m^{-3} of bulk density), $q_{p,net,ar}$ represents the LHV as received (MJ·kg^{-1}), the bulk density, i.e., volume weight of the biofuel as received is given by BD_{ar} (kg·m^{-3} bulk volume), and 1/3600 represents the conversion factor for the energy units (MJ to MWh).

For the calculations of energy density after harvesting we used for branches and needles of maritime pine, an average bulk density of 285 Kg·m^{-3} [38] for green bundles of logging residues. For wood stem and top of tree, wood density was used in the calculations.

2.2.6. Determination of Ash Elemental Metals

After the combustion of biofuel, following CEN/TS 14775:2004, the ash elemental metals—iron (*Fe*), aluminium (*Al*), arsenic (*As*), cadmium (*Cd*), lead (*Pb*), cobalt (*Co*), copper (*Cu*), chromium (*Cr*), manganese (*Mn*), zinc (*Zn*) and nickel (*Ni*)—were measured by following the standard CEN/TS 15290:2006 Solid Biofuels—Determination of major elements, and by following the standard CEN/TS 15297:2006 Solid biofuels—Determination of minor elements.

The detection of the major elements (*Fe* and *Al*) was performed by using flame atomic absorption spectrometry, known as FAAS, and the detection of the minor elements (*As*, *Pb*, *Cd*, *Co*, *Cr*, *Cu*, *Mn*, *Zn* and *Ni*) was carried out by means of graphite furnace atomic absorption spectroscopy, known as

GF-AAS, after the ash digestion with nitric acid—HNO_3 (65%), hydrofluoric acid—HF (40%), hydrogen peroxide—H_2O_2 (30%) and boric acid—H_3BO_3 (4%).

It should be noted, as concluded by [23], that the method used in the assessment of major and minor ash-forming elements in solid biofuels can give values that are significantly different.

3. Results and Discussion

3.1. Pine Stand Characteristics

The biometrical characteristics of pine stands where the biomass samples were collected for fuel and ash analysis are presented in Table 1.

Table 1. Descriptive statistics of 16 maritime pine plots.

Stands Plots (n)	N (trees·ha^{-1})	T (year)	d_{bh} (cm)	H (m)	BA (m^2·ha^{-1})	SI (m)	B_{dry} (ton·ha^{-1})
1	700	52	39.8	21.3	59.2	19	420.2
2	800	40	28.6	19.4	55.8	24	260.9
3	600	51	31.5	19.9	49.7	19	232.5
4	600	44	32.0	19.6	50.5	20	231.1
5	600	42	29.5	18.6	43.4	18	193.1
6	600	54	29.4	19.1	43.4	19	194.2
7	620	26	23.6	14.9	30.9	20	123.4
8	460	39	30.6	13.6	38.6	15	158.1
9	900	20	14.1	10.6	16.7	16	58.4
10	620	20	15.0	9.0	15.4	14	53.2
11	460	51	27.4	20.6	29.3	19	132.7
12	2020	27	13.3	12.9	30.1	15	106.3
13	460	60	33.9	22.8	66.9	25	406.2
14	520	48	24.0	16.5	43.1	20	203.7
15	980	26	17.9	11.2	36.7	18	140.1
16	720	35	22.0	15.8	31.0	20	130.1
Mean	729	40	26	17	40	19	190
Min	460	20	13	9	15	14	53
Max	2020	60	40	23	67	25	420
SD	375.6	12.7	7.7	4.2	14.4	3.0	105.3

Where N is the number of stems ha^{-1}; t is the stand mean age; h is the mean height; d_{bh} is the mean diameter at breast height; BA is the basal area; SI is the site index and B_{dry} represents the dry biomass (ton·ha^{-1}).

The maritime pine stands are pure self-thinned and even-aged, with ages ranging between 20 and 60 years. The younger plants with a stand mean age of 20 years (plots 9 and 10) showed, as expected, lower heights of 9 m and 10.6 m, lower d_{bh} of 14.1 cm and 15 cm, lower basal area and lower biomass dry estimates of 58.4 ton·ha^{-1} and 53.2 ton·ha^{-1}, respectively. The maximum tree density was found for plot 12, with an average age of 27 years and 106.3 ton·ha^{-1} biomass dry yield.

Plots 1 and 13 corresponded to 52 and 60-year-old trees, with dry biomass yields of 420.2 ton·ha^{-1} and 406.2 ton·ha^{-1}, respectively. The highest site index (SI) of 25 was found for plot 13, with average 60-year-old trees, and $SI = 20$ was found for plots 4, 7 and 14 with tree densities of 600, 620 and 520 trees ha^{-1} and average ages 44, 26 and 48 years, respectively. Plot 7 showed a peculiar biomass potential, given its relatively high SI of 29, good biomass productivity taking into account its young age, and its relatively small diameter (23.6 cm).

3.2. Proximate Analysis and Physical Properties

The final proximate analysis, moisture content and density of woody biomass of pine tree components are shown in Table 2.

Table 2. Proximate analysis and basic density of maritime pine woody biomass elements.

	Proximate Analysis (wt %)			D_{db} (Kg·m^{-3})
	Ash	VM	FC	
Wood stem	0.22 (±0.0011)	85.7 (±0.0096)	14.0 (±0.0095)	476.4 (±0.039)
Top	0.24 (±0.0035)	84.2 (±0.0031)	15.6 (±0.0058)	332.7 (±0.040)
Branches	1.00 (±0.0023)	80.8 (±0.0061)	18.2 (±0.0058)	419.6 (±0.047)
Needles	1.97 (±0.0022)	79.4 (±0.0065)	18.7 (±0.0074)	358.6 (±0.050)

Where the proximate and ultimate analyses are in terms of wt. % dry biomass basis. The ash is represented as percentage (%), VM is the Volatile Matter (%), FC is the Fixed Carbon (%), wt % is the moisture content wet basis (%) and the basic density (kg·m^{-3}) is D_{db}. Numbers in parentheses are the standard deviations.

Ash content ranged from 0.22 to 1.97% for stem and needles, respectively. These are favourable values, insofar that they are lower than the data quoted by the literature. For example, in [39], values of 3% were found for maritime pine foliage, and in [40], values between 1.16 and 2.55% are shown for *Pinus pinaster* dead meddles in Mediterranean countries [40].

In a study comparing the calorific value and chemical composition of maritime pine woody biomass infected and not infected with pine nematode [41], the average ash weight percent was provided as 0.66%, and ranged between 0.25% and 1.04% for healthy maritime pine wood [41].

In [42], lower values are reported for Spain, ranging between 0.31 and 0.69% for pine needles, with values for branches ranging from 0.05 to 0.026%. In other work [43], values between 0.71 and 1.04% are shown for pine aboveground biomass ash, and in [41,44], 0.25% and 0.2% were the values found, respectively, for *Pinus pinaster* wood samples in Portugal.

Other references, such as [45], give the ash weight percentage as 0.6%, and in [46], ash amounts of 1.3% were obtained for maritime pine shells. The digital database of the Biomass Energy Foundation (BEF) [47] reports ash values of 1.31% for yellow pine and 0.29% for ponderosa pine. In [22], weight percent data of 2.7% for pine chips and 3.5% for ash was reported for a set of 28 woody biomass species.

For the Fixed Carbon (FC) in [45], values around 26% were obtained for maritime pine bark biomass, while in [46], values of 26% were reported for pine shells, and on the BEF digital database [47], FC reaches around 26.1% for pine needles. As a comparison, Ponderosa pine has FC of about 17%. In [22], values of 21.6% for pine chips, 14% for stem, 15.6% for top, 18.2% for branches and 18.7% for needles were found.

Volatile matter (VM) percentage for maritime pine woody biomass reaches 63.7%, according to [48] and 59% for pine shells, according to [46], and 72% for pine needles. In [22] values of 72.4% for pine chips and 78% for wood residues were reported for a set of 28 wood species. For maritime pine, VM for stems (85.7%), tops (84.2%), branches (80.8%) and needles (79.4%) are a bit higher than the cited values for other pine species. VM for ponderosa pine biomass reaches 82.5%.

The basic densities (D_{db}) obtained for maritime pine biomass tree components (Table 2) were 476 kg·m^{-3}, 419 kg·m^{-3}, 358.6 kg·m^{-3} and 332.7 kg·m^{-3}, for stem, branches, tops and needles, respectively. Our results are in the range of the average values reported in the literature for *P. pinaster* biomass tree components and total AGB [49–51].

3.3. Ultimate Analysis

Ultimate analysis results for tree biomass components of maritime pine are shown in Table 3.

Elementary carbon varied slightly, with values of 46.5%, 48.4%, 48.6%, and 48.2% for stem, top, branches and needles, respectively.

As for similar studies, in [45], an elementary carbon value of 46.9% was also obtained for maritime pine woody biomass. In [46], a similar value of 47.8% was reported for maritime pine shells, and in [22], values such as 44.3% for stem, 50.8% for branches and 47.5% for needles are given for maritime pine biomass components. In [41], an average of 49.32% elementary carbon is reported in Portuguese maritime pine wood samples. Finally, in [42], slightly higher values are given for maritime pine in NW Spain, ranging from 49.40% to 55.23% for needles, and from 48.33% to 54.92%, for branches.

Table 3. Ultimate analysis maritime pine woody biomass components.

	Ultimate Analysis (wt. %)				
	C	**H**	**O**	**N**	**S**
Wood Stem	46.5 (±0.107)	6.7 (±0.082)	46.4 (±0.026)	0.13 (±0.025)	0.029 (±0.005)
Top	48.4 (±0.120)	6.8 (±0.070)	43.8 (±0.060)	0.21 (±0.010)	0.050 (±0.002)
Branches	48.6 (±0.319)	6.8 (±0.050)	43.3 (±0.408)	0.28 (±0.013)	0.032 (±0.010)
Needles	48.2 (±0.225)	6.9 (±0.063)	42.1 (±0.184)	0.74 (±0.063)	0.86 (±0.026)

Where ultimate analyses are in terms of wt. % dry biomass basis. Numbers in parentheses represent the Standard Deviations.

Elementary carbon from the BEF digital database [47] was 49.3% and 52.6% for yellow pine and ponderosa pine woody biomass, respectively. In [22], values of 52.8% of elementary carbon for pine wood species and an average of 52% elementary carbon are also reported for a set of 28 woody biomass species.

The elementary hydrogen weight contents achieved were 6.7%, 6.8%, 6.8% and 6.9% for maritime pine stem, top, branches and needles, respectively. These values are of the same order of magnitude as those quoted in literature. Indeed, in [45], 6.3% hydrogen for maritime pine wood was found, while in [52], an amount of 6.1% hydrogen was reported in pine woody biomass, with an average of 5.4% for a set of 28 woody biomass species. In [41], an average value of 6.3% was reported for maritime pine woody biomass, with values ranging between 6% and 6.78%, while in [46], a hydrogen amount of 4.3% was provided, and the digital database of BEF [47] provided 6% and 7% for the hydrogen amounts in ponderosa pine and yellow pine, respectively.

The elementary oxygen weight percentage values, although lower than elementary carbon, are higher than those quoted by the literature. Indeed, the authors in [45] reached an amount of 45.7% for maritime pine woody biomass. In [52], the authors determined 41% for pine chips, and the digital database of BEF [47] provided information of the order of magnitude of 44.36% of oxygen for Ponderosa pine and 40.1% for yellow pine. In [46], 32.4% was found for oxygen percentage in maritime pine shells. The research in [41], on maritime pine, was that with values of elementary oxygen weight amounts closest to those obtained in this work, with an average of 42.62%, ranging between 38.69% and 45.30%.

The sulphur content was under the 0.1% threshold fixed in [53] for minimising sulphur-related risk of corrosion in combustion of biomass in boilers, which is comparable to the mean values of woody fuels. This elementary sulphur amount is slightly higher in [45] for maritime pine crown, bark and wood. Table 4 shows the calculated atomic ratios (O/C) and (H/C), which have values in the usual range for biomass [54].

Table 4. Ratios O/C and H/C for maritime pine biomass components.

	O/C (Atomic)	**H/C (Atomic)**
Wood stem	0.75	1.73
Top	0.68	1.69
Branches	0.67	1.68
Needles	0.66	1.72

3.4. Higher and Lower Heating Values

Higher and lower heating values (HHV and LHV) of biomass tree components of maritime pine, as well as the basic density in dry basis (D_{db}), of fuelwood are presented in Table 5 and Figure 1.

In Table 5, the HHV is the Higher Heating Value (MJ·kg^{-1} at dry basis) and basic density. The numbers in parentheses give the Standard Deviations.

Table 5. Higher and lower heating values (HHV and LHV) and basic density of maritime pine biomass components.

	HHV (MJ·kg^{-1})	D_{db} (Kg·m^{-3})
Wood stem	21.60 (±0.541)	476.4
Top	20.65 (±0.60)	332.7
Branches	20.92 (±0.555)	419.6
Needles	21.61 (±0.492)	358.6

The HHV analysed for stem, top, branches and needles resulted in values of 21.6 MJ·kg^{-1}, 20.65 MJ·kg^{-1}, 20.92 MJ·kg^{-1} and 21.61 MJ·kg^{-1}, respectively. Similar HHV values are presented in [42] for maritime pine biomass tree components in NW Spain for needles (20.42 MJ·kg^{-1} to 21.7 MJ·kg^{-1}) for thin branches (20.18 MJ·kg^{-1} to 21.40 MJ·kg^{-1}) and for thick branches (19.09 MJ·kg^{-1} and 20.26 MJ·kg^{-1}), and the conclusions reached in [41] present HHV values for woody maritime pine in Central Portugal ranging between 19.48 MJ·kg^{-1} and 20.69 MJ·kg^{-1}.

Data compiled in [55] for maritime pine showed a HHV of 21.2 MJ·kg^{-1} for needles in France, and 21.49 MJ·kg^{-1} and 21.3 MJ·kg^{-1} for live and dead needles in Spain, respectively; for dead twigs with dimensions 0–2 mm, 2–6 mm and 6.25 mm, 22.071 MJ·kg^{-1}, 21.37 MJ·kg^{-1} and 21.03 MJ·kg^{-1}, respectively; for dead cones, 20.7 MJ·kg^{-1}. However, in [45], lower calorific values of 19 MJ·kg^{-1} were reported for maritime pine woody biomass, as well as 19.8 MJ·kg^{-1} for maritime pine bark and around 20 MJ·kg^{-1} for maritime pine crown biomass. Also, the HHV value for maritime pine cones is assessed in [46] and is equal to 18.82 MJ·kg^{-1}.

Comparing the achieved values with other pine species, these are in the same range. In [55], for Aleppo pine dead needles and dead twigs, values of HHV of 22.08 MJ·kg^{-1} and 20.6 MJ·kg^{-1} were reported, respectively, in Castilla y León. For stone pine dead needles and twigs, an HHV of 21.8 MJ·kg^{-1} was reported, and for cones, an HHV of 20.1 MJ·kg^{-1}. For Monterey pine live and dead needles, HHVs of 21.4 MJ·kg^{-1} and 22.5 MJ·kg^{-1}, respectively, were reported, and for dead cones an HHV of 20.1 MJ·kg^{-1}.

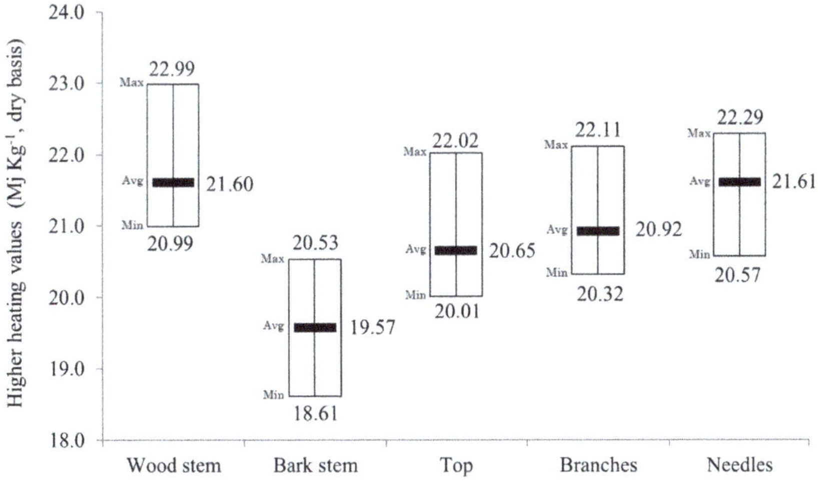

Figure 1. Average higher heating values of maritime pine woody biomass (Max, Min and Avg stand for Maximum, Minimum and Average).

3.5. Fuelwood Value Index (FVI) and Energy Density (E$_d$)

Fuelwood value index is a valid parameter for evaluating quality, insofar as woody biomasses with higher FVI exhibit good fuel qualities. Several studies point clearly to that conclusion. Indeed, from its very definition, FVI for woody biomasses should be correlated with high energy content, high wood density, and low ash and moisture content. Consequently, FVI has even been considered as a standard parameter for identifying suitable tree species for fuelwood production [56,57].

The results of FVI for the four tree components of the maritime pine woody biomass—stem, top, branches and needles—are shown in Table 6, wherein it can be observed that there is a significant decrease from stem (4658), to top (2861), to branches (880), to needles (394).

Table 6. Higher and lower heating values (HHV and LHV), ash maritime pine woody biomass Fuelwood value index (FVI), and energy density (E$_d$).

	HHV (MJ·kg^{-1})	D_{db} (kg·m^{-3})	Ash (%)	FVI	E_d (GJ·m^{-3})
Wood stem	21.60	476.4	0.22	4658.0	8.99
Top	20.65	332.7	0.24	2861.8	6.31
Branches	20.92	419.6	1.00	880.4	2.08
Needles	21.61	358.6	1.97	394.2	2.06

By comparing the calculated FVI for maritime pine biomass tree components with other species, it is possible to obtain a better picture of the quality of maritime pine fuelwood; Ref. [56] presents, in India, a FVI ranging between 792.6 for *Quercus serrata* and 1109.7 for *Castanopsis indica* (Chinquapin). In [57], a longitudinal study of FVI with *Melia dubia*, an Indian fast growing hardwood, was conducted, showing that this parameter increases with age from 1 to 5 years of age, with values varying from 1540 to 4125 [57]. Also, ref. [37] reported experimental work with a set of small branch cuttings from 25 Indian woody biomass species with values of FVI ranging between 343 and 1434, compared with the value of FVI of 880.4 for maritime pine branches obtained in this work.

Energy density (E_d), as mentioned above, is an important variable for addressing the costs of storage and transport and biomass, which became cheaper and more efficient with higher energy density.

In Table 6, it can be seen that energy density exhibits the same variability as FVI insofar that it also decreases from woody stem with 8.99 GJ·m^{-3}, top with 6.31 GJ·m^{-3}, branches with 2.08 GJ·m^{-3} and needles with 2.06 GJ·m^{-3}. Ash amount increases in the maritime pine biomass components with a decrease of FVI and E_d.

The E_d values found in this work for maritime pine biomass components (Table 6) compare well with values cited in the literature, such as: grass in high pressed bales (2.74 GJ·m^{-3}), woodchips of softwood (2.8 GJ·m^{-3}), straw (winter wheat) (1.74 GJ·m^{-3}), and triticale (cereals) (1.92 GJ·m^{-3}). If this biomass is processed into pellets for household usage, the bulk density is higher, ≈550–650 kg·m^3 [58,59], so the energy density also increases, e.g., from 8.9 to 11.5 GJ·m^{-3} [58].

Densification of biomass through chipping, bundling, torrefaction or pelletising maximises E_d and FVI. Small branches and leaves need greater space for transporting and storing, so the low energy density is an evident problem associated with this wood fuel. The compression into bundles varies according to the type of biomass residues and to the type of machinery utilised [60–62].

3.6. Ash Elemental Metals

The analysis results for heavy metal—*Fe, Al, As, Cd, Pb, Co, Cu, Cr, Mn, Zn*, and *Ni*—concentrations present in ash (mg·kg^{-1}) generated in the combustion process are presented in Table 7. These results are then compared to the limit values instituted by legislation on the utilisation ash as a fertiliser in forestry and agriculture in several European countries [63], such as Denmark, Finland, UK, Sweden and Austria. The presently existing Portuguese legislation, with restraining limits for such types of elements, applies to the sludge from water treatment plants, which is used for agriculture, as established in the Decree-Law n° 118/2006 [64].

Table 7. Concentrations (mg·kg^{-1}) of heavy metals in ash of *Pinus pinaster* and limit values in European countries' legislation.

Pinus pinaster	*Fe*	*Al*	*As*	*Cd*	*Pb*	*Co*	*Cu*	*Cr*	*Mn*	*Zn*	*Ni*
Wood stem	29,375.0	5884.4	3.8	0.9	13.4	17.1	926.3	109.4	9750.0	1250.0	37.5
Bark stem	9794.0	8231.2	3.0	2.1	0.4	26.3	85.1	8.5	7239.0	1618.1	21.3
Top	4897.2	10,980.2	5.7	9.7	10.2	9.0	791.2	10.8	8605.1	2863.7	21.0
Branches	7522.3	8782.3	3.8	4.7	3.9	18.7	333.7	88.8	7346.8	2557.6	20.1
Needles	7758.9	3253.7	2.4	0.7	1.5	5.6	43.2	113.5	21,623.2	737.7	25.4
Country (application)											
Portugal (Agriculture) *				20	750		1000	1000		2500	300
Denmark (Agriculture/Forestry)				15	120			100			30
Finland (Agriculture)			25	1.5	100		600	300		1500	100
Finland (Forestry)			30	17.5	150		700	300		4500	150
Sweden (Forestry)			30	30	300		400	100		7000	300
Austria (Field and grassland)			20	8	100	100	250	250		1500	100
Spain (Soils with $p_h < 7$)				20	750		1000	1000		2500	300
Spain (Soils with $p_h > 7$)				40	1200		1750	1500		4000	400

* Limit values of concentration of heavy metals in sludge from water treatment plants.

The measured contents of ash *Fe*, *As*, *Al*, *Pb*, *Cd*, *Cu*, *Co*, *Cr*, *Mn*, *Ni*, and *Zn* are close to the usual ranges commonly found in the ash of biomass tree components (wood, bark, leaves, etc.) for such elements, as conveyed in many research publications [12,65–70] such as in the most popular wood and ash properties databases already discussed previously.

Some exceptions were observed in Al content for pine needles, values for which were lower than the reported values in the literature. In maritime pine wood stem, the *Cu* content was higher than that mentioned in the literature, but in [71], an even higher content value for this parameter was presented in *P. pinaster* measured in NW Spain.

The results of the assessed values happened to be comfortably below the limits instituted by the current Portuguese legislation for all of the studied elements (except for Zn content in pine branches and tops), and also below the limits of the majority of the legislations of several European countries, as can be observed in Table 7. However, some of the observed values in biomass ash would not meet the most restrictive limits enforced by some Central and North European countries for Cu (<250 mg·kg^{-1}), Zn (<1500 mg·kg^{-1}), Cd (<1.5 mg·kg^{-1}), Cr (<100 mg·kg^{-1}) and Ni (<1500 mg·kg^{-1}), and could therefore potentially induce volatilisation problems in thermo-chemical conversion, requiring special filtering systems, or in soil contamination.

Despite the potential soil contamination by some elements present in the ash, given their low concentration, especially in the ash-slag, from the results of this work, the constraints placed on the use of maritime pine biomass ash as fertiliser on soil should be minimal. However, the authors in [67] call attention to the fact that, in cases in which an application of ash is made in large quantities, even if the typical trace element contents that are found in the ash of forest residues do not present any real risk, as long as the fly ash is not utilised, the variation of *Cu*, *Ni* and *Zn* in the bottom ash represents a risk of surpassing the allowed thresholds [67].

Even though the application of ash has an effect on the surrounding environment, it is still safe to apply ash up to 10 t·ha^{-1} from ordinary boilers, since the results in heavy metal soil levels are still 2 levels below the United States Environmental Protection Agency (USEPA) guidelines. However, the environmental impact should still be a target of exhaustive research, as highlighted in [52], given that long-term study on the ecological impact of the ash is fairly limited [56].

4. Conclusions

Biomass samples of maritime pine (*Pinus pinaster Aiton.*) species were assessed with respect to the contents of carbon, nitrogen, hydrogen, oxygen and sulphur, determined by ultimate analysis and by the ash content of volatile matter and fixed carbon, as determined by proximate analyses. The calorific values were measured, and the Energy Density and the Fuelwood Value Index of biomass tree components were calculated. The metal elements of the ash were measured and compared to the restricted values enforced by the legislation of several European countries on ash utilisation in forestry and agriculture as a fertiliser.

The high calorific value of biomass tree components (19.57 to 21.61 MJ·kg^{-1}) and calculated E_d of 2.06 to 8.9 GJ·m^{-3} reveal the considerable potential of this biomass to be utilised as an important source of energy. The FVI ranks, from higher to lower quality, the pine wood stem (4658) and top (2861.8) and pine needles (394.2).

The elemental analysis indicated that high amounts of carbon content (46.5 to 49.3%) are stored in the biomass of pinewood components. Such information has an underlining significance for future studies that aim to quantify the importance of this species in the global cycle of carbon.

A phenomenon that typically occurs in the use of biomass in wood-fired power plants is the employment of ash from combustion. As for the chemical composition of the ash, the analysis indicated that the ash can be employed on forest or agricultural soils, since the analysis showed that the heavy metals were below the legal maximum allowable quantities. A reparation for the loss of nutrients could be achieved from the harvesting location by recycling ash as a soil improving agent or as a fertiliser, thus offering environmental and, in some cases, economic benefits.

Author Contributions: H.F.d.S.V. performed the writing and original draft preparation. A.M.R. performed parts of the literature review. R.G. handled the writing and editing of the manuscript and contributed with parts of the literature review. J.C.d.O.M. and L.J.R.N. supervised, revised and corrected the manuscript.

Funding: This work was supported by European Investment Funds by FEDER/COMPETE/POCI-Operational Competitiveness and Internationalisation Programme, under Project POCI-01-0145-FEDER-006958 and National Funds by FCT—Portuguese Foundation for Science and Technology, under the project UID/AGR/04033/2013 and by SFRH/PROTEC/49626/2009 grant provided to the first author. We thank Cristóvão Santos from Mechanical Department of the University of Trás-os-Montes and Alto Douro, where calorific analysis was done, and Rui Rocha from Leco laboratory for the assistance in elemental analysis. This work was also supported by Fundação para a Ciência e Tecnologia (FCT), under the project UID/EMS/00151/2013 C-MAST, with reference POCI-01-0145-FEDER-007718. This research was also funded by the following project: Centro-01-0145-FEDER-000017-EMaDeS-Energy, Materials and Sustainable Development, co-financed by the Portugal 2020 Program (PT 2020), comprised of the Regional Operational Program of the Center (CENTRO 2020) and the European Union through the European Regional Development Fund (ERDF).

Conflicts of Interest: The authors declare no conflicts of interest.

References

1. Dias, A.C.; Arroja, L. Environmental impacts of eucalypt and maritime pine wood production in Portugal. *J. Clean. Prod.* **2012**, *37*, 368–376. [CrossRef]
2. Viana, H.; Cohen, W.B.; Lopes, D.; Aranha, J. Assessment of forest biomass for use as energy. GIS-based analysis of geographical availability and locations of wood-fired power plants in Portugal. *Appl. Energy* **2010**, *87*, 2551–2560. [CrossRef]
3. Gómez-González, S.; Ojeda, F.; Fernandes, P.M. Portugal and Chile: Longing for sustainable forestry while rising from the ashes. *Environ. Sci. Policy* **2018**, *81*, 104–107. [CrossRef]
4. Parente, J.; Pereira, M.G.; Amraoui, M.; Tedim, F. Negligent and intentional fires in Portugal: Spatial distribution characterization. *Sci. Total Environ.* **2018**, *624*, 424–437. [CrossRef] [PubMed]
5. Sá, A.C.L.; Turkman, M.A.A.; Pereira, J.M.C. Exploring fire incidence in Portugal using generalized additive models for location, scale and shape (GAMLSS). *Model. Earth Syst. Environ.* **2018**, *4*, 199–220. [CrossRef]
6. Godina, R.; Nunes, L.J.R.; Santos, F.M.B.C.; Matias, J.C.O. Logistics cost analysis between wood pellets and torrefied Biomass Pellets: The case of Portugal. In Proceedings of the 2018 7th International Conference on Industrial Technology and Management (ICITM), Oxford, UK, 7–9 March 2018; pp. 284–287.

7. Ribeiro, J.M.C.; Godina, R.; de Oliveira Matias, J.C.; Nunes, L.J.R. Future Perspectives of Biomass Torrefaction: Review of the Current State-Of-The-Art and Research Development. *Sustainability* **2018**, *10*, 2323. [CrossRef]
8. Alén, R. Structure and chemical composition of wood. In *Forest Products Chemistry*; Stenius, P., Ed.; Gummerus Printing: Jyväskylä, Finland, 2000; pp. 11–57.
9. Kilpeläinen, A.; Peltola, H.; Ryyppö, A.; Sauvala, K.; Laitinen, K.; Kellomäki, S. Wood properties of Scots pines (*Pinus sylvestris*) grown at elevated temperature and carbon dioxide concentration. *Tree Physiol.* **2003**, *23*, 889–897. [CrossRef] [PubMed]
10. Viana, H.; Vega-Nieva, D.J.; Ortiz Torres, L.; Lousada, J.; Aranha, J. Fuel characterization and biomass combustion properties of selected native woody shrub species from central Portugal and NW Spain. *Fuel* **2012**, *102*, 737–745. [CrossRef]
11. Sjostrom, E. *Wood Chemistry: Fundamentals and Applications*; Elsevier: New York, NY, USA, 2013; ISBN 978-0-08-092589-9.
12. Saarela, K.E.; Harju, L.; Rajander, J.; Lill, J.O.; Heselius, S.J.; Lindroos, A.; Mattsson, K. Elemental analyses of pine bark and wood in an environmental study. *Sci. Total Environ.* **2005**, *343*, 231–241. [CrossRef] [PubMed]
13. Vassilev, S.V.; Baxter, D.; Andersen, L.K.; Vassileva, C.G.; Morgan, T.J. An overview of the organic and inorganic phase composition of biomass. *Fuel* **2012**, *94*, 1–33. [CrossRef]
14. Vassilev, S.V.; Baxter, D.; Andersen, L.K.; Vassileva, C.G. An overview of the composition and application of biomass ash. Part 1. Phase-mineral and chemical composition and classification. *Fuel* **2013**, *105*, 40–76. [CrossRef]
15. Nunes, L.J.R.; Matias, J.C.O.; Catalão, J.P.S. A review on torrefied biomass pellets as a sustainable alternative to coal in power generation. *Renew. Sustain. Energy Rev.* **2014**, *40*, 153–160. [CrossRef]
16. Sarvaramini, A.; Assima, G.P.; Larachi, F. Dry torrefaction of biomass—Torrefied products and torrefaction kinetics using the distributed activation energy model. *Chem. Eng. J.* **2013**, *229*, 498–507. [CrossRef]
17. Van der Stelt, M.J.C.; Gerhauser, H.; Kiel, J.H.A.; Ptasinski, K.J. Biomass upgrading by torrefaction for the production of biofuels: A review. *Biomass Bioenergy* **2011**, *35*, 3748–3762. [CrossRef]
18. Rodrigues, A.; Loureiro, L.; Nunes, L.J.R. Torrefaction of woody biomasses from poplar SRC and Portuguese roundwood: Properties of torrefied products. *Biomass Bioenergy* **2018**, *108*, 55–65. [CrossRef]
19. Grigiante, M.; Brighenti, M.; Antolini, D. A generalized activation energy equation for torrefaction of hardwood biomasses based on isoconversional methods. *Renew. Energy* **2016**, *99*, 1318–1326. [CrossRef]
20. Núñez-Regueira, L.; Proupín-Castiñeiras, J.; Rodríguez-Añón, J.A. Design of an experimental procedure for energy evaluation from biomass. *Thermochim. Acta* **2004**, *420*, 29–31. [CrossRef]
21. Moya, R.; Tenorio, C. Fuelwood characteristics and its relation with extractives and chemical properties of ten fast-growth species in Costa Rica. *Biomass Bioenergy* **2013**, *56*, 14–21. [CrossRef]
22. Vassilev, S.V.; Baxter, D.; Vassileva, C.G. An overview of the behaviour of biomass during combustion: Part II. Ash fusion and ash formation mechanisms of biomass types. *Fuel* **2014**, *117*, 152–183. [CrossRef]
23. Baernthaler, G.; Zischka, M.; Haraldsson, C.; Obernberger, I. Determination of major and minor ash-forming elements in solid biofuels. *Biomass Bioenergy* **2006**, *30*, 983–997. [CrossRef]
24. James, A.K.; Thring, R.W.; Helle, S.; Ghuman, H.S. Ash Management Review—Applications of Biomass Bottom Ash. *Energies* **2012**, *5*, 3856–3873. [CrossRef]
25. Pöykiö, R.; Mäkelä, M.; Watkins, G.; Nurmesniemi, H.; Dahl, O. Heavy metals leaching in bottom ash and fly ash fractions from industrial-scale BFB-boiler for environmental risks assessment. *Trans. Nonferrous Met. Soc. China* **2016**, *26*, 256–264. [CrossRef]
26. Kalembkiewicz, J.; Chmielarz, U. Ashes from co-combustion of coal and biomass: New industrial wastes. *Resour. Conserv. Recycl.* **2012**, *69*, 109–121. [CrossRef]
27. Rivero-Huguet, M.; Huertas, R.; Francini, L.; Vila, L.; Darré, E. Concentrations of As, Ca, Cd, Co, Cr, Cu, Fe, Hg, K., Mg, Mn, Mo, Na, Ni, Pb, and Zn in Uruguayan rice determined by atomic absorption spectrometry. *At. Spectrosc.* **2006**, *27*, 48–55. [CrossRef]
28. Bonanno, G.; Cirelli, G.L.; Toscano, A.; Lo Giudice, R.; Pavone, P. Heavy metal content in ash of energy crops growing in sewage-contaminated natural wetlands: Potential applications in agriculture and forestry? *Sci. Total Environ.* **2013**, *452–453*, 349–354. [CrossRef] [PubMed]
29. Nieminen, M.; Piirainen, S.; Moilanen, M. Release of mineral nutrients and heavy metals from wood and peat ash fertilizers: Field studies in Finnish forest soils. *Scand. J. For. Res.* **2005**, *20*, 146–153. [CrossRef]

30. Michelsen, O.; Solli, C.; Strømman, A.H. Environmental impact and added value in forestry operations in Norway. *J. Ind. Ecol.* **2008**, *12*, 69–81. [CrossRef]
31. Verma, M.; Loha, C.; Sinha, A.N.; Chatterjee, P.K. Drying of biomass for utilising in co-firing with coal and its impact on environment—A review. *Renew. Sustain. Energy Rev.* **2017**, *71*, 732–741. [CrossRef]
32. TruSpec Micro. Available online: https://www.leco.com/products/analytical-sciences/carbon-hydrogen-nitrogen-protein-sulfur-oxygen-analyzers/truspec-micro (accessed on 23 July 2018).
33. ISO/TC 44/SC 8. *ISO 15296:2004, Gas Welding Equipment—Vocabulary—Terms Used for Gas Welding Equipment*; Multiple; American National Standards Institute: Washington, DC, USA, 2007.
34. TC 107/WG 1, I. *ISO 14918:1998, Thermal Spraying—Approval Testing of Thermal Sprayers*; Multiple; American National Standards Institute: Washington, DC, USA, 2007.
35. Bhatt, B.P.; Tomar, J.M.S. Firewood properties of some Indian mountain tree and shrub species. *Biomass Bioenergy* **2002**, *23*, 257–260. [CrossRef]
36. Nirmal Kumar, J.I.; Patel, K.; Kumar, R.N.; Bhoi, R.K. An evaluation of fuelwood properties of some Aravally mountain tree and shrub species of Western India. *Biomass Bioenergy* **2011**, *35*, 411–414. [CrossRef]
37. Bhatt, B.P.; Tomar, J.M.S.; Bujarbaruah, K.M. Characteristics of some firewood trees and shrubs of the North Eastern Himalayan region, India. *Renew. Energy* **2004**, *29*, 1401–1405. [CrossRef]
38. Parikka, M. Global biomass fuel resources. *Biomass Bioenergy* **2004**, *27*, 613–620. [CrossRef]
39. Vázquez, G.; Antorrena, G.; González, J.; Freire, S. Studies on the composition of Pinus pinaster foliage. *Bioresour. Technol.* **1995**, *51*, 83–87. [CrossRef]
40. Keane, R.E. *Wildland Fuel Fundamentals and Applications*; Springer International Publishing: New York, NY, USA, 2015; ISBN 978-3-319-09014-6.
41. Reva, V.; Fonseca, L.; Lousada, J.L.; Abrantes, I.; Viegas, D.X. Impact of the pinewood nematode, Bursaphelenchus xylophilus, on gross calorific value and chemical composition of Pinus pinaster woody biomass. *Eur. J. For. Res.* **2012**, *131*, 1025–1033. [CrossRef]
42. Núñez-Regueira, L.; Rodríguez-Añón, J.; Proupín, J.; Romero-García, A. Energy evaluation of forest residues originated from pine in Galicia. *Bioresour. Technol.* **2003**, *88*, 121–130. [CrossRef]
43. Núñez-Regueira, L.; Rodríguez, J.; Proupín, J.; Mouriño, B. Forest waste as an alternative energy source. *Thermochim. Acta* **1999**, *328*, 105–110. [CrossRef]
44. Telmo, C.; Lousada, J.; Moreira, N. Proximate analysis, backwards stepwise regression between gross calorific value, ultimate and chemical analysis of wood. *Bioresour. Technol.* **2010**, *101*, 3808–3815. [CrossRef] [PubMed]
45. Álvarez-Álvarez, P.; Pizarro, C.; Barrio-Anta, M.; Cámara-Obregón, A.; Bueno, J.L.M.; Álvarez, A.; Gutiérrez, I.; Burslem, D.F.R.P. Evaluation of Tree Species for Biomass Energy Production in Northwest Spain. *Forests* **2018**, *9*, 160. [CrossRef]
46. Ye, T.H.; Azevedo, J.; Costa, M.; Semião, V. Co-Combustion of Pulverized Coal, Pine Shells, and Textile Wastes in a Propane-Fired Furnace: Measurements and Predictions. *Combust. Sci. Technol.* **2004**, *176*, 2071–2104. [CrossRef]
47. Biomass Energy Foundation: Woodgas Home Page. Available online: http://drtlud.com/BEF/proximat.htm (accessed on 26 July 2018).
48. Du, S.; Yang, H.; Qian, K.; Wang, X.; Chen, H. Fusion and transformation properties of the inorganic components in biomass ash. *Fuel* **2014**, *117*, 1281–1287. [CrossRef]
49. Countryman, C.M.; Philpot, C.W. *Physical Characteristics of Chamise as a Wildland Fuel*; Res. Paper PSW-RP-66; Pacific Southwest Forest & Range Experiment Station, Forest: Berkeley, CA, USA, 1970; 16p.
50. Núez-Regueira, L.; Proupín-Castieiras, J.; Rodríguez-Aón, J.A. Energy evaluation of forest residues originated from Eucalyptus globulus Labill in Galicia. *Bioresour. Technol.* **2002**, *82*, 5–13. [CrossRef]
51. Bert, D.; Danjon, F. Carbon concentration variations in the roots, stem and crown of mature *Pinus pinaster* (Ait.). *For. Ecol. Manag.* **2006**, *222*, 279–295. [CrossRef]
52. Vassilev, S.V.; Baxter, D.; Andersen, L.K.; Vassileva, C.G. An overview of the chemical composition of biomass. *Fuel* **2010**, *89*, 913–933. [CrossRef]
53. Obernberger, I.; Biedermann, F.; Widmann, W.; Riedl, R. Concentrations of inorganic elements in biomass fuels and recovery in the different ash fractions. *Biomass Bioenergy* **1997**, *12*, 211–224. [CrossRef]
54. McGroddy, M.E.; Daufresne, T.; Hedin, L.O. Scaling of C:N:P stoichiometry in forests worldwide: Implications of terrestrial redfield-type ratios. *Ecology* **2004**, *85*, 2390–2401. [CrossRef]

55. Antolín, G.; Irusta, R.; Velasco, E.; Carrasco, J.; González, E.; Ortíz, L. Biomass as an energy resource in Castilla y León (Spain). *Energy* **1996**, *21*, 165–172. [CrossRef]

56. Meetei, S.B.; Singh, E.J.; Das, A.K. Fuel wood properties of some oak tree species of Manipur, India. *J. Environ. Biol.* **2015**, *36*, 1007–1010. [PubMed]

57. Saravanan, V.; Parthiban, K.T.; Kumar, P.; Anbu, P.V.; Pandian, P.G. Evaluation of Fuel Wood Properties of Melia dubia at Different Age Gradation. *Res. J. Agric. For. Sci.* **2013**, *1*, 8–11.

58. Obernberger, I.; Thek, G. Physical characterisation and chemical composition of densified biomass fuels with regard to their combustion behaviour. *Biomass Bioenergy* **2004**, *27*, 653–669. [CrossRef]

59. Van Loo, S.; Koppejan, J. *The Handbook of Biomass Combustion and Co-Firing*; Earthscan: London, UK, 2008; Volume 1.

60. Felix, M. Future prospect and sustainability of wood fuel resources in Tanzania. *Renew. Sustain. Energy Rev.* **2015**, *51*, 856–862. [CrossRef]

61. McKendry, P. Energy production from biomass (part 1): Overview of biomass. *Bioresour. Technol.* **2002**, *83*, 37–46. [CrossRef]

62. Spinelli, R.; Magagnotti, N. Logging residue bundling at the roadside in mountain operations. *Scand. J. For. Res.* **2009**, *24*, 173–181. [CrossRef]

63. Guideline for Classification Classification of Ash from Solid Biofuels and Peat Utilised for Recycling and Fertilizing in Forestry and Agriculture (NT TR 613)—Nordtest.info. Available online: http://www. nordtest.info/index.php/technical-reports/item/guideline-for-classification-classification-of-ash-from-solid-biofuels-and-peat-utilised-for-recycling-and-fertilizing-in-forestry-and-agriculture-nt-tr-613.html (accessed on 26 July 2018).

64. Decreto-Lei 118/2006, 2006–2006-21. Available online: https://dre.pt (accessed on 26 July 2018).

65. Misra, M.K.; Ragland, K.W.; Baker, A.J. Wood ash composition as a function of furnace temperature. *Biomass Bioenergy* **1993**, *4*, 103–116. [CrossRef]

66. Steenari, B.M.; Lindqvist, O. Stabilisation of biofuel ashes for recycling to forest soil. *Biomass Bioenergy* **1997**, *13*, 39–50. [CrossRef]

67. Pitman, R.M. Wood ash use in forestry—A review of the environmental impacts. *Forestry* **2006**, *79*, 563–588. [CrossRef]

68. Augusto, L.; Bakker, M.R.; Meredieu, C. Wood ash applications to temperate forest ecosystems—Potential benefits and drawbacks. In *Plant and Soil*; Springer: New York, NY, USA, 2008; Volume 306, pp. 181–198.

69. Sánchez-Rodríguez, F.; Rodríguez-Soalleiro, R.; Espaol, E.; López, C.A.; Merino, A. Influence of edaphic factors and tree nutritive status on the productivity of *Pinus radiata* D. Don plantations in Northwestern Spain. *For. Ecol. Manag.* **2002**, *171*, 181–189. [CrossRef]

70. Haraldsen, T.K.; Pedersen, P.A.; Grønlund, A. Mixtures of Bottom Wood Ash and Meat and Bone Meals as NPK Fertilizer. In *Recycling of Biomass Ashes*; Springer: New York, NY, USA, 2011; pp. 33–44.

71. Núñez-Regueira, L.; Rodríguez Añón, J.A.; Proupín Castiñeiras, J. Calorific values and flammability of forest species in Galicia. Coastal and hillside zones. *Bioresour. Technol.* **1996**, *57*, 283–289. [CrossRef]

 *sustainability*

Article

Technological Innovation in Biomass Energy for the Sustainable Growth of Textile Industry

Leonel Jorge Ribeiro Nunes [1,2,*], Radu Godina [3] and João Carlos de Oliveira Matias [1,2]

[1] DEGEIT—Department of Economics, Management, Industrial Engineering and Tourism, University of Aveiro, 3810-193 Aveiro, Portugal; jmatias@ua.pt

[2] GOVCOPP—Research Unit on Governance, Competitiveness and Public Policies, University of Aveiro, 3810-193 Aveiro, Portugal

[3] Research and Development Unit in Mechanical and Industrial Engineering (UNIDEMI), Department of Mechanical and Industrial Engineering, Faculty of Science and Technology (FCT), New University of Lisbon, 2829-516 Caparica, Portugal; rd@ubi.pt

* Correspondence: leonelnunes@ua.pt; Tel.: +351-232-446-600

Received: 15 December 2018; Accepted: 16 January 2019; Published: 20 January 2019

Abstract: The growing increase in world energy consumption favors the search for renewable energy sources. One of the existing options for the growth and sustainable development of such types of sources is through the use of biomass as an input. The employment of biomass as solid fuel is widely studied and is no longer a novelty nor presents any difficulty from the technical point of view. It presents, however, logistic obstacles, thus not allowing their direct dissemination in every organization that is willing to replace it as an energy source. Use of biomass can be rewarding due to the fact that it can bring significant economic gains attained due to the steadiness of the biomass price in Portugal. However, the price may rise as predicted in the coming years, although it will be a gradual rising. The main goal of this study was to analyze whether biomass in the case of the Portuguese textile industry can be a viable alternative that separates the possibility of sustainable growth from the lack of competitiveness due to high energy costs. The study showed that biomass can be a reliable, sustainable and permanent energy alternative to more traditional energy sources such as propane gas, naphtha and natural gas for the textile industry. At the same time, it can bring savings of 35% in energy costs related to steam generation. Also, with new technology systems related to the Internet of Things, a better on-time aware of needs, energy production and logistic chain information will be possible.

Keywords: energy from biomass; textile industrial sector; alternative energy; SWOT analysis; energy costs; Internet of Things

---

1. Introduction

The energy sector, with particular emphasis on renewable energy, has witnessed an increased interest of scientific research on the production of energy for the manufacturing industry, particularly in the form of thermal energy [1]. Global demand for increased energy efficiency, partly driven by climate changes, but mainly by a relentless pursuit of lower energy costs, has influenced and motivated the introduction of a holistic perspective in the analysis of thermal systems. Supported by the Internet of Things (IoT), the development of new concepts and tools for the intelligent management of energy systems, with particular emphasis on thermal systems, has become one of the most pressing current demands [2].

Biomass is considered a renewable energy source that could decidedly support the mitigation of climate change [3–5] and can be used in the production of energy from processes such as the combustion of organic material produced and accumulated in an ecosystem [6]. It can be distinguished from different energy sources with considerable energy potential: wood (and waste), agricultural waste, solid municipal waste, animal waste, food waste, aquatic plants and algae [7]. Compared with fossil fuels such as petroleum products, these wastes generate fewer greenhouse gas emissions [8]. Therefore, biomass is considered a sustainable type of energy [9–11]. Energy security is an element of significant evidence and is crucial for the sustainable economic growth of every country [12]. As indicated by the European Commission, the goal is to reduce the greenhouse gas (GHG) emissions in the EU-27 by at least 80% in 2050 when compared to the emissions in 1990 [13,14]. Also, worldwide energy necessities could double in 2050, largely due to the quick development of emerging nations [15]. A different study confirms a similar tendency, with a rise of approximately 50% of energetic needs by 2030, of which as much as 70% are part of India, China and other emerging nations [16–18].

Although biomass is less utilized than hydro and wind energy for the production of electricity in Portugal, it has the advantage of being able to be stored and used only when demand justifies it. Thus, in this way, although biomass is commonly dependent on the growth, production of wastes and agriculture, therefore, it can be highly impacted on by the weather, production rate and economic parameters, which could lead to raw material shortage, it allows a constant supply, unlike other renewable energy sources that are intermittent and dependent on the atmospheric conditions in the case of wind and the level of water in dams in the case of water, making the production of energy through these sources dependent and uncertain [19].

As a result, one of the solutions to reduce Portugal's energy dependence from the exterior is to replace the use of fossil fuels, one of the most promising alternatives being biomass, which includes forest, agricultural residues, and all biodegradable organic waste originating from man and animals. This energy source is very flexible and can be used in its pure form or processed to produce biofuels. According to Eurostat [20,21], the energy produced from biomass and solid biofuels sources in Portugal in the last 10 years shows a significant decline in consumption between the years 2009 and 2012, as can be seen in Figure 1. This decrease is due mainly to the exportation of biomass products, mainly in the form of wood pellets, to northern Europe, instead of national use due to the higher prices paid by such countries.

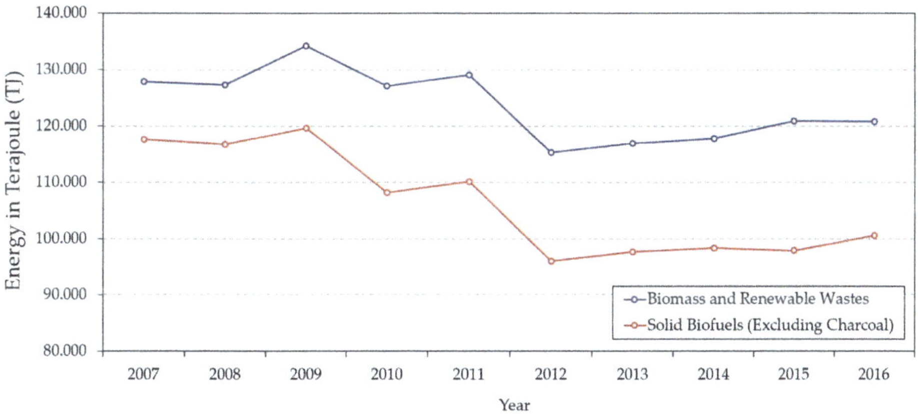

Figure 1. The energy produced from biomass and solid biofuels sources in Portugal from 2007 to 2016.

This new perspective on the use of biomass is also due to the discovery of this type of energy by industries. Large companies that produce their own energy are replacing the fossil fuels that moved their plants through alternative sources. Among the main benefits is achieving the emissions reduction targets [22]. Biomass has started to become a more attractive source thanks to the advancement of its processing technology. With the improvement of the equipment used in combustion, the efficiency of the process has increased. One of the main challenges of the industry is to make the most of biomass, since the conditions of humidity and conservation of waste have a direct impact on the generation of energy [23].

The main objective of this work is to research into the energy efficiency management and logistics associated with the use of biomass as a source of thermal energy in the Portuguese textile industry, since the increasing consumption of this energy will require the creation of information networks and data analysis that will support decision making, contributing to the development of the collective awareness systems.

This work is organized as follows. In Section 2, the literature review is performed. In Section 3, the energy sources used in Portugal by the textile dye Industry are addressed. A comparison between biomass and fuels of fossil origin is made in Section 3. In Section 4, a SWOT analysis of the use of biomass energy in the textile industry is made. Finally, the results are analyzed in Section 4, and conclusions are drawn in Section 5.

2. State-of-The-Art

The textile industry is considered one of the most complex industries due to the variety of processes, machinery and components used throughout the process, as well as all types of fibres and yarns, production methods, finishing, preparation, dyeing and coating, amongst others which are also important. The number of enterprises that are part and are operating in this industry in Portugal can be seen in Figure 2, where the turnover in millions of euros that these enterprises generate when combined can also be observed [24]. Energy is one of the main components of this industry, and consequently, energy consumption costs are especially high during a period of high price volatility. One of the main interests of the stakeholders of this industry is energy efficiency.

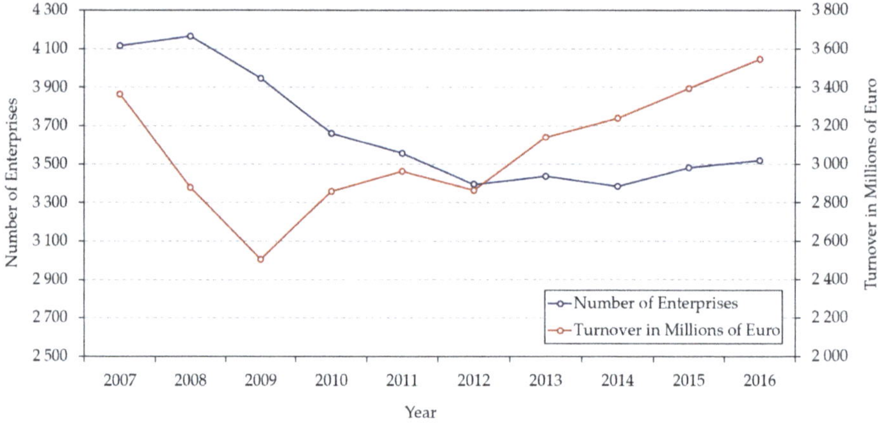

Figure 2. Number of Portuguese enterprises operating in the textile industry vs the turnover in millions of euro.

The textile industry is considered a significant element of the global economy, which suffered profound changes in the last few decades due to the technological advancements that followed during the same years [25]. This industry, in Portugal, is up against difficult competitive challenges, and the reason is elevated costs of production, which in turn are provoked by increasing costs of energy as a part of the European energy cost trend [26]. However, this industry has been resilient, as represented in Figure 3, where it can be seen that the turnover has been steadily increasing since the 2007–2008 financial crisis [24]. However, this industry is responsible for a substantial share of energy consumption, as can be observed in Figure 4 [24]. The textile industry is known for being very diverse, with several distinct production processes. However, within this industry, the textile dyeing sector is recognized as a considerable energy consumer, since its manufacturing process requires a substantial quantity of steam for it to operate properly [27].

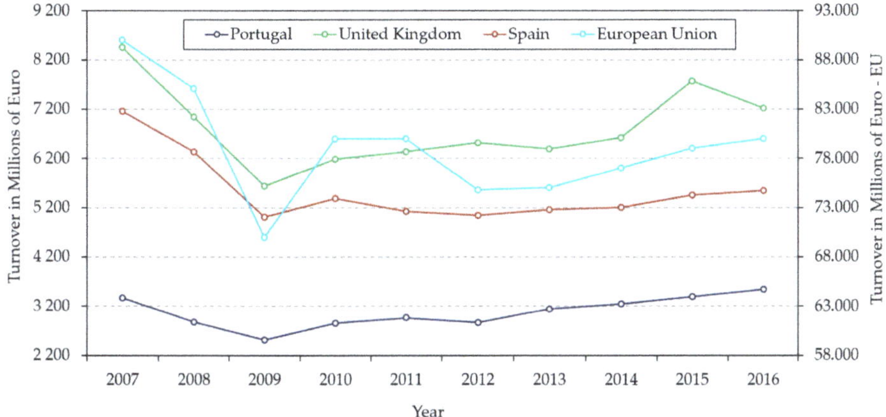

Figure 3. The turnover of the textile industry in EU, Portugal, United Kingdom and Spain.

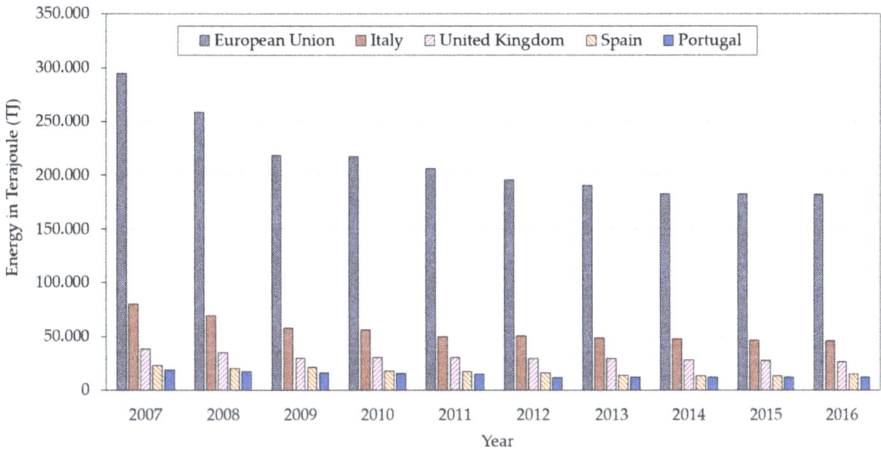

Figure 4. The energy consumed by the textile industry in EU, Portugal, United Kingdom and Spain.

The Portuguese textile dyeing industry relies mostly on natural gas as fuel. However, in some instances, the use of naphtha and propane gas still occurs, not by the choice of prices being more competitive, but due to logistical obstacles or due to a lack of access to the natural gas supply chain [4]. The price of utilising steam boilers for steam production, usually at 10 bar and at 90 °C, makes up as much as 60% of the textile dyeing industry's energy costs. Therefore, the opportunity to use biomass as an alternative energy source brings many advantages for this industry [28,29].

Researchers have been steadily increasing their focus and attention on the use of biomass in the textile industry. A study focusing on the analysis of the energy mix profile and energy efficiency of the Brazilian dairy industry is presented in Reference [30], in which, biomass sources for thermal energy generation are addressed. In Reference [4], the environmental and economic benefits of utilising textile waste for thermal energy generation are investigated and a comparative economic assessment with distinct fuels is made. A study of biomass employment as an energetic substitute for the textile dyeing industry of Portugal is made in Reference [25]; however, a more detailed consumption of steam production in m^3 is lacking.

A decrease of the adsorbed dyes' volume by utilizing the aerobically-activated sludge process in the discharged sludge originating from an industrial textile wastewater treatment plant is proposed in Reference [31]. By selecting a performant biomass, the authors argue that microorganisms are capable of generating an adjusted conglomerate capable of lowering toxic dye molecules. In the author's analyses, the energy footprints of the textile industry, namely textile manufacturing, utilization, and recycling stages [32]. In Reference [33], the potential of future energy saving of the textile industry of China is predicted by analysing the energy substitution effect of technological progress through the use of a macroeconomics approach while the authors in Reference [34] conducted a comprehensive study and analysis of the GHG emissions in the textile industry of China and then assessed the nature of the emission. The results showed that coal was the main source of GHG emissions. Since biomass pellets or torrefied biomass pellets are a direct replacement of coal, here lies an opportunity to significantly decrease the emissions of the textile industry. Authors in Reference [35] also identified potential cost-effective opportunities to improve the energy use and energy efficiency in this industry. Also, in Reference [36], such types of opportunities and potential are discussed, and the author identifies the compressor, steam boiler and lighting as the elements taking a substantial slice of the overall energy consumption. Finally, the assessment and the capability of a photovoltaic/thermal-energy system of cogeneration equipped with a storage device applicable to the textile industry is proposed in Reference [37].

3. Materials and Methods

The textile dyeing industry operating in Portugal has a significant share of the Portuguese economic activity, which translates into having substantial importance. The major industrial units and organizations are all situated in and around the proximate regions of Vale do Cávado and Vale do Ave, located in the north of Portugal. Thus, for this paper, these industrial units were selected, specifically in the municipalities of Barcelos, Famalicão and Fafe, and the city of Guimarães, which are the areas were the Portuguese textile industry is mostly concentrated and which have been operating for more than 150 years [27,38].

The textile industry in Portugal suffered a steep decline a few years ago, as can be deduced from Figure 2, a period during which many industrial units closed doors. In the aftermath of the 2007–2008 financial crisis, many other organizations suffered profound changes and drastically reduced their workforce with the purpose of keeping the operation active and trying to survive [25]. Several of the abovementioned industrial units were willing to modernise through the implementation of distinct and original requalification policies and procedures for their production processes. Thus, these industrial units implemented several measures with the purpose of achieving a higher energy efficiency, which also meant reducing the production costs and liberating more income for other more urgent expenses [39].

Three types of fuel in steam boilers were used: natural gas, propane gas and naphtha; they can be observed in Figure 5. The last two are mainly used in smaller units without direct access to the natural gas distribution grid.

Not even one industrial unit considered in this study was employing biomass for their energetic input. Thus, an opportunity arose to assess the use of this source in this industry. The steam boilers that were most frequently met in the sought industrial units are usually equipped with capacities of steam generation between 8 and 12 ton/h at 10 bar pressure, as can be observed in Figure 6.

A modern biomass boiler (for pellets, chips, wood briquettes and wood residues) represents an ecological and convenient solution, alternatively or by integration, of traditional heating systems for fossil fuels. The transition from fossil fuels to biomass also means altering and upgrading the equipment, as can be observed in Figure 7, where the first steam boiler in a textile dyeing industry was assembled. This can frequently be very challenging to accomplish due to logistical and spatial order obstacles. This happens mainly due to the fact that these units have been operating unaltered for decades and are occupying obsolete facilities. These installations have been established to satisfy the production needs of that period. Thus, in facilities adapted for biomass, they must have space for storage and for the filter system, as can be observed in Figures 8 and 9, respectively.

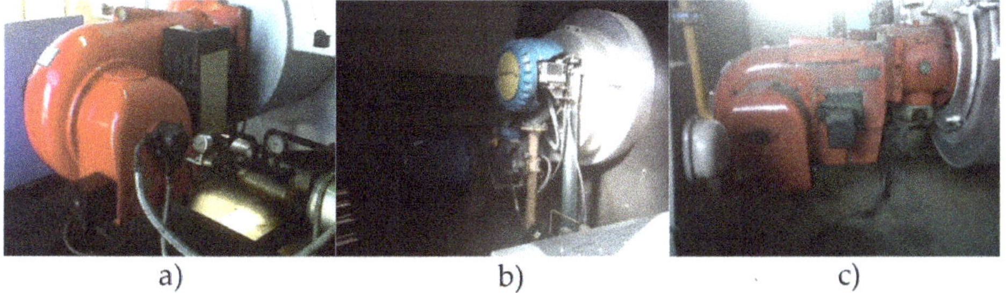

Figure 5. Example of (**a**) naphtha, (**b**) propane gas and (**c**) natural gas burners connected to steam boilers.

Figure 6. An example of a steam boiler that produces 10 tonnes per-hour of steam.

Figure 7. Example of a biomass burner in a steam boiler of a textile production plant.

(a)   (b)

Figure 8. Biomass warehouse (**a**) with a burner automatic feeding system (**b**).

Figure 9. Biomass steam boiler filtering system.

4. Results and Discussion

By being utilised as a substitute fuel to fossil fuels, the biomass can support the reduction of GHG emissions [40]. However, every biofuel positively impacts the environment. The production of biomass and transportation techniques could have a negative effect on the environment as well as in certain areas of the world the production of biofuel threatens agricultural crops [41].

Natural gas is the most utilized fuel in steam boilers while propane gas and naphtha are only used as a last resort in cases when the natural gas is not readily available. Thus, in this paper, is only made reference to natural gas and biomass comparing.

Previous consumption data from the top 10 textile dyeing industrial units in the regions of Vale do Cávado and Vale do Ave were gathered and compiled by the authors, thus presenting in Table 1 the average of the annual consumptions for the year 2016, assuming that of the energy total amounts calculated, only 60% is related to the production of steam. For the comparative study, woodchips of pine wood were used as biomass form, which is considered to be a product containing particles with an average moisture content of 40%, a size around $40 \times 40 \times 20$ mm and a lower heating value (LHV) of 3.50 kWh [42], which is in opposition to natural gas that has a LHV of 9.16 kWh [43].

Table 1. Natural Gas Consumption of the top 10 textile dyeing industrial units in 2016.

	Monthly Average Values	Annual Average Values
Consumption of steam production in m^3 (60% of the entire consumption)	175.000 m^3	1.925.000 m^3
Consumption of steam production in kWh (60% of the entire consumption)	2.250.000 kWh	24.750.000 kWh
Steam production costs (60% of the entire consumption)	75.000 €	825.000 €
The cost of kWh	0.034 €/kWh	

Thus, in order to achieve a similar result, it is required to consume a monthly mass of around 645,000 kg of woodchips, which is overall 7,095,000 kg per year, as can be observed in Table 2. In this study, a yearly reference value of 11 months was utilized, due to the reason that August is usually vacation month and is used for maintenance. The 11-month year was used as the reference for the present calculations, the highest quotation accessed from several local suppliers of pine woodchips, and that was 75 €/t. All values used are presented free of VAT.

Table 2. Biomass (woodchips) consumption—estimated annual average values.

	Monthly Average Values	Annual Average Values
Consumption of steam production in m^3 (60% of the entire consumption)	645.000 kg	7.095.000 kg
Consumption of steam production in kWh (60% of the entire consumption)	2.250.000 kWh	24.750.000 kWh
Steam production costs (60% of the entire consumption)	48.375 €	532.125 €
The cost of kWh	0.024 €/kWh	

By comparing the annual steam production costs of the natural Gas and biomass, using pine woodchips, it is possible to asses from the results that real savings in energy costs for steam production of about 35% can be achieved, as can be observed in Table 3.

Table 3. Yearly savings if biomass is chosen instead of natural gas.

	Natural Gas	Biomass (Pine Woodchips)
Annual kWh consumption for steam production	24.750.000 kWh	
Annual steam production costs	825.000 €	532.125 €
Annual total savings	±35%	

The SWOT analysis, as can be observed in Table 4, is a technique capable of identifying weaknesses, strengths, threats and opportunities for a given goal. In the current study, employing biomass as a different and more sustainable energy source for the textile industry helps achieve an optimized operational strategy [44]. As can be observed, the use of IoT technologies in monitoring energetic consumption and logistic control of solid fuel's supply is identified as a potential opportunity.

Table 4. Employing Biomass Energy in Textile Industry—a SWOT Analysis.

Strengths	Weaknesses
<ul><li>Energy efficiency.</li><li>The rise in competitiveness achieved by reducing the energy costs.</li><li>The rise of textile-dependent industries and suppliers, meaning job creation.</li><li>An important role in decreasing the imports of natural gas, coal, etc.</li></ul>	<ul><li>Biomass logistics and supply.</li><li>The necessity for initial investment.</li><li>Spaces adaptation for the updated equipment.</li><li>Recurrent price variation and certain physical traits of the biomass.</li><li>Absence of a clear successful example that could validate biomass as an alternative.</li></ul>
Opportunities	**Threats**
<ul><li>Positive impact in the economic development of the region.</li><li>Opportunity to implement a resource preservation policy.</li><li>Increase the use of autochthone energy resources.</li><li>Development and modernization of the forest management and industry.</li><li>Use of IoT technologies in monitoring energetic consumption and logistic control of solid fuels supply.</li></ul>	<ul><li>Difficult to penetrate into the energy market.</li><li>Lack of a steady national policy that could promote sustainable exploitation of renewable energy in industry.</li><li>Doubt of decision makers regarding the potential of biomass.</li></ul>

Among the different sources of renewable energy, the importance of Biomass stands out. There is wide use of Biomass in energy production, namely Biomass Energy Forestry and Biomass Residual Forest (surplus exploitation). The biomass sector for energy purposes has been undergoing strong development, with an increase in the production of electricity at the Portuguese national level. By looking at Table 4, it is possible to assess the general outline of the energy mix of the textile dyeing industry in Portugal and its potential to grow and develop, since it is an important sector for the entire Portuguese economy and currently is going through a very difficult period as a result of very high costs of production, which in turn is caused mainly by the ever-rising energy taxes and costs.

Because it is a renewable resource, the utilization of biomass as the most important source for the production of steam encourages the textile industry to consume an autochthone energy source, thus reducing the dependence on imported energy of Portugal, while additionally being more environmentally friendly.

Regardless of all the aforementioned listed benefits, replacing fossil fuels is not an easy task, since it requires high investing efforts in order to upgrade the current equipment used in this industry, or replace it entirely. However, replacing the equipment is not enough, significant changes of the entire supply chain need to occur and how to store the biomass must be investigated into.

One of the most significant benefits that the use of biomass could bring is the incentive for improved economic development of rural areas through job creation, thus decreasing the rural exodus and strengthening the local industry.

However, despite the benefits achieved in this study, the substitution of conventional fossil fuels is not a sure bet, since it requires substantial investments for upgrading the existing equipment or replacing it entirely. Such a transition will also signify a transformation in the entire process of storage, movement and supply chain of biomass. Yet, this transition could be strengthened by employing IoT. Thus, as soon as great service-based, distributed energy infrastructure with IoT is implemented, it will be possible to develop innovative cost-effective concepts that could empower the textile industry with more efficient capabilities and tools to solve old problems. However, in order for this technology to be successfully implemented, several challenges have to be overcome through research and real test bed implementation.

5. Conclusions

The Portuguese textile industry is of great importance for the country's exports and economy. However, in this industry, it has been reported that the energy consumption reaches up to 60% of the total production cost. The aim of this study was to explore the potential of energy efficiency management and logistics associated with the use of biomass as a source of thermal energy in the Portuguese textile industry. Given that the increasing consumption of this type of energy source will require the creation of information networks and data analysis that will support decision making, the study performed in this paper showed that Biomass could be a reliable, sustainable and permanent substitute for fossil fuels for the textile industry. Thus, it could make the energy consumed at the industrial units more cost-effective, which is detrimental to other more traditional energy sources such as propane gas, naphtha and natural gas. Such a transition could bring substantial savings by the order of 35% in energy costs related to steam generation, which is essential to the industrial process, thus increasing the overall competitiveness of this industrial sector.

Author Contributions: L.J.R.N. conducted the study and performed the writing and original draft preparation. R.G. handled the writing and editing of the manuscript and contributed with parts of the literature review. J.C.d.O.M. supervised, revised and corrected the manuscript.

Funding: Radu Godina would like to acknowledge financial support from Fundação para a Ciência e Tecnologia (grant UID/EMS/00667/2013).

Conflicts of Interest: The authors declare no conflict of interest.

References

1. Zuberi, M.J.S.; Bless, F.; Chambers, J.; Arpagaus, C.; Bertsch, S.S.; Patel, M.K. Excess heat recovery: An invisible energy resource for the Swiss industry sector. *Appl. Energy* **2018**, *228*, 390–408. [CrossRef]
2. Lee, I.; Lee, K. The Internet of Things (IoT): Applications, investments, and challenges for enterprises. *Bus. Horiz.* **2015**, *58*, 431–440. [CrossRef]
3. Ko, S.; Lautala, P.; Handler, R.M. Securing the feedstock procurement for bioenergy products: A literature review on the biomass transportation and logistics. *J. Clean. Prod.* **2018**, *200*, 205–218. [CrossRef]
4. Nunes, L.J.R.; Godina, R.; Matias, J.C.O.; Catalão, J.P.S. Economic and environmental benefits of using textile waste for the production of thermal energy. *J. Clean. Prod.* **2018**, *171*, 1353–1360. [CrossRef]
5. Perea-Moreno, A.-J.; Perea-Moreno, M.-Á.; Hernandez-Escobedo, Q.; Manzano-Agugliaro, F. Towards forest sustainability in Mediterranean countries using biomass as fuel for heating. *J. Clean. Prod.* **2017**, *156*, 624–634. [CrossRef]
6. Ioannou, K.; Tsantopoulos, G.; Arabatzis, G.; Andreopoulou, Z.; Zafeiriou, E. A Spatial Decision Support System Framework for the Evaluation of Biomass Energy Production Locations: Case Study in the Regional Unit of Drama, Greece. *Sustainability* **2018**, *10*, 531. [CrossRef]
7. Athari, H.; Soltani, S.; Rosen, M.; Mahmoudi, S.; Morosuk, T. Thermodynamic Analysis of a Power Plant Integrated with Fogging Inlet Cooling and a Biomass Gasification. *Sustainability* **2015**, *7*, 1292–1307. [CrossRef]
8. Han, G.; Martin, R. Teaching and Learning about Biomass Energy: The Significance of Biomass Education in Schools. *Sustainability* **2018**, *10*, 996. [CrossRef]
9. Filipe dos Santos Viana, H.; Martins Rodrigues, A.; Godina, R.; Carlos de Oliveira Matias, J.; Jorge Ribeiro Nunes, L. Evaluation of the Physical, Chemical and Thermal Properties of Portuguese Maritime Pine Biomass. *Sustainability* **2018**, *10*, 2877. [CrossRef]
10. Yub Harun, N.; Parvez, A.; Afzal, M.; Yub Harun, N.; Parvez, A.M.; Afzal, M.T. Process and Energy Analysis of Pelleting Agricultural and Woody Biomass Blends. *Sustainability* **2018**, *10*, 1770. [CrossRef]
11. Li, M.; Luo, N.; Lu, Y.; Li, M.; Luo, N.; Lu, Y. Biomass Energy Technological Paradigm (BETP): Trends in This Sector. *Sustainability* **2017**, *9*, 567. [CrossRef]

12. Ren, J.; Dong, L. Evaluation of electricity supply sustainability and security: Multi-criteria decision analysis approach. *J. Clean. Prod.* **2018**, *172*, 438–453. [CrossRef]

13. Hübler, M.; Löschel, A. The EU Decarbonisation Roadmap 2050—What way to walk? *Energy Policy* **2013**, *55*, 190–207. [CrossRef]

14. Mousa, E.; Wang, C.; Riesbeck, J.; Larsson, M. Biomass applications in iron and steel industry: An overview of challenges and opportunities. *Renew. Sustain. Energy Rev.* **2016**, *65*, 1247–1266. [CrossRef]

15. Fragkos, P.; Tasios, N.; Paroussos, L.; Capros, P.; Tsani, S. Energy system impacts and policy implications of the European Intended Nationally Determined Contribution and low-carbon pathway to 2050. *Energy Policy* **2017**, *100*, 216–226. [CrossRef]

16. Outlook, Southeast Asia Energy. *World Energy Outlook 2013*; Organization for Economic: Paris, France, 2013; ISBN 978-92-64-20130-9.

17. Outlook, Southeast Asia Energy. *Energy Balances of Non-OECD Countries 2013*; OECD Publishing: Paris, France, 2013; ISBN 978-92-64-20306-8.

18. Heubaum, H.; Biermann, F. Integrating global energy and climate governance: The changing role of the International Energy Agency. *Energy Policy* **2015**, *87*, 229–239. [CrossRef]

19. Guilhermino, A.; Lourinho, G.; Brito, P.; Almeida, N. Assessment of the Use of Forest Biomass Residues for Bioenergy in Alto Alentejo, Portugal: Logistics, Economic and Financial Perspectives. *Waste Biomass Valoriz.* **2018**, *9*, 739–753. [CrossRef]

20. Database—Eurostat. Available online: https://ec.europa.eu/eurostat/web/environmental-data-centre-on-natural-resources/data/database (accessed on 9 October 2018).

21. Eurostat Energy Balance Sheets 2016 DATA. Available online: https://ec.europa.eu/eurostat/web/products-statistical-books/-/KS-EN-18-001?inheritRedirect=true (accessed on 19 January 2019).

22. Godina, R.; Nunes, L.J.R.; Santos, F.M.B.C.; Matias, J.C.O. Logistics cost analysis between wood pellets and torrefied Biomass Pellets: The case of Portugal. In Proceedings of the 2018 7th International Conference on Industrial Technology and Management (ICITM), Oxford, UK, 7–9 March 2018; pp. 284–287. [CrossRef]

23. Proskurina, S.; Heinimö, J.; Schipfer, F.; Vakkilainen, E. Biomass for industrial applications: The role of torrefaction. *Renew. Energy* **2017**, *111*, 265–274. [CrossRef]

24. Eurostat—Data Explorer. Available online: http://appsso.eurostat.ec.europa.eu/nui/show.do?dataset=sbs_na_ind_r2&lang=en (accessed on 8 October 2018).

25. Nunes, L.J.R.; Matias, J.C.O.; Catalão, J.P.S. Analysis of the use of biomass as an energy alternative for the Portuguese textile dyeing industry. *Energy* **2015**, *84*, 503–508. [CrossRef]

26. Gotzens, F.; Heinrichs, H.; Hake, J.-F.; Allelein, H.-J. The influence of continued reductions in renewable energy cost on the European electricity system. *Energy Strategy Rev.* **2018**, *21*, 71–81. [CrossRef]

27. Nunes, L.J.R.; Matias, J.C.O.; Catalão, J.P.S. Economic evaluation and experimental setup of biomass energy as sustainable alternative for textile industry. In Proceedings of the 2013 48th International Universities' Power Engineering Conference (UPEC), Dublin, Ireland, 2–5 September 2013; pp. 1–6.

28. Florin, N.H.; Harris, A.T. Enhanced hydrogen production from biomass with in situ carbon dioxide capture using calcium oxide sorbents. *Chem. Eng. Sci.* **2008**, *63*, 287–316. [CrossRef]

29. Kobayashi, N.; Guilin, P.; Kobayashi, J.; Hatano, S.; Itaya, Y.; Mori, S. A new pulverized biomass utilization technology. *Powder Technol.* **2008**, *180*, 272–283. [CrossRef]

30. De Lima, L.P.; de Deus Ribeiro, G.B.; Perez, R. The energy mix and energy efficiency analysis for Brazilian dairy industry. *J. Clean. Prod.* **2018**, *181*, 209–216. [CrossRef]

31. Haddad, M.; Abid, S.; Hamdi, M.; Bouallagui, H. Reduction of adsorbed dyes content in the discharged sludge coming from an industrial textile wastewater treatment plant using aerobic activated sludge process. *J. Environ. Manag.* **2018**, *223*, 936–946. [CrossRef] [PubMed]

32. Palamutcu, S. 2—Energy footprints in the textile industry. In *Handbook of Life Cycle Assessment (LCA) of Textiles and Clothing*; Muthu, S.S., Ed.; Woodhead Publishing Series in Textiles; Woodhead Publishing: Sawsto, UK, 2015; pp. 31–61. ISBN 978-0-08-100169-1.

33. Lin, B.; Chen, Y.; Zhang, G. Impact of technological progress on China's textile industry and future energy saving potential forecast. *Energy* **2018**, *161*, 859–869. [CrossRef]

34. Huang, B.; Zhao, J.; Geng, Y.; Tian, Y.; Jiang, P. Energy-related GHG emissions of the textile industry in China. *Resour. Conserv. Recycl.* **2017**, *119*, 69–77. [CrossRef]

35. Hasanbeigi, A.; Price, L. A review of energy use and energy efficiency technologies for the textile industry. *Renew. Sustain. Energy Rev.* **2012**, *16*, 3648–3665. [CrossRef]
36. Çay, A. Energy consumption and energy saving potential in clothing industry. *Energy* **2018**, *159*, 74–85. [CrossRef]
37. Ben Youssef, W.; Maatallah, T.; Menezo, C.; Ben Nasrallah, S. Assessment viability of a concentrating photovoltaic/thermal-energy cogeneration system (CPV/T) with storage for a textile industry application. *Sol. Energy* **2018**, *159*, 841–851. [CrossRef]
38. Nunes, L.J.R.; Matias, J.C.O.; Catalão, J.P.S. Application of biomass for the production of energy in the Portuguese textile industry. In Proceedings of the 2013 International Conference on Renewable Energy Research and Applications (ICRERA), Madrid, Spain, 20–23 October 2013; pp. 336–341.
39. Douglas, P. Woodward Presidential Address: Industry Location, Economic Development Incentives, and Clusters. *Rev. Reg. Stud.* **2012**, *42*, 5–23.
40. Ribeiro, J.M.C.; Godina, R.; de Matias, J.C.; Nunes, L.J.R. Future Perspectives of Biomass Torrefaction: Review of the Current State-Of-The-Art and Research Development. *Sustainability* **2018**, *10*, 2323. [CrossRef]
41. Busato, P.; Sopegno, A.; Berruto, R.; Bochtis, D.; Calvo, A.; Busato, P.; Sopegno, A.; Berruto, R.; Bochtis, D.; Calvo, A. A Web-Based Tool for Energy Balance Estimation in Multiple-Crops Production Systems. *Sustainability* **2017**, *9*, 789. [CrossRef]
42. Mokhatab, S.; Poe, W.A.; Mak, J.Y. *Handbook of Natural Gas Transmission and Processing: Principles and Practices*, 3rd ed.; Gulf Professional Publishing: Amsterdam, The Netherlands, 2015; ISBN 978-0-12-801499-8.
43. McKendry, P. Energy production from biomass (part 1): Overview of biomass. *Bioresour. Technol.* **2002**, *83*, 37–46. [CrossRef]
44. Carneiro, P.; Ferreira, P. The economic, environmental and strategic value of biomass. *Renew. Energy* **2012**, *44*, 17–22. [CrossRef]

 *sustainability*

Article

An Integrated Multi-Criteria Decision Making Model and AHP Weighting Uncertainty Analysis for Sustainability Assessment of Coal-Fired Power Units

Dianfa Wu *, Zhiping Yang, Ningling Wang, Chengzhou Li and Yongping Yang

National Research Center for Thermal Power Engineering and Technology, North China Electric Power University, Changping District, Beijing 102206, China; zhiping_yang@ncepu.edu.cn (Z.Y.); 50201940@ncepu.edu.cn (N.W.); chengzhou_li@ncepu.edu.cn (C.L.); yyp@ncepu.edu.cn (Y.Y.)
* Correspondence: wudian508@ncepu.edu.cn; Tel.: +86-10-6177-2000

Received: 3 April 2018; Accepted: 18 May 2018; Published: 23 May 2018

Abstract: The transformation of the power generation industry from coal-based to more sustainable energy sources is an irreversible trend. In China, the coal-fired power plant, as the main electric power supply facility at present, needs to know its own sustainability level to face the future competition. A hybrid multi-criteria decision making (MCDM) model is proposed in this paper to assess the sustainability levels of the existing Chinese coal-fired power units. The areal grey relational analysis (AGRA) method is involved in the hybrid model, and a combined weighting method is used to determine the priorities of the criteria. The combining weight fuses the fuzzy rough set (FRS) and entropy objective weighting method together with the analytic hierarchy process (AHP) subjective weighting method by game theory. Moreover, an AHP weighting uncertainty analysis using Monte Carlo (MC) simulation is introduced to measure the uncertainty of the results, and a 95 percent confidence interval (CI) is defined as the uncertainty measurement of the alternatives. A case study about eight coal-fired power units is carried out with a criteria system, which contains five aspects in an operational perspective, such as the flexibility, economic, environmental, reliability and technical criterion. The sustainability assessment is performed at the unit level, and the results give a priority rank of the eight alternatives; additionally, the uncertainty analysis supplies the extra information from a statistical perspective. This work expands a novel hybrid MCDM method to the sustainability assessment of the power generation systems, and it may be a benefit to the energy enterprises in assessing the sustainability at the unit level and enhance its ability in future sustainable development.

Keywords: areal grey relational analysis; fuzzy rough set; game theory; AHP; uncertainty analysis; coal-fired power unit

1. Introduction

Nowadays, people have realized the importance of the sustainability of energy utilization with the depletion of fossil fuel and global warming [1,2]. The expected way of future energy utilization is most likely mainly to be powered by renewable energy sources like hydropower, bioenergy, solar energy, wind power, geothermal power [3–8], etc. The sustainable development of the power industry plays an important role in the sustainable development of energy. Despite the rapid growth of renewable energy, coal-fired power generation still retains a large proportion of the whole power generation, especially in China. By the end of 2016, coal-fired power generation accounted for about 57 percent of the total installed power capacity [9], and the plants are still the main suppliers of electricity. Thus, the sustainability of the coal-fired power facility has an important impact on the sustainable utilization of energy.

In recent years, the coal-fired power plants in China have encountered some new challenges such as the generation share decreasing and utilization hours' reduction caused by the transition to renewable energy and the irrational investment of commercial capital [10,11]. For example, the average utilization hours of the coal-fired power generation equipment (600 megawatts and above) were only 3786 h in the year 2016 [9]. That means the main power units have in the long-term a low-load operation condition. Actually, the level of operation and management of the coal-fired power plants has risen up to a relatively high standard today. Thus, the potential of reaching further energy savings and emission reductions, which will enhance the sustainability of the power generation facilities, is becoming smaller than ever before. Even so, some efforts have been made to improve the situation of sustainable development in the technical aspects, which contain the operational optimization of the unsteady state [12], deep heat recovery from boiler flue gas [13], emission control technologies [14–18], etc. However, the guiding role of the government's policy is essential in the power generation alternation from coal-based to coal-free mode, such as the supply-side structural reform [19] and the Chinese 13th five-year energy development plan [20]. Therefore, it is necessary to make a reasonable assessment of the sustainability level of the coal-fired power plants. The appropriate policy suitable for the current level of sustainable development can reduce the unnecessary cost of the energy sustainable development process, on the one hand, such as the power over-investment [10,11], and on the other hand, it can optimize the sustainability of the overall energy generation system, such as enhancing the flexibility of the coal power units to accommodate more capacity of wind power and photovoltaic power [21,22].

There are several conventional methodologies that can be employed for the sustainability assessment, such as exergy [23,24], emergy [25,26], life cycle assessment (LCA) [27] and the multi-criteria decision making (MCDM) methods [28–35]. The MCDM method, which is particularly suitable for the evaluation or decision making of a complex system with multiple indicators, is widely used in the energy field. Many prevalent specific methods are available for sustainability assessment of thermal power plants, which include TOPSIS [36–38], AHP [39–42], the entropy method [43], the fuzzy method [41,44], the grey method [40,45], the rough set [46–48], set pair analysis (SPA) [26], etc. Sometimes, we use a hybrid model to take advantage of each classical theory. Grey relational analysis (GRA) is one of the branches of grey theory, which was introduced by Deng in 1982 [45]. The merit of the GRA is its insensitivity to the amount of raw data that usually has insufficient and poor information, and it can derive an unbiased estimate by determining the relationship of the data sequences. However, the conventional GRA only considers the linear relationship of alternatives of the same indicators using the grey relational coefficient number, and it is unable to measure the relationships between indicators. Thus, an improved evaluation method, areal grey relational analysis (AGRA), was developed to cope with the demand of the areal grey relational coefficient number of indices [49–51].

It is important to determine the weights of the criteria in the process of an MCDM method. Generally, the weighting methods can be divided into several categories such as subjective, objective and combination methods. The subjective weight needs the knowledge of experts. Except for the Delphi [30] method, AHP has been widely used as a subjective weighting method benefiting from its ability to quantify the qualitative factors by the hierarchical decomposition process [39–41]. For the objective weight determination, the entropy method is a typical principle to cope with the uncertainty of the index [41,44,52], and the CRITIC method can handle the relationship between indicators by introducing the linear correlation coefficient [53]. However, in the real world, the relationship between the data sequences is usually nonlinear and complex. Then, the rough set theory and fuzzy set theory can be employed to treat the nonlinear problem from different perspectives [46–48,54]. We usually use an integration weight based on various principles to achieve practical results. The weighting combination methods mainly utilize the rules of addition, multiplication [26], game theory [50,55,56], etc. Game theory is a practical way to get the optimum equilibrium solution among different kinds of weights with conflicts.

When performing the AHP method to determine the importance degree of criteria, the result is strongly affected by the expert's knowledge and may vary significantly with different decision makers. One of the reasons is that the experts may be unfamiliar with some specific indices, so they cannot make a precise judgment in the pairwise comparison process. In order to overcome these problems, the Monte Carlo (MC) method was introduced in many works to promote the performance of the classical AHP method [57–61]. The stochastic-based AHP approach can give us more decision information from a statistical perspective.

For the sustainability assessment of an energy system, we usually establish the evaluation system considering the technical, economic, environmental and social factors [4,31,33,62]. There is no doubt that it is significant to discuss the sustainability of the energy system from a broad perspective with the consideration of the interaction with human society, but it is also worthy to research the sustainability of specific power generation systems from a smaller perspective, such as the sustainability evaluation of the coal-fired power unit at the equipment level. In order to obtain a reasonable evaluation of the power units, some performance indices such as economic, environmental and technical can be selected under the premise of device safety. For the existing coal-fired power units of China, improving their operational flexibility is the most practical option at present for the systematic sustainable energy transition [4,21,22].

In this paper, we develop a hybrid MCDM model to evaluate the sustainability of the coal-fired power units. The proposed model uses the AGRA method integrating a combined weight. We employ the AHP method to gather the knowledge of experts in the thermodynamic field, and we use the entropy technique to measure the uncertainties of the criteria. Moreover, the FRS is used to get an additional objective weight, and ultimately, the three weights (AHP, entropy, FRS-based) are combined with the optimizing method based on game theory. We also establish a multi-criteria evaluation system to perform the sustainability assessment at the power unit level, which considers five criterion categories including flexibility, economic, environmental, reliability and the technical index and with sub-criteria under every criterion category, respectively. Furthermore, in order to detect the uncertainty of the assessment results, a stochastic simulation work is performed based on the AHP method. A case study is also carried out with the sustainability assessment of eight coal-fired power units of a similar design generation capacity. We believe this work is beneficial to the sustainability of the total energy system and may provide some valuable references for the policy-making sectors.

The rest of the paper is organized as follows: In Section 2, we introduce the hybrid model that combines the AGRA method and some selected weighting methods (AHP, entropy and FRS weight). Section 3 introduces the MC-AHP method to develop the weight uncertainty analysis. Section 4 discusses a case study to illustrate the effectiveness of the evaluation model for the sustainability assessment of power units. Section 5 shows the results and discussions about the case research, and Section 6 draws out some conclusions. Additionally, the symbols used in this paper are listed in the Nomenclature Section.

2. The Integrated MCDM Model

This section introduces the proposed hybrid MCDM model used in the paper for the sustainability evaluation of coal-fired power units. The integrating model involves the areal-based grey theory, AHP, entropy, FRS and game theory. Different methods have their own advantages in dealing with the raw data from the different point of views. The details are briefly introduced as follows.

2.1. Areal Grey Relational Analysis Method

Grey system theory has been widely used in various fields because of its advantages of treating complex systems that have various interrelated indicators [45]. Although Deng's GRA method, which was developed from grey system theory, is prevalent in multi-objective decision-making problems, it still can be improved in some aspects [28,40,63,64].

Suppose there are m kinds of evaluated criteria $a_j (1 \leq j \leq m)$ and n kinds alternatives $f_i (1 \leq i \leq n)$; they form a sample matrix $X = (x_{ij})_{n \times m}$, and X can be written as:

$$X = (x_{ij})_{n \times m} = \begin{array}{c} \\ f_1 \\ f_2 \\ \vdots \\ f_n \end{array} \begin{array}{cccc} a_1 & a_2 & \cdots & a_m \\ \left[\begin{array}{cccc} x_{11} & x_{12} & \cdots & x_{1m} \\ x_{21} & x_{22} & \cdots & x_{2m} \\ \vdots & \vdots & \ddots & \vdots \\ x_{n1} & x_{n2} & \cdots & x_{nm} \end{array} \right] \end{array} \tag{1}$$

Suppose there are two alternative sequences f_i, $f_k (1 \leq i,k \leq n)$ with the same criteria $a_j (1 \leq j \leq m)$; the classical GRA method calculates the grey relational degree (GRD) of the two sequences by the absolute difference of the corresponding index values, in the format of $\left| x_{ij} - x_{kj} \right|$. However, for the AGRA, one of the methods derived from the GRA [49–51], its GRD is calculated not only considering the absolute difference of the criteria, but also taking into account the geometric area between the adjacent indices. Generally speaking, AGRA theory determines the areal correlation degrees according to the similarity among the sequence curves like the GRA method. That means, the more similarity between two sequences, the higher the correlation degree will be. We assume an optimal criterion sequence as a reference, and then, we can determine the similarity of the indices between the selected alternatives and the reference one. The more the similarity between the two index sequences, the better the comprehensive performance of the alternatives investigated. The detailed procedure is introduced as follows.

2.1.1. Normalizing the Criteria

The criteria usually have different dimensions and magnitudes. Therefore, the normalizing procedure should be performed first. For the sample matrix X described in Equation (1), the following formulas can be used. For the criterion the bigger the better (or benefit attribute), it can be normalized as:

$$y_{ij} = \frac{x_{ij} - \min\limits_{i}(x_{ij})}{\max\limits_{i}(x_{ij}) - \min\limits_{i}(x_{ij})}, (1 \leq i \leq n, 1 \leq j \leq m) \tag{2}$$

and for the criterion the smaller the better (or cost attribute), it can be normalized as:

$$y_{ij} = \frac{\max\limits_{i}(x_{ij}) - x_{ij}}{\max\limits_{i}(x_{ij}) - \min\limits_{i}(x_{ij})}, (1 \leq i \leq n, 1 \leq j \leq m) \tag{3}$$

where in Equations (2) and (3), $\max\limits_{i}(x_{i,j})$ and $\min\limits_{i}(x_{i,j})$ respectively mean the maximum and the minimum value of the row i, and $Y = (y_{ij})_{n \times m}$ is the linear scale standardized matrix.

2.1.2. Calculating the Areal Grey Relational Coefficient Number

Take Figure 1 as an example: to calculate the areal grey relational coefficient values, the key step is to determine the area values between the alternative sequences (for instance, the solid line) and the reference one (for instance, the dotted line). The geometric figure made up by the two adjacent criteria of the two sequences may have the following cases: (A) triangle, (B) two lines intersecting, (C) trapezoid and (D) synthesized to one line. Obviously, the smaller the correlation area between the sequences is, the closer the two curves are, and then, the greater the correlation degree is, and vice versa. The extreme situation is the case of (D), that is to say, in that case, the two points of a criterion totally coincide with each other, and the grey relational degree is up to the biggest value of one.

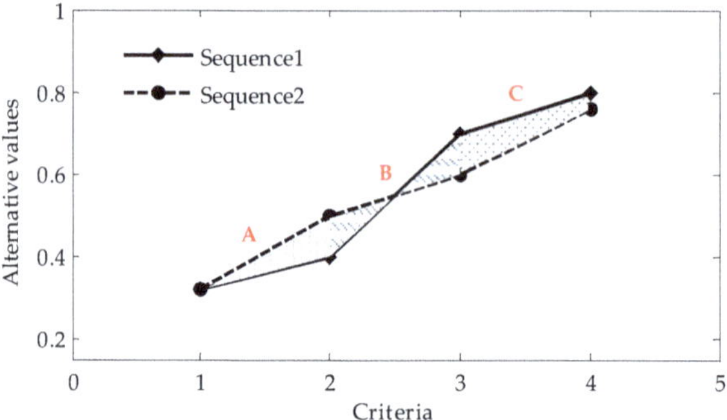

Figure 1. Graphical representation of possible relations between two sequences.

For the normalized matrix $Y = (y_{ij})_{n \times m}$, set $y_0 = \{y_0(1), y_0(2), \ldots, y_0(m)\}$ as the reference sequence, where $y_0(j) = \max(y_j(i))$, $(1 \leq i \leq n, 1 \leq j \leq m)$, and the series $y_i, (1 \leq i \leq n)$ is the comparison one. Set the horizontal distance of two adjacent criterion equal to one, then the areas s_{ij} can be calculated with several cases depicted in the following steps.

First, we need to define a flag function $f_i(j) = F(z_i(j))$, where $z_i(j) = y_0(j) - y_i(j)$. This function is used to judge the relative positions of the two curves, and it has the possible values:

$$\begin{cases} f_i(j) > 0, & z_i(j) > 0 \\ f_i(j) = 0, & z_i(j) = 0 \\ f_i(j) < 0, & z_i(j) > 0 \end{cases} , (1 \leq i \leq n, 1 \leq j \leq m) \tag{4}$$

According to the geometric relationship, the calculation methods about the different areal shapes shown in Figure 1 can be summarized as six formulas:

$$s_{ij} = \begin{cases} |(y_0(2) - y_i(2)) \times 1|/2, & f_i(j) = f_i(1) = 0 \\ |(y_0(j) - y_i(j) + y_0(j+1) - y_i(j+1)) \times 1|/2, & f_i(j) \times f_i(j+1) > 0 \\ (|(y_0(j) - y_i(j)| \times (y_i(j) - j) \\ \quad + |y_0(j+1) - y_i(j+1)| \times ((j+1) - y_i(j)))/2, & f_i(j) \times f_i(j+1) < 0 \\ 0, & f_i(j) = f_i(j+1) = 0 \\ |(y_0(j+1) - y_i(j+1)) \times 1|/2, & f_i(j) = 0 \ \& \ f_i(j+1) \neq 0 \\ |(y_0(j) - y_i(j)) \times 1|/2, & f_i(j) \neq 0 \ \& \ f_i(j+1) = 0 \end{cases} \tag{5}$$

Second, for the j-th index of the i-th alternative, after the areas s_{ij} obtained with Equation (5), the areal grey relational coefficient matrix can be obtained with the elements formulated as:

$$\zeta_{ij} = \frac{\min\limits_{i} \min\limits_{j} \{s_{ij}\} + \rho \max\limits_{i} \max\limits_{j} \{s_{ij}\}}{s_{ij} + \rho \max\limits_{i} \max\limits_{j} \{s_{ij}\}}, \ (1 \leq i \leq n, 1 \leq j \leq m) \tag{6}$$

where ζ_{ij} is the areal grey relational coefficient of the j-th index of the i-th alternative. The factor $\rho \in [0, 1]$ is the distinguishing coefficient like the GRA method, and we set it as 0.5.

2.1.3. Calculating the Areal Grey Relational Result Vector

Based on the methodology of AGRA and the indices' weights $W = (w_1, w_2, \ldots, w_m)^T$, the final calculation model can be deduced as:

$$R = \zeta \cdot W = (r_1, r_2, \ldots r_n)^T \tag{7}$$

where $r_i = \sum_{j=1}^{m} \xi_i(j) \cdot w_j, (1 \leq i \leq n, 1 \leq j \leq m)$, the vector R is the result of the evaluated alternatives, ξ is the areal grey relational coefficients matrix of the criteria and W is the weight of the evaluated criteria. According to the principle of maximum correlation, the evaluation alternative can be sorted, and the larger the r_i, the better the alternative.

2.2. Selected Weighting Methods

In this subsection, some weighting methods are briefly introduced, including the FRS, entropy principle and AHP method, and we use game theory to combine these three weights to perform a trade-off of their benefits.

2.2.1. Fuzzy Rough Set Objective Weight

The datasets we face are usually incomplete, imprecise and inaccurate. To settle these problems, the rough set theory was developed by Pawlak in 1982, which can describe the rough data and dig out useful knowledge just based on the raw data themselves [65]. The rough set theory is usually used as a powerful tool of information reduction in the machine learning and data mining fields. However, the classical rough set is good at dealing with symbolic variables, and for the continuous numerical values, a key step of data discretization is needed. The problem is that the discrete methods are not uniform, and they may cause additional loss of information of the research data. Then, fuzzy set theory, proposed by Zadeh, was introduced to the classical rough set to overcome the weakness, and it forms the branch of the fuzzy rough set (FRS) method [46–48,54,66].

First, we introduce some basic concepts about the fuzzy equivalence relation according to the reference paper [66]. Let U be a nonempty universe of discourse and $F(U \times U)$ be the fuzzy power set on $U \times U$. R is the fuzzy relation on $U \times U$, if $R \epsilon F(U \times U)$, where $R(x,y)$ measures the strength of the relationship between $x \epsilon U$ and $y \epsilon U$. Then, the fuzzy relation R has the following properties. R is reflexive if $R(x,x) = 1$ for any $x \epsilon U$, and R is T-transitive if $R(x,y) \geq T(R(x,z), R(y,z))$ for a triangular norm T and any $x, y, z \epsilon U$. Furthermore, R is called a T-similarity relation if R is reflexive, symmetric and T-transitive. Specially, if $T = min$, then R is called a fuzzy equivalence relation.

Then, suppose a knowledge representation system $S = (U, A, V, f)$, where U is the collection of objects, A is the collection of properties, $V = \cup_{a \in A} V_a$ is the value range and $f : U \times A \to V$ is the information function. For a fuzzy set X, we define the lower and upper approximation operators based on a T-similarity relation R as:

$$\left\{ \begin{array}{l} \underline{R}X(x) = \inf_{y \in U} \max\{1 - R(x,y), X(y)\} \\ \overline{R}X(x) = \sup_{y \in U} \min\{R(x,y), X(y)\} \end{array} \right. \tag{8}$$

where $x \epsilon U$. The lower and upper approximation operators are used to determine the certainty degree and possibility degree of x that belong to X, respectively. Let $P, Q \in A$ be a pair of fuzzy equivalence relations, then the positive region of Q set about P set is denoted as:

$$POS_P(Q) = \cup_{X \subseteq U/Q} \underline{P}X \tag{9}$$

For a criterion $a_j \in A$, we define the dependence degree of a_j on the criteria set A as:

$$\gamma_{a_j}(A) = \left| pos_{a_j}(A) \right| / \left| U \right| \tag{10}$$

In order to measure the importance of the index a_j, we first remove this criteria, then we can calculate the dependence degree of the rest criteria set $A - \{a_j\}$ on A, the significance degree of the measured index can be obtained with the equation:

$$sig_{A-\{a_j\}}(a_j) = 1 - \gamma_{A-\{a_j\}}(A) = 1 - \left| pos_{A-\{a_j\}}(A) \right| / \left| U \right| \tag{11}$$

Finally, we normalize the significance degree, and the weight based on the FRS method is formed as:

$$w_j^{frs} = \frac{sig_{A-\{a_j\}}(a_j)}{\sum\limits_{j=1}^{m} sig_{A-\{a_j\}}(a_j)} \tag{12}$$

2.2.2. Entropy Objective Weight

The entropy principle is usually used to measure the irreversible phenomenon of the motion state, and later, it was introduced into the information field to calculate the uncertainty of data series. The entropy weight we used in this work is according to some published literature [41,44,52]. Generally, the greater the amount of information, the less the uncertainty and the entropy will be, and vice versa. The steps of the method are introduced in the following briefly.

For the sample dataset with n kinds of alternatives and m kinds of criteria, which form as:

$$X = (x_{ij})_{n \times m}, \ i = 1, 2, \ldots, n; \ j = 1, 2, \ldots, m \tag{13}$$

First, we need to normalize the matrix X with the equation:

$$p_{ij} = x_{ij} / \sum\limits_{i=1}^{n} x_{ij}, \ i = 1, 2, \ldots, n; \ j = 1, 2, \ldots, m \tag{14}$$

Second, we calculate the entropy of each criterion with the formula:

$$e_j = \frac{\sum\limits_{i=1}^{n} p_{ij} \ln p_{ij}}{\ln n}, \ j = 1, 2, \ldots, m \tag{15}$$

Finally, we get the entropy-based weigh of the criteria by using the following formula:

$$w_j^{ent} = \frac{1 - e_j}{\sum\limits_{j=1}^{m} (1 - e_j)}, \ j = 1, 2, \ldots, m \tag{16}$$

2.2.3. AHP Subjective Weight

The AHP method is becoming a more and more popular tool in dealing with the MCDM problems in recent years [39–42]. Although it has been disputed and has some insufficiencies in the calculation process, there is no doubt that it is a powerful approach to cope with the real-world complex evaluations or decision matters due to the ability to quantify some qualitative questions. Generally, the method reflects the procedures of people in working out a complex or messy scientific problem; that is, firstly decomposition, then judgment and finally synthesizing. Except for being used as an MCDM method to draw out the priority of the candidate alternatives, it can also be used as a weighting method just to give the prioritization of the criteria. The latter application we used in this paper is briefly introduced as follows in several steps according to the previous literature.

Step 1: For the established hierarchical decision system with m kinds of criteria, we use the classical Saaty's comparison scale of 1–9 (see Table 1) to determine the pairwise comparison matrix $A = \left(a_{jk}\right)_{m \times m}$ for the decision makers.

Table 1. Saaty's comparison scale of AHP.

Intensity of Weight	Definition
1	Equal importance
3	Moderate importance
5	Strong importance
7	Very strong importance
9	Absolute (extreme) importance
2, 4, 6, 8	Intermediate values

Step 2: We standardize the comparison matrix A using the equation:

$$\overline{a}_{jk} = a_{jk} / \sum_{k=1}^{m} a_{jk}, \ (j, k = 1, 2, \cdots, m) \tag{17}$$

Then, we sum the elements of the matrix $\overline{A} = \left(\overline{a}_{jk}\right)_{m \times m}$ as columns, and get a vector $\overline{w}_j$:

$$\overline{w}_j = \sum_{k=1}^{m} \overline{a}_{jk}, \ (i = 1, 2, \cdots, m) \tag{18}$$

Furthermore, we normalize the vector $\overline{w}_j$, the criteria weight based on the AHP method terms, as:

$$w_j^{ahp} = \overline{w}_j / \sum_{j=1}^{m} \overline{w}_j, j = 1, 2, \cdots, m \tag{19}$$

Step 3: We verify the weight we obtained in Step 2. To validate the decision matrix of the AHP, the maximum feature value λ_{max} is calculated by the formula:

$$\lambda_{\max} \cdot w^{ahp} = A \cdot w^{ahp} \tag{20}$$

then, the consistency ratio CR is used to check the consistency by comparing the random index values listed in Table 2, using the equation:

$$CR = CI/RI \tag{21}$$

where $CI = (\lambda_{max} - m)/(m - 1)$ (m is the dimension of the comparison matrix) and RI is the random index. Finally, if the value CR is smaller than 0.1, the results are assumed to pass the consistency check. Otherwise, we need to adjust the pairwise comparison matrix for the decision makers and recalculate it until the results pass the consistency check.

Table 2. Random index values.

m	3	4	5	6	7	8	9	10
RI	0.58	0.9	1.12	1.24	1.32	1.41	1.45	1.49

2.2.4. Weighting Combination by Game Theory

Game theory is a branch of modern mathematics and is a popular method in weight aggregation [50,55,56]. The key idea of the method is to minimize the deviation between weights by the mathematical programming method. In this work, we use this theory to get the optimum equilibrium solution among the selected criteria weights. The calculation steps can be summarized as follows.

Step 1: For l kinds of weights, they form a weight set $W = \{w_1, w_1, \cdots w_l\}$. Then, a possible weight vector w with the form of an arbitrary linear combination may be expressed as:

$$w = \sum_{j=1}^{l} \alpha_j w_j^{\mathrm{T}} \ (\alpha_j > 0) \tag{22}$$

where α_j is the weight coefficient number.

Step 2: Calculate the weight coefficient α_j with the following game theory model:

$$\min || \sum_{j=1}^{l} \alpha_j w_j^T - w_i^T ||^2 \ (i = 1, 2, \cdots, l) \tag{23}$$

where l is the number of selected weights and w_j is the possible weight of the weight set. Then, we solve the equation with the first-order derivation and have the formula as:

$$\sum_{j=1}^{l} \alpha_j w_i w_j^{\mathrm{T}} = w_i w_i^{\mathrm{T}} \tag{24}$$

The equation can also be written in the expanded way:

$$\begin{bmatrix} w_1 \cdot w_1^{\mathrm{T}} & w_1 \cdot w_2^{\mathrm{T}} & \cdots & w_1 \cdot w_l^{\mathrm{T}} \\ w_2 \cdot w_1^{\mathrm{T}} & w_2 \cdot w_2^{\mathrm{T}} & \cdots & w_2 \cdot w_l^{\mathrm{T}} \\ \cdots & \cdots & \cdots & \cdots \\ w_l \cdot w_1^{\mathrm{T}} & w_l \cdot w_2^{\mathrm{T}} & \cdots & w_l \cdot w_l^{\mathrm{T}} \end{bmatrix} \begin{bmatrix} \alpha_1 \\ \alpha_2 \\ \cdots \\ \alpha_l \end{bmatrix} = \begin{bmatrix} w_1 \cdot w_1^{\mathrm{T}} \\ w_2 \cdot w_2^{\mathrm{T}} \\ \cdots \\ w_l \cdot w_l^{\mathrm{T}} \end{bmatrix} \tag{25}$$

Step 3: Then, the weight coefficient α_j can be calculated by the Equation (24), and usually, the norm of vector α is not constantly equal to one; thus we need to normalize the vector using the formula:

$$\alpha_j^* = \alpha_j / \sum_{j=1}^{l} \alpha_j \tag{26}$$

Step 4: Finally, we obtain the integrated weight by the equation:

$$w^* = \sum_{j=1}^{l} \alpha_j^* w_j^{\mathrm{T}} \tag{27}$$

2.3. The Framework of the Integrated MCDM Model

Based on the methods we introduced above, we construct a hybrid MCDM model in this subsection. The sustainability assessment problems usually can be depicted using the hierarchical indicator system shown in Figure 2. The problem we faced can be divided into several categories, and we can describe every category with a criterion (or factor). Every criterion may have several sub-criteria (or indicators), and so on. The criteria are usually selected relying on the experiences of decision makers or estimators and are adjusted dynamically in practice. The proposed integrating MCDM model is illustrated in the following based on the indicator system in Figure 2.

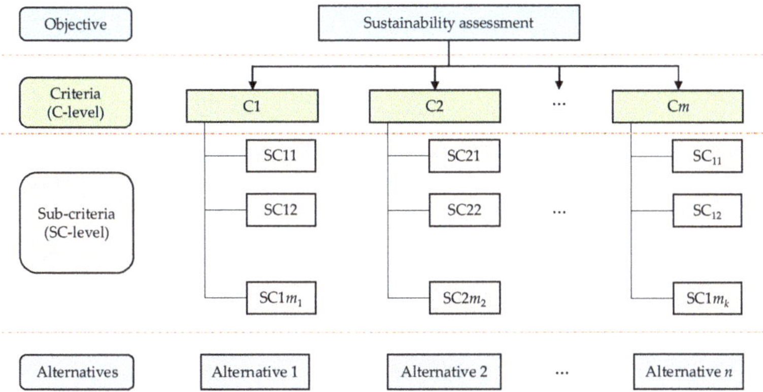

Figure 2. The hierarchy of the criteria system for sustainability assessment.

The calculating method mainly includes two steps. The first step is to synthesize the information of n kinds of alternatives represented by the respective sub-criteria (SC-level). Specifically, for each criterion category (m kinds), we use the FRS weighting method and entropy weighting method to get the objective weights and then use game theory to combine them. Then, based on the combined weight, we use the AGRA method to obtain the assessment results. This procedure is repeated for each criterion until we get m group sustainability results corresponding to each criterion. Then, the m group results form a new matrix (denoted as R_C). In the second step, we use the matrix obtained from the first step to calculate the final evaluation results of the criteria (C-level). Specifically, we use the FRS and entropy method to calculate the objective weights, and additionally, in this step, we performed the AHP method by experts to obtain the subjective weight, then we integrated the three weights by game theory again. At last, we reach the objective with the AGRA method synthesizing the R_C and the corresponding weight factor. The flowchart of the whole algorithm is shown in Figure 3.

In the process of the two-fold calculation for sustainability assessment, we only use the AHP weight in the second step, i.e., when calculating the sub-criteria weights, we just used the objective weighting methods without the subjective experiences. We take this strategy mainly for the following reasons. The AHP process usually needs much expert knowledge to fulfill the pairwise comparison matrix and occasionally the subsequent adjustments if the matrix has not passed the consistency check. These additional difficulties will be obvious if there is a large number of the sub-criteria in each criterion category. Another reason is that, actually, some experts more easily give a reasonable decision value to the high-level index than to the bottom one. For example, they may easily make a reasonable judgment about the priority of the economic factor versus the environmental factor, but they may confuse the specific indicators subordinate to one of the criteria. One of the reasons is that sometimes, they are not familiar with all of the indicators due to their work usually focusing on different perspectives. Based on the above reasons, we only fuse the expert knowledge by AHP weights in the criteria-level calculation process. In this way, expert preference is ultimately reflected in the final result.

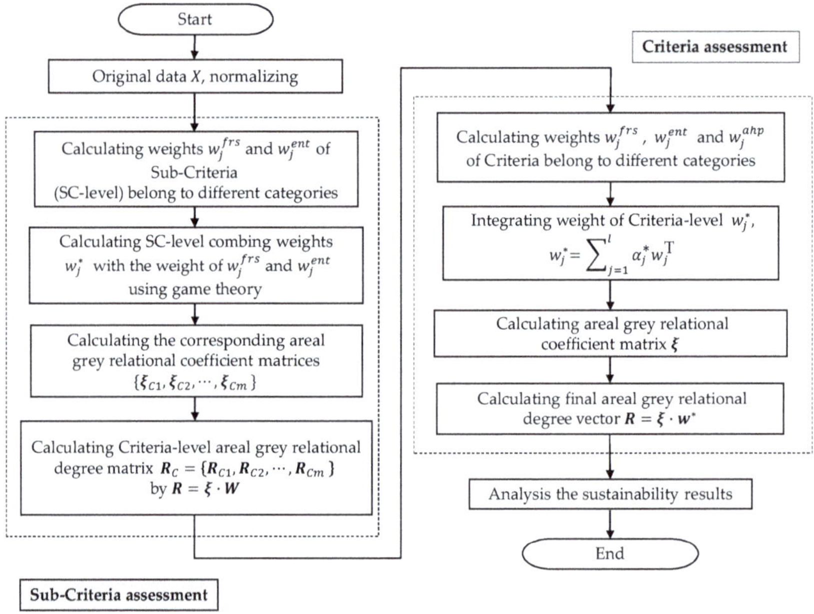

Figure 3. The framework of the hybrid MCDM algorithm.

3. AHP Weighting Uncertainty Analysis

In this section, we introduce the weighting uncertainty analysis with the stochastic AHP method. Usually, the MC method is used in the AHP process to treat the imprecise or incomplete pairwise comparison matrix, which is the work that ought to be completed by the experts in the classical AHP method [57,58,60,67]. However, the purpose we introduced the MC-AHP method in this paper is mainly to illustrate the sensitivity of the assessment results with the stochastic AHP weights from a statistical perspective. The algorithm is shown in Figure 4 and it is described as follows.

Step 1: Suppose there are n kinds of alternatives with m kinds of criteria; if we perform the AHP method to assess the alternatives, we need $m(m-1)/2$ independent elements to complete the pairwise comparison matrix, and the possible element is one of the integers between one and nine and their reciprocals, i.e., the optional element set C ought to be constructed as:

$$C = \left\{ \frac{1}{9}, \frac{1}{8}, \frac{1}{7}, \frac{1}{6}, \frac{1}{5}, \frac{1}{4}, \frac{1}{3}, \frac{1}{2}, 1, 1, 2, 3, 4, 5, 6, 7, 8, 9 \right\} \tag{28}$$

Step 2: Creating a pairwise comparison matrix $A_{m \times m}$, we select every element a_{ij} ($2 \leq i \leq m$, $1 \leq j \leq m-1$) from the set C randomly with equal probabilities, and after that, we complement the reciprocals with $a_{ji} = 1/a_{ij}$ correspondingly. Then, we calculate the AHP weight with the method mentioned in Section 2.2.3.

Step 3: We check the AHP weight with the consistency rules shown in Table 2. If the consistency check failed, we repeat Step 2 to create a new weight vector.

Step 4: Repeating the process of Step 2 and Step 3, we generate B group's AHP weight vector.

Step 5: We perform the sustainability assessment with the hybrid MCDM model that we proposed in Section 2 (see Figure 3) B times using the B group's AHP weights generated from Step 4 correspondingly. Then, we get B group's assessment results about the n kind alternatives, that is the results matrix has a dimension of $n \times B$.

Step 6: We analyze the result vectors of each alternative with the statistical methods, i.e., giving out the points of 0.025 fractile and 0.975 fractile, the width of the 95% confidence intervals (CI) and the probability distributions.

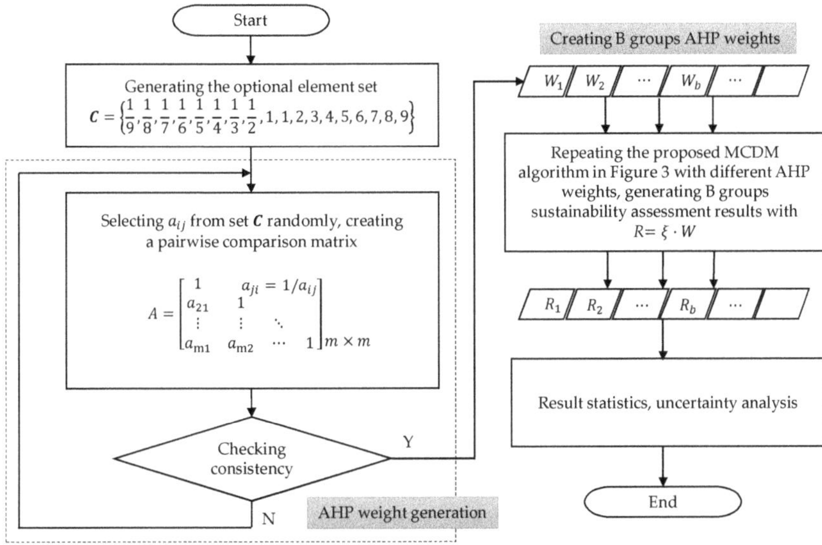

Figure 4. The flowchart of the AHP uncertainty analysis procedures.

Besides, in order to calculate the proposed model and implement the uncertainty analysis introduced in this section, we developed a series of programs in MATLAB (v 8.3.0.532) using a personal computer equipped with an Intel(R) Core(TM) i5-2400 CPU @ 3.10 GHz configuration and a 64-bit Windows 7 system.

4. A Case Study of the Sustainability Assessment of Power Plants

In this section, we performed a case study about the power units selected from some power plants. Firstly, we construct a hierarchical criteria system with the detailed indicators and also give some necessary data in the following subsection.

The schematic of a conventional coal-fired power unit (plant) is shown in Figure 5. The unit includes several key devices: boiler, steam turbine, generator, steam condenser and the flue gas purification equipment such as selective catalytic reduction (SCR), electrostatic precipitator (ESP) and the flue gas desulfurization (FGD). The system is briefly introduced as follows: the fuel (coal) is sent to the boiler to heat the water into steam, and the steam with a high temperature and high pressure is expanded in the turbine component. Then, the turbine drives the generator to generate the power, which is sent to the power grid, and the exhausted steam from the turbine is condensed by the condenser, which is usually cooled by water or air. Then, the water is sent to the boiler again as feedwater, and they complete a thermodynamic cycle finally. Simultaneously, the exhaust flue gas from the boiler is released into the atmosphere through the purification facilities.

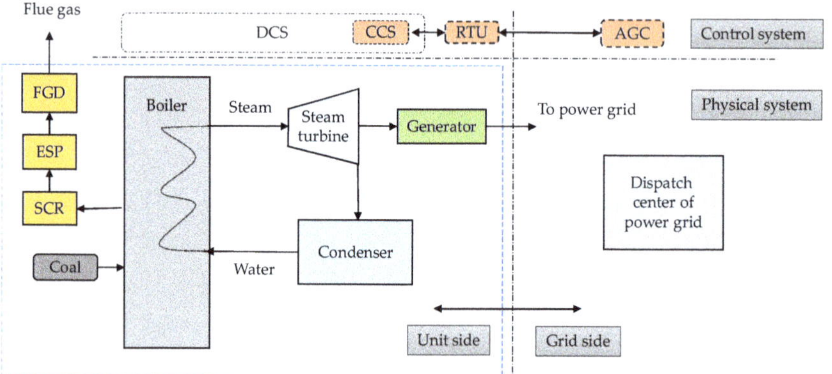

Figure 5. Schematic of a coal-fired power plant accessed by the power grid (adapted from [33]; FGD, flue gas desulfurization; ESP, electrostatic precipitator; SCR, selective catalytic reduction; DCS, distributed control system; CCS, coordinate control system; RTU, remote terminal unit; AGC, automatic generation control).

The physical device system has been briefly introduced above, and for the control system, the whole coal-fired power unit is under the control of the distributed control system (DCS). The coordinate control system (CCS), one of the subsystems of the DCS, has the ability to respond to power load instructions sent out by the automatic generation control (AGC) system through the remote terminal unit (RTU) device, which is on the power plant side. However, the RTU is a functional part of the AGC system, which is one of the control systems of the power grid dispatch center.

4.1. Criteria Selection

As we mentioned in the Introduction, the coal-fired power plant is the most important power generation facility of the national power grid of China. Unlike the generalized energy system sustainable development analysis, in this paper, we just focus on the specific coal-fired power generation systems. That is, we performed the sustainability assessment just at the unit level and from the operational perspective.

For the coal-fired power units, we select the criteria that can reflect the operational sustainability considering the rules that the indicator set should be integrated, independent, available, etc. Referencing the literature [32,33,63,68], we selected five criteria categories with their own sub-criteria, that is flexibility (C1), economic (C2), environmental (C3), reliability (C4) and the technical criterion (C5). The detailed information is introduced as follows.

4.1.1. Flexibility Criterion

In recent years, for the existing coal-fired power units of China, what the electric grid has wanted is the operational flexibility to make the national energy system more sustainable, other than the power generating ability, as in the early years. Considering the technical availability, we select five sub-criteria from the AGC system, which can be obtained from the DCS system of the power unit. The indicators include: AGC availability ratio (SC11), which reflects the available state of the unit's AGC function and is the ratio of the statistical time when the device maintains an available state to the total effective time during the statistical period. The effective time of the AGC system refers to the statistical period removing the unavailable time, which is not caused by the power plant itself, but usually by the maintenance reason or channel fault, etc. The AGC regulation rate factor (SC12) is the ratio of the regulating rate to the standard regulating rate of the electric power generation unit. The regulating rate is the average of the rising rate and the declining rate in the statistical period.

The AGC regulation precision factor (SC13) is the ratio of the regulation deviation to the allowable deviation, where the regulation precision means the difference between the actual output to the AGC load instruction output. The regulation deviation is the value between the AGC load instruction with the actual stable output of the generator responding. The AGC response time factor (SC14) is the ratio of actual response time to the standard response time. The AGC response time is used to break through the regulation dead zone in the same adjustment direction based on the original generator output point. The AGC adjustable capacity (SC15) refers to the ratio of the adjustable unit capacity for the power grid to the design load rate.

4.1.2. Economic Criterion

For a power unit, the economic operation is the ability of resource savings and mainly reflects the efficiency of the facilities. The economic concept we used here is in techno-economic [63,68], which is a narrow sense of the conventional economic conception. These indicators can be gathered from the department of production management of the power plant.

The indicators include: The net coal consumption rate (SC21) is one of the most important indices representing power plant performance, which represents the amount of standard coal consumed per kWh and can reflect the operating status, maintenance quality and management level of a power plant. It can be deduced from the net efficiency of the power unit with the equation:

$$b = 3600/(H_{std}\eta_{net}) \tag{29}$$

where b is the net coal consumption rate, g/kWh; H_{std} is the heat value of standard coal with the value of 29.308 kJ/g; and η_{net} is the net efficiency of the unit.

The auxiliary power ratio (SC22) reflects the electric power consumption of the power unit itself in the process of power generation by the auxiliary devices such as fans and pumps. The oil consumption rate (SC23) is caused by the unit startup and sundown and the low-load combustion stability. The water consumption rate (SC24) reflects the water saving benefits and usually has a big difference in the different cooling conditions (water cooling or air cooling).

4.1.3. Environmental Criterion

Pollutant emissions are a major contributor to the environmental deterioration of China. To reflect the environment protection ability of the power unit operation, we mainly consider the gaseous pollutants in coal-fired boiler flue gas exhaust according to the current level of technology. The indicators include SO_2 emission concentration (SC31), NO_X emission concentration (SC32) and the dust emission concentration (SC23), which can be obtained by an online monitoring platform integrated with the DCS.

It should be noted that CO_2 is not considered in this paper like [63] does. On the one hand, we argue that the emission level of CO_2 mainly depends on the features of the input coal and the efficiency of the unit facility. However, the composition of the coal depends on its source and has little relationship with the sustainability level of the unit itself. Moreover, the efficiency of the unit is reflected by the C21 indicator (net coal consumption rate) indirectly. On the other hand, emission removal equipment has not been widely used in China, that is we have no effective control measures. The other pollutants like Hg and Cu [16,17] are not considered either for a similar reason. Otherwise, the original design of the power plants was based on the principle of water saving, and the optimization of the water balance system is carried out during the operation period. Thus, most of the wastewater is reused, and the remaining small amount of wastewater which is difficult to recover is comprehensively utilized in another way such as wetting dry ash and coal dust suppression. The power plants are discharged without any waste liquid [69]. Meanwhile, the coal ash is comprehensively utilized in a cyclic way. Thus, we do not consider the indicators of sewage.

4.1.4. Reliability Criterion

Sustainability is based on the premise of the reliability of the devices. The reliability reflects the technical level and management level of the power plant at present. We selected three comprehensive indicators to represent the reliability of the coal-fired power unit. Previously, we introduce some typical states of the unit possibly used by the reliability assessment: the status includes the active and inactive situation, and the active status contains available and unavailable states. When the unit is in an available state, it may be in service or in reserve. While the unit is in the service, in reserve or unavailable state, it can be divided into two situations, that is planned or unplanned, respectively. This information is shown in Figure 6.

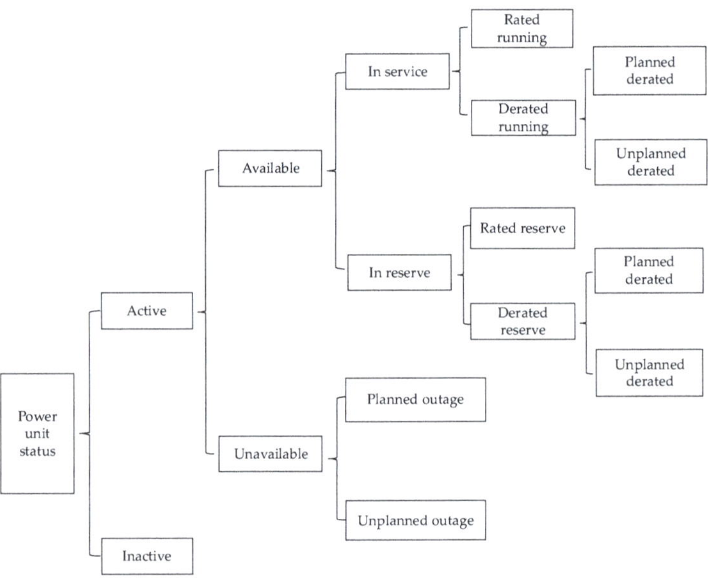

Figure 6. Basic status classification of a coal-fired power unit.

The equivalent available factor (SC41) is a coefficient of the equivalent available hours to the statistical hours, where the equivalent available hours are equal to the available hours subtracting the equivalent outage hours during derated available hours. The operation exposure rate (SC42) is the ratio of the service time to the available time of a unit, and the available time is the summation of the service time and reserve time. This index reflects the reserve ratio indirectly. The equivalent forced outage rate (SC43) reflects the situation of an unplanned outage in a year. It is the ratio of the forced outage hours and the equivalent outage hours of the unplanned derated hours to the total statistical hours. These indicators can be accessed by the statistics of the production department of the power plants, and we take a year as a statistical period.

4.1.5. Technical Criterion

The power plant is a complex industrial system, which includes many subsystems as functional components, as illustrated in Figure 5. To ensure the continuous operation of the whole system, many subsystems have their own technical control index. However, in this paper, we aimed to assess the sustainability of the power generation system at the unit level. Thus, we were not concerned with the specific parameters such as the main steam pressure and temperature, whether or not they are beyond their technical allowable standards, and we only take the unit-level comprehensive indicators as the technique supervisory sub-criterion to assess the operational safety of the unit.

They include the desulphurization system input rate (SC51) and denitration system input rate (SC52) to assess the utilization of the environmental purification equipment, which refers to the ratios of the operational time of purification facilities to the total running time of coal-fired units during the statistical periods, respectively. The qualified steam and water quality ratio (SC53) refers to the ratio of qualification times to the total measurement times. The thermal protection system input rate (SC54) is the ratio of the amount of the in-service protection devices to the total amount protection devices of a unit. The relay protection system correct action rate (SC55) is the ratio of the number of protection system correct actions to the total number of actions in the statistical periods.

The five criterion categories we selected above (flexibility, economic, environmental, reliability and technical) and their sub-criteria construct the sustainability assessment system of the coal-fired power unit from different perspectives. The flexibility criteria reflect the ability to adapt to the future power grid with a high penetration rate of renewable energy generation. The economic criteria reflect the ability to save resources. The environmental criterion reflects the ability to be in harmony with the environment. The reliability and technical criteria reflect the precondition guarantee of the unit's stable operation. Moreover, the criteria system is listed in Table 3.

Table 3. Criteria hierarchy for sustainability assessment of coal-fired power units.

Objective	Criteria	Sub-Criteria	Unit	Attribute *
Sustainability assessment of coal-fired power units	C1: flexibility	SC11: AGC availability ratio	%	(+)
		SC12: AGC regulation rate factor	-	(+)
		SC13: AGC regulation precision factor	-	(−)
		SC14: AGC response time factor	-	(−)
		SC15: AGC adjustable capacity	%	(+)
	C2: economic	SC21: net coal consumption rate	g/kWh	(−)
		SC22: auxiliary power ratio	%	(−)
		SC23: oil consumption rate	T/a	(−)
		SC24: water consumption rate	kg/(kWh)	(−)
	C3: environmental	SC31: SO_2 emission concentration	mg/Nm^3	(−)
		SC32: NOx emission concentration	mg/Nm^3	(−)
		SC33: dust emission concentration	mg/Nm^3	(−)
	C4: reliability	SC41: equivalent availability factor	%	(+)
		SC42: operating exposure rate	%	(+)
		SC43: equivalent forced outage rate	%	(−)
	C5: technical	SC51: desulphurization system input rate	%	(+)
		SC52: denitration system input rate	%	(+)
		SC53: qualified steam and water quality ratio	%	(+)
		SC54: thermal protection system input rate	%	(+)
		SC55: relay protection correct action rate	%	(+)

* (+) represents benefit attributes, the bigger the better, and (−) represents cost attributes, the smaller the better.

4.2. Data Collection

The case study was performed with the data of eight 600-megawatt subcritical condensing power units, which are located in mid-west of the Inner Mongolia Autonomous Region of China, and the basic information is listed in Table 4. The basic configuration in the design mode is as follows.

Table 4. Basic data of the selected power units *.

Sub-Criteria	Unit 1	Unit 2	Unit 3	Unit 4	Unit 5	Unit 6	Unit 7	Unit 8
SC11	100	99.2	100	100	100	100	100	100
SC12	0.95	0.82	0.97	1.2	1.1	1.08	1.18	1.3
SC13	1.44	1.39	1.48	1.56	1.53	1.21	1.45	1.42
SC14	0.83	0.78	0.87	1.02	0.92	1.15	1.09	1.2
SC15	50	50	50	55	55	50	50	51
SC21	315.7	301.64	300	312.72	320.99	318.28	319.74	328.75
SC22	4.72	4.78	4.45	4.23	4.75	5.02	4.58	4.44
SC23	44.54	99.45	37.01	21.15	61.44	54.01	112.94	24.22
SC24	0.89	0.47	1.98	0.63	1.55	2.62	2.31	1.35
SC31	18.29	84.1	18.12	19.14	79.65	26.1	47.59	18.95
SC32	20.32	116.17	12.93	26.38	47.16	27.15	54.08	14.76
SC33	2.54	17.08	2.32	1.51	21.72	3.93	15.62	2.65
SC41	100	94.86	100	98.2	100	100	99.91	96.09
SC42	97.86	100	95.98	100	88.84	80.15	100	97.29
SC43	0	0	0	0	0	0	0	0
SC51	99.71	100	100	100	99.6	99.89	100	100
SC52	100	100	100	100	98.7	98.79	100	100
SC53	99.86	99.97	100	99.92	99.93	99.93	100	99.94
SC54	100	100	100	100	100	100	100	100
SC55	100	100	100	100	100	100	100	100

* Data information: the data were collected by the authors from some plants located in the mid-west of the Inner Mongolia Autonomous Region of China at the end of 2016; we treat the specific unit names as anonymous according to the requirements of some related enterprises.

The boilers of Unit 1 and Unit 2 use forced circulation and tangential combustion, and the other boilers use natural circulation and opposed firing. The feeding coal is bitumite with the supply mode of straight blowing, and all the boilers use plasma ignition mode. The design efficiencies of the boilers are 93.95% (Units 1–2), 93.43% (Units 3–4) and 94.36% (Units 5–8), respectively. Correspondingly, the designed heat consumption rates of steam turbine units are 7762 kJ/kWh (Units 1–2), 7773 kJ/kWh (Units 3–4) and 8153 kJ/kWh (Units 5–8), respectively. The condensers of Units 1–4 are cooled by water and the other four by air. All the units are equipped with a flue gas purification system, i.e., the flue gas desulfurization (FGD) system for SO_2, the selective catalytic reduction (SCR) system for NO_X and the electrostatic precipitator (ESP) for dust removal. Additionally, based on the design configuration, some units except Units 2 and 5 have completed the low-NO_X combustion retrofit coupling with SCR to meet the rigorous ultra-lower emission requirements of China, and the other improvements such as the high-frequency power source retrofit of the ESP system and the upgrading of the desulfurization system were also performed.

The data in Table 4 were gathered by the authors at the end of the year 2016 as the annual average level and supplied by the electric production department of each plant, and the data were also checked by each superior of the power enterprises.

5. Results and Discussion

5.1. Calculation and Sustainability Assessment Results

The sustainability assessment was performed according to the procedures introduced in Section 2. Firstly, we obtained the objective weights of each sub-criteria (SC-level) of the five criteria categories, i.e., flexibility (C1), economic (C2), environmental (C3), reliability (C4) and the technical (C5). In this step, the weights of FRS and the entropy method were calculated by the corresponding methods for the SC-level, and furthermore, the hybrid weights of each criterion categories were calculated by game theory. The weight results are listed in Table 5. The objective weights of the sub-criteria SC43, SC54 and SC55 scored zero mainly because the samples of the alternatives in these sub-criteria have the same data and the weighting methods were data-driven. Thus, they cannot be used to distinguish the alternatives. However, this does not mean that these sub-criteria can be neglected in the assessment processes.

Table 5. Weight information of sub-criteria.

Weight	SC11	SC12	SC13	SC14	SC15	SC21	SC22	SC23	SC24	SC31
FRS	0.1903	0.2679	0.3499	0.0718	0.1201	0.422	0.0762	0.1437	0.3581	0.3053
Entropy	0.0632	0.1029	0.1566	0.1356	0.5417	0.2604	0.2256	0.2357	0.2782	0.4037
Hybrid	0.1120	0.1663	0.2309	0.1111	0.3797	0.4535	0.0471	0.1258	0.3736	0.3386

Weight	SC32	SC33	SC41	SC42	SC43	SC51	SC52	SC53	SC54	SC55
FRS	0.5423	0.1524	0.4999	0.5001	0	0.5475	0.2683	0.1842	0	0
Entropy	0.2295	0.3668	0.5551	0.4449	0	0.3049	0.3951	0.3001	0	0
Hybrid	0.4366	0.2248	0.5551	0.4449	0	0.5126	0.2865	0.2008	0	0

With the hybrid weights shown in Table 5 and the AGRA method, we obtained five groups of sustainability assessment values of the eight power units under the five criteria, respectively. The results are listed in Table 6. The result shows that Unit 5 has the highest score with the criterion C1, and Unit 2 has the lowest. For the criterion C2, Unit 4 has the biggest score, followed by Unit 3. The two units are obviously better than the others in economic operation, and Unit 7 ranks last with the C1 criterion. For the environmental aspect, Unit 3 obtained the biggest sustainability score followed by Unit 8. Conversely, Units 2 and 5 rank at the end. The low scores of the last two are mainly because Units 2 and 5 have not completed the ultra-low emission retrofit like the other six units by the time we obtained data. Unit 7 has the highest reliability score, followed by Unit 1, with Unit 8 and Unit 2 ranking last. For the technical criterion, Units 3 and 7 have the maximum score due to all the assessed alternatives having the same data (the maximum possible value) under the two criteria, and Unit 5 has the lowest score.

Table 6. Evaluation values (AGRA score) under five criterion groups.

Criteria	Unit 1	Unit 2	Unit 3	Unit 4	Unit 5	Unit 6	Unit 7	Unit 8
Flexibility(C1)	0.5068	0.4229	0.4923	0.6028	0.6273	0.5387	0.5126	0.5706
Economic (C2)	0.5923	0.6743	0.7497	0.7926	0.4836	0.4577	0.4253	0.5562
Environmental (C3)	0.8974	0.2493	0.9833	0.8445	0.3028	0.7084	0.4109	0.9376
Reliability (C4)	0.8940	0.2564	0.8354	0.4961	0.7242	0.6692	0.9517	0.2966
Technical (C5)	0.3167	0.9247	1.0000	0.8764	0.2170	0.3953	1.0000	0.8905

The five groups' sustainability values of the eight power units (in Table 6) performed with the sub-criteria data formed a new matrix. For this matrix, we calculated the weights using the FRS and entropy methods again, and also, in this step, we implemented the AHP procedures to get a subjective weight with the knowledge of experts (with a consistency ratio of 0.031 smaller than 0.1). The results are shown in Table 7. The AHP-based weight reflects that the flexibility criterion (C1) and the environmental (C3) criterion have obtained much more attention from experts, while the economic (C2) and reliability (C4) criteria have a similar priority score, and the technical (C5) criterion obtained the lowest attendance in the sustainability assessment. The high AHP weight score of C1 is based on the fact that it is a hot topic to improve the operational flexibility of the conventional coal-based power units in China at present, which is also considered to be one of the most promising ways to solve the current contradictions, such as the overcapacity of power generation and the high permeability ratio, which is caused by the rapid growth of renewable energy generation. On the other hand, the strict emission standard for power plants has been widely accepted and has gained continued attention, causing C2 to have the highest priority value, while C2 and C4 turn out to be relatively unimportant ones.

Table 7. Weight information on five criteria categories.

Criteria	C1	C2	C3	C4	C5
FRS	0.1500	0.3090	0.0551	0.2508	0.2350
Entropy	0.0206	0.0716	0.3052	0.2565	0.3461
AHP *	0.3160	0.1495	0.3433	0.1395	0.0516
Hybrid **	0.1603	0.1501	0.2686	0.2105	0.2105

* The consistency ratio $CR = 0.031 < 0.1$; the consistency test is passed. ** With the weight coefficient vector $\alpha = [0.2612, 0.5254, 0.4890]$.

Different from the AHP method, the FRS and entropy methods obtained the weighting values from different perspectives correspondingly by exploiting the inherent data structure. The respective results reflect that C2 is the most important factor for sustainability assessment according to the FRS method, and C3 is the least important one; while for the entropy weight, C3 and C5 are obviously more important than the C1 and C2 criteria in the evaluation process. However, the final weight combines the three weights with a reasonable compromise by game theory, and the result shows that C3 is still the most important one as the AHP weight, while the C4 and C5 factors are becoming more important than C1and C2, so it is reasonable that C4 and C5 are the premises of the operational sustainability of power units.

Using the AGRA method with the data of Table 6 and the hybrid weight listed in Table 7, we get the final sustainability scores as follows:

$$R = \begin{bmatrix} 0.5716 & 0.3225 & 0.6543 & 0.6332 & 0.4292 & 0.4764 & 0.4990 & 0.5065 \end{bmatrix} \tag{30}$$

Finally, we sort the scores (areal grey relational degree) of the eight coal-fired power units from small to large, and the corresponding rank result is:

$$Uint\ 2 \prec Uint\ 5 \prec Uint\ 6 \prec Uint\ 7 \prec Uint\ 8 \prec Uint\ 1 \prec Uint\ 4 \prec Uint\ 3 \tag{31}$$

The ranking result shows that Unit 3 has the highest score of sustainability from the operational perspective of the power unit, while Unit 2 ranks last. Unit 3 ranks first mainly owing to the good performance in the environmental and the technical aspects (see Table 6), and it also performed relatively well in the economic and reliability criteria. Unit 4 ranks second mainly due to the high scores in the economic and flexibility aspects. However, the performance of the rest of the criteria of Unit 4 is not outstanding among the other alternatives. The sustainability score of Unit 2 is the lowest one among the evaluated power units mainly because of the poor performance ranks in the flexibility, environmental and the reliability aspects. Unit 5 is just a little better than Unit 2 in the ranking list, which has poor performance on the technical and environmental aspects, and the other criteria are not prominent either. As mentioned in Section 4.2, Unit 2 and Unit 5 have poor environmental sustainability scores mainly due to the incomplete ultra-lower emission retrofit.

As analyzed above, the sustainability assessment using the proposed MCDM model has synthesized different sustainable features of the coal-fired power units from the operational aspects. The assessment framework is beneficial to the sustainable development of the existing power generation facilities.

5.2. The AHP Uncertainty Analysis

The MCDM methods can drive out a deterministic assessment result. However, sometimes, we need to know the result robustness information with the criteria weights that contain subjective factors. One of the optional methods is to implement sensitivity analysis on the weights. In this paper, we conducted this work in a statistical way, that is we used the probabilistic method to explore the features of the sustainability results caused by the process of the AHP weight generating.

We use the uncertainty analysis procedures mentioned in Section 3, generating 5000 group AHP weights that passed the consistency check. After that, we ran the proposed hybrid MCDM assessment method with the 5000 group AHP weights. Finally, we obtain 5000 group assessment results of the power units' sustainability. The approximate statistical probability distribution is illustrated in Figure 7.

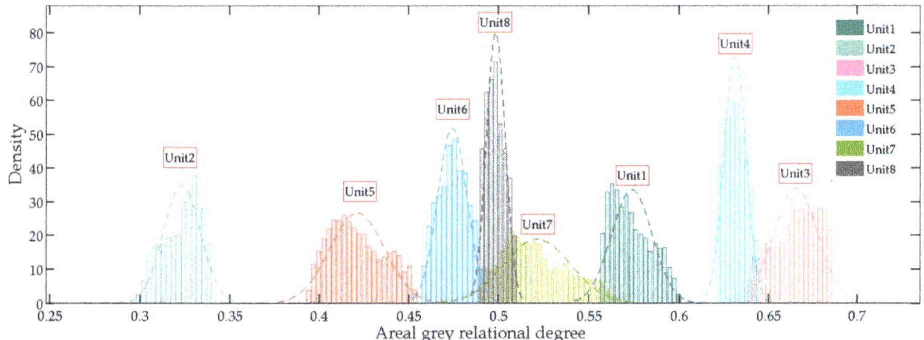

Figure 7. Result distribution of each power unit.

The distribution of the sustainability results of the eight power units shows that most of them are irregular in comparison with the corresponding normal distributions. Additionally, the range of the distribution intervals has the possibility of intersecting with the adjacent alternatives, such as Unit 3 with Unit 4, Unit 8 with Unit 6 and Unit 7 and Unit 7 with Unit 1. To express the uncertainty information quantitatively, we carried out some statistical information about the results series of the eight power units, which is listed in Table 8, including the mean value, standard deviation, 2.5% and 95.5% fractile and the interval range between them. Due to the possible asymmetry of these distributions, we define the width of the 95% confidential interval (CI) range (the last column in Table 8) other than the standard deviation to measure the alternatives' uncertainties caused in the AHP procedures.

Table 8. Statistical information of AHP uncertainty analysis.

Alternative	Mean Value	Standard Deviation	2.5% Fractile	97.5% Fractile	95% Confidential Interval Range
Unit 1	0.5734	0.0112	0.5707	0.5724	0.0016
Unit 2	0.3233	0.0102	0.3247	0.3260	0.0012
Unit 3	0.6650	0.0124	0.6656	0.6674	0.0018
Unit 4	0.6306	0.0061	0.6299	0.6310	0.0010
Unit 5	0.4208	0.0151	0.4173	0.4194	0.0021
Unit 6	0.4743	0.0079	0.4739	0.4750	0.0011
Unit 7	0.5221	0.0191	0.5181	0.5204	0.0023
Unit 8	0.4981	0.0053	0.4972	0.4979	0.0007

The results in Table 8 show that Unit 7 and Unit 5 have larger uncertainty values, and Unit 8 has the smallest value followed by Unit 4. That means Unit 8 and Unit 4 are more robust than the other alternatives with the possible weighting changes incurred by the AHP procedures, and the results are consistent with Figure 7.

However, the intersections of the distribution interval range such as Unit 3 and Unit 4 shown in Figure 7 do not mean the ranks of the two power units will be certainly swapped with each other. That is to say, Unit 3 may rank ahead of Unit 4 always. This is because the distribution curves shown in Figure 7 are only the representation of the alternatives in an independent way, and they are used just to illustrate the uncertainty ranges caused by the AHP method. In order to obtain more information, the statistical ranks are shown in Table 9, as [58,67] did.

Table 9. Statistical information on alternative ranks.

Order	#1	#2	#3	#4	#5	#6	#7	#8
Unit 1	0	0	4989	11	0	0	0	0
Unit 2	0	0	0	0	0	0	0	5000
Unit 3	5000	0	0	0	0	0	0	0
Unit 4	0	5000	0	0	0	0	0	0
Unit 5	0	0	0	0	0	0	5000	0
Unit 6	0	0	0	0	0	5000	0	0
Unit 7	0	0	11	4286	703	0	0	0
Unit 8	0	0	0	703	4297	0	0	0
Total	5000	5000	5000	5000	5000	5000	5000	5000

As we discussed above, with 5000 times simulation, the result shows that Unit 3 always ranks first and Unit 4 second, and the others with a similar situation are Unit 2, Unit 5 and Unit 6. The numerical test also shows that the rank of Unit 7 may be changed with Unit 1 and Unit 8, with a probability of 0.22% and 14.06%, respectively. Obviously, Unit 7 is probabilistically superior to Unit 8 with a probability of 85.94% (Unit 7 and Unit 8 rank fourth and fifth probabilistically, respectively). However, the sustainability assessment result we performed in Section 5.1 shows that the sustainability score of Unit 7 is less than Unit 8. The contrast shows that the stochastic method can obtain extra information sometimes, and Unit 7 may have a bigger sustainability score than Unit 8 with different experts.

5.3. The Distribution of the AHP Weight Components

In this subsection, we discuss some other complementary information about the AHP uncertainty analysis process. As mentioned in Section 3, we simulate the process in performing the pairwise comparison matrix with equal probability under the classical scales (see Table 1), which should be the work of the experts in the conventional AHP procedures. We display the probability distributions of the AHP weight components (w.c.) with the dimensions (criterion number) from 3–6 in Figure 8. The curves are discrete because the AHP weight components are discontinuous with the simulation method.

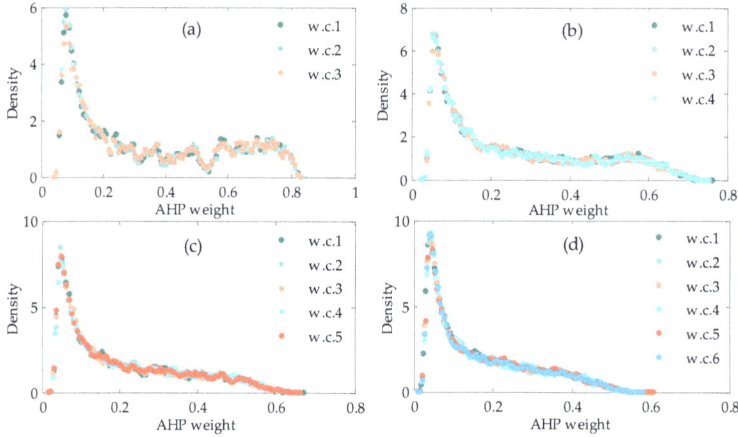

Figure 8. Distributions of AHP weight components (w.c.) with different index dimensions: (**a**) three dimensions, (**b**) four dimensions, (**c**) five dimensions and (**d**) six dimensions.

Figure 8 shows that every weight component has almost the same distribution and distribution curves with different dimensions, and it has a similar shape except for the main value interval of every component. The main value intervals become smaller as the dimension number increases from 3–6. The results also reflect that the asymmetrical alternative distributions in Figure 7 may not only be due to the nonlinearity of the MCDM model itself, but also may be related to the features of the input AHP weights.

On the other hand, in the process of calculating the AHP weight by creating a pairwise comparison matrix with its elements selected randomly, the time cost increases sharply to obtain the right AHP weight that can pass through the consistency check. This is because the combination of the matrix elements have $18^{m(m-1)/2}$ possible values, where the m is the criteria number of the AHP weight. Some corresponding information is listed in Table 10 with the weight varying from 3–6. The mean denied number refers to the average failed times with a consistency check passed weight, and the total time consumption means the cumulative time we used to generate 5000 group available weights. The result shows that the time increased to 30,681 s (about 8.5 h) when the weight dimension is six. The details reflect that it is not a good idea to use the uncertainty analysis method introduced in this work on a personal computer when the MCDM problem has many criteria (such as greater than six).

Table 10. Stochastic AHP with different weight dimensions with 5000 times *.

Weight Dimension	3	4	5	6
Mean denied number	3.96	31.82	443.39	15,692
Totally time consumption (s)	6.88	48.11	690.8	30,681

* Computing configuration: MATLAB (v 8.3.0.532), Intel(R) Core(TM) i5-2400 CPU @ 3.10 GHz and a 64-bit Windows 7 system.

5.4. Limitations

In this paper, we developed a hybrid MCDM model and introduced the AHP weight uncertainty analysis with stochastic simulation method, and after that, we implement a case study to illustrate the sustainability assessment of coal-fired power units from an operational perspective. However, it some limitations of our work should be noted.

The proposed MCDM approach using AGRA and a hybrid weight was only validated by the theoretical analysis of the case study results, and it is not verified in a practical way at present due to the current technology situation. That is, in order to figure out whether the assessment results work well, a long-term tracking study may be needed. For example, if the reward and punishment policies published by the electric plant management based on the sustainability assessment are widely accepted by most of the candidate power units, then the method is verified to be effective for the problem of coal-fired power plants.

In the process of AHP weight calculation, we invited three experts (two professors from the university and one engineer from the power plant) to complete the pairwise comparison matrix. The elements of the matrix (priorities of the five criteria) were determined by compromising their arguments, and the element adjustment work was done in the same way when the consistency check failed. However, the AHP weight result would be more reliable if more experts were involved, with the aggregation methods of group decision [42] or statistics [60]. That is, more experts coming from different subfields may balance each other's prejudices and the incompleteness of their knowledge.

Additionally, in the process of AHP weight uncertainty analysis, we selected the elements of the comparison matrix from Equation (28) in Section 3 only with a uniform distribution by the MC simulation method. However, the distribution may vary with the specific evaluation problems due to the decision makers who always have their own preferences for some kind of criteria. Anyway, the stochastic method we used in this work is mainly to measure the possibly varying range of the assessment result caused by the uncertainty of the AHP method. However, the distribution of the sampling can be replaced by a custom one if necessary, for example the normal distribution [60].

Diverse distributions are also allowable with different elements. On the other hand, the uncertainty measurement of the attribute value is obviously another point worthy of study, which was not carried out in this paper. The MC method may be used for this question if we have the uncertainty ranges and the determinate distributions for all the attribute values of specific indicators, and we will focus on this related issue in future work.

It is a complex project to assess the sustainability of the existing coal-fired power units with the MCDM method. The non-uniqueness of the assessment result is not only due to the variation of the optional MCDM approach but also due to the difficulty in establishing an appropriate criteria system. For the assessment of the power unit's sustainability, we only focus on the operational aspect at the device level in this paper. Additionally, the economic indicators we selected are closer to the techno-economic aspects that mainly focus on the consumption of economic resources like fuel, water and power. However, the criteria such as the overall levelized cost of electricity generation and the marginal operating cost will be more meaningful in the future for sustainability purposes because they are more comprehensive indicators with the consideration of the fixed assets' depreciation, the operation cost, maintenance cost, financial expenses, tax factor, etc.

6. Conclusions

In this paper, a hybrid MCDM model was developed for the sustainability assessment of coal-fired power units. The model integrates AGRA with a combined criteria weight. As a case study, sustainability assessment of the eight coal-fired power units with five criteria categories (flexibility, economic, environmental, reliability and technical) was performed subsequently.

We can conclude that it is feasible to use the proposed hybrid model to evaluate the sustainability of the coal-fired power plant from an operational perspective. The AGRA method has the inherent features of considering the relationship between the adjacent indicators, while the combined weight fuses the objective and subjective weights together. This work has extended a novel hybrid MCDM model used for sustainability measurement. Moreover, the AHP weighting uncertainty analysis can provide extra information in the process of sustainability assessment. The uncertainty analysis results can reflect the sensitivity of the evaluation results to the subjective criteria weight from a statistical perspective. The 95% confidential interval range of the result distributions obtained by the stochastic AHP method can be used as an uncertainty measure index. However, the amount of calculation grows sharply with the increase of the criteria dimension.

This work provided a method of sustainability assessment for the existing coal-fired power generation facilities by focusing on the operational performances. It will benefit the coal-based power generation enterprises of China to make their own sustainability level clear. The enterprises, which are encountering the development dilemma caused by the energy transition from fossil to sustainable energy sources, can take corresponding measures to enhance their sustainability level in the future based on a reasonable assessment.

Author Contributions: D.W. and Y.Y. proposed the comprehensive evaluation method. D.W. and Z.Y. conducted the analysis of various evaluation methods. N.W. and C.L. conducted the case study and wrote the paper. D.W. developed the main program codes and contributed to the main work reported.

Funding: This research was funded by the National Basic Research Program (973 Program) (2015CB251505).

Acknowledgments: We thank Degang Chen for providing subprogram codes of the fuzzy rough set for the weighting calculation.

Conflicts of Interest: The authors declare no conflict of interest.

Nomenclature

Abbreviations

AGC	automatic generation control
AGRA	areal grey relational analysis
AHP	analytic hierarchy process
CCS	coordination control system
CI	consistency index in AHP, confidence interval in statistics
CR	consistency ratio in the AHP method
CRITIC	criteria importance through inter-criteria correlation
DCS	distributed control system
ESP	electrostatic precipitator
FGD	flue gas desulfurization
GRA	grey relational analysis
GRD	grey relational degree
LCA	life cycle assessment
MCDM	multi-criteria decision making
MC	Monte Carlo
RTU	remote terminal unit
SCR	selective catalytic reduction
SPA	set pair analysis
TOPSIS	technique for order preference by similarity to an ideal solution

Symbols

ζ	the areal grey relational coefficient
ρ	the distinguishing coefficient of GRA
λ_{max}	maximum eigenvalue of the AHP comparison matrix
η_{net}	net efficiency of the power unit
a, x, y, z, p	element of matrix
b	net coal consumption rate
a_j	alternative
$\bar{a}, \bar{z}$	mean value
e_j	entropy value
$f_i(j)$	flag function
max	maximum
min	minimum
r	linear correlation coefficient
s_{ij}	areal elements in AGRA method
sig	significance degree
A, X, Y, U	matrix
B	simulation times
$F(U \times U)$	fuzzy power set
H_{std}	the heat value of the standard coal
P, Q	fuzzy equivalence relation
$\underline{R}X(x), \overline{R}X(x)$	lower and upper approximation operators
R, r	grey relational degree, fuzzy relation
S	area, knowledge representation system
W, w	weight vector and element

Subscripts and Superscripts

m, n, l	dimension of matrix
i, j, h, k, b	index of vector or matrix
0	reference
frs, ent, ahp	based on FRS, entropy, AHP method
T	transpose symbol of the vector, similarity relation

References

1. Farfan, J.; Breyer, C. Structural changes of global power generation capacity towards sustainability and the risk of stranded investments supported by a sustainability indicator. *J. Clean. Prod.* **2017**, *141*, 370–384. [CrossRef]

2. Li, J.; Geng, X.; Li, J. A Comparison of Electricity Generation System Sustainability among G20 Countries. *Sustainability* **2017**, *8*, 1276. [CrossRef]

3. Pambudi, N.A.; Itoi, R.; Jalilinasrabady, S.; Gürtürk, M. Sustainability of geothermal power plant combined with thermodynamic and silica scaling model. *Geothermics* **2018**, *71*, 108–117. [CrossRef]

4. Sun, X.; Zhang, B.; Tang, X.; Mclellan, B.; Höök, M. Sustainable Energy Transitions in China: Renewable Options and Impacts on the Electricity System. *Energies* **2016**, *9*, 980. [CrossRef]

5. Beires, P.; Vasconcelos, M.H.; Moreira, C.L.; Lopes, J.A.P. Stability of autonomous power systems with reversible hydro power plants: A study case for large scale renewables integration. *Electr. Power Syst. Res.* **2018**, *158*, 1–14. [CrossRef]

6. Phillips, J. Determining the sustainability of large-scale photovoltaic solar power plants. *Renew. Sustain. Energy Rev.* **2013**, *27*, 435–444. [CrossRef]

7. Zhao, X.; Cai, Q.; Zhang, S.; Luo, K. The substitution of wind power for coal-fired power to realize China's CO_2 emissions reduction targets in 2020 and 2030. *Energy* **2016**, *120*, 164–178. [CrossRef]

8. Stougie, L.; Tsalidis, G.A.; Kooi, H.J.V.D.; Korevaar, G. Environmental and exergetic sustainability assessment of power generation from biomass. *Renew. Energy* **2017**. [CrossRef]

9. Council, C.E. *Annual Development Report of China Electric Power Industry 2017*, 1st ed.; China Market Press: Beijing, China, 2017.

10. Yuan, J.; Li, P.; Wang, Y.; Liu, Q.; Shen, X.; Zhang, K.; Dong, L. Coal power overcapacity and investment bubble in China during 2015–2020. *Energy Policy* **2016**, *97*, 136–144. [CrossRef]

11. Zeng, M.; Zhang, P.; Yu, S.; Liu, H.; Zeng, M.; Zhang, P.; Yu, S.; Liu, H.; Zeng, M.; Zhang, P. Overall review of the overcapacity situation of China's thermal power industry: Status quo, policy analysis and suggestions. *Renew. Sustain. Energy Rev.* **2017**, *76*, 768–774.

12. Hübel, M.; Meinke, S.; Andrén, M.T.; Wedding, C.; Nocke, J.; Gierow, C.; Hassel, E.; Funkquist, J. Modelling and simulation of a coal-fired power plant for start-up optimisation. *Appl. Energy* **2017**. [CrossRef]

13. Liu, M.; Zhang, X.; Ma, Y.; Yan, J. Thermo-economic analyses on a new conceptual system of waste heat recovery integrated with an S-CO_2 cycle for coal-fired power plants. *Energy Convers. Manag.* **2018**, *161*, 243–253. [CrossRef]

14. Hu, Q.; Li, X.; Lin, A.; Qi, W.; Li, X.; Yang, X.J. Total emission control policy in China. *Environ. Dev.* **2017**. [CrossRef]

15. Ma, Z.; Deng, J.; Li, Z.; Li, Q.; Zhao, P.; Wang, L.; Sun, Y.; Zheng, H.; Pan, L.; Zhao, S. Characteristics of NOx emission from Chinese coal-fired power plants equipped with new technologies. *Atmos. Environ.* **2016**, *131*, 164–170. [CrossRef]

16. Li, R.; Li, J.; Cui, L.; Wu, Y.; Fu, H.; Chen, J.; Chen, M. Atmospheric emissions of Cu and Zn from coal combustion in China: Spatio-temporal distribution, human health effects, and short-term prediction. *Environ. Pollut.* **2017**, *229*, 724–734. [CrossRef] [PubMed]

17. Li, C.; Duan, Y.; Tang, H.; Zhu, C.; Li, Y.N.; Zheng, Y.; Liu, M. Study on the Hg emission and migration characteristics in coal-fired power plant of China with an ammonia desulfurization process. *Fuel* **2017**, *211*, 621–628. [CrossRef]

18. Ji, X.; Li, G.; Wang, Z. Impact of emission regulation policies on Chinese power firms, reusable environmental investments and sustainable operations. *Energy Policy* **2017**, *108*. [CrossRef]

19. Cai, F. *China's Economic New Normal and Supply-Side Structural Reform*, 1st ed.; Foreign Languages Press: Beijing, China, 2016.

20. Nuer, B. *Counseling Reader for the 13th Five-Year Energy Development Plan*, 1st ed.; China Electric Power Press: Beijing, China, 2017.

21. Kubik, M.L.; Coker, P.J.; Barlow, J.F. Increasing thermal plant flexibility in a high renewables power system. *Appl. Energy* **2015**, *154*, 102–111. [CrossRef]

22. Kopiske, J.; Spieker, S.; Tsatsaronis, G. Value of power plant flexibility in power systems with high shares of variable renewables: A scenario outlook for Germany 2035. *Energy* **2017**. [CrossRef]

23. Kai, W.; Carmona, L.G.; Sousa, T. A review of the use of exergy to evaluate the sustainability of fossil fuels and non-fuel mineral depletion. *Renew. Sustain. Energy Rev.* **2017**, *76*, 202–211.

24. Bilgen, S.; Sarıkaya, İ. Exergy for environment, ecology and sustainable development. *Renew. Sustain. Energy Rev.* **2015**, *51*, 1115–1131. [CrossRef]

25. Pan, H.; Geng, Y.; Jiang, P.; Dong, H.; Sun, L.; Wu, R. An emergy based sustainability evaluation on a combined landfill and LFG power generation system. *Energy* **2018**, *143*. [CrossRef]

26. Su, M.R.; Yang, Z.F.; Chen, B.; Ulgiati, S. Urban ecosystem health assessment based on emergy and set pair analysis—A comparative study of typical Chinese cities. *Ecol. Model.* **2009**, *220*, 2341–2348. [CrossRef]

27. Atilgan, B.; Azapagic, A. Assessing the Environmental Sustainability of Electricity Generation in Turkey on a Life Cycle Basis. *Energies* **2016**, *9*, 31. [CrossRef]

28. Elena Arce, M.; Saavedra, Á.; Míguez, J.L.; Granada, E. The use of grey-based methods in multi-criteria decision analysis for the evaluation of sustainable energy systems: A review. *Renew. Sustain. Energy Rev.* **2015**, *47*, 924–932. [CrossRef]

29. Cinelli, M.; Coles, S.R.; Kirwan, K. Analysis of the potentials of multi criteria decision analysis methods to conduct sustainability assessment. *Ecol. Indic.* **2014**, *46*, 138–148. [CrossRef]

30. Zhao, H.; Li, N. Optimal Siting of Charging Stations for Electric Vehicles Based on Fuzzy Delphi and Hybrid Multi-Criteria Decision Making Approaches from an Extended Sustainability Perspective. *Energies* **2016**, *9*, 270. [CrossRef]

31. Diaz-Balteiro, L.; González-Pachón, J.; Romero, C. Measuring systems sustainability with multi-criteria methods: A critical review. *Eur. J. Oper. Res.* **2017**, *258*, 607–616. [CrossRef]

32. Claudia Roldán, M.; Martínez, M.; Peña, R. Scenarios for a hierarchical assessment of the global sustainability of electric power plants in México. *Renew. Sustain. Energy Rev.* **2014**, *33*, 154–160. [CrossRef]

33. Škobalj, P.; Kijevčanin, M.; Afgan, N.; Jovanović, M.; Turanjanin, V.; Vučićević, B. Multi-Criteria sustainability analysis of thermal power plant Kolubara-A unit 2. *Energy* **2017**, *125*. [CrossRef]

34. Cobuloglu, H.I.; Büyüktahtakın, İ.E. A stochastic multi-criteria decision analysis for sustainable biomass crop selection. *Expert Syst. Appl.* **2015**, *42*, 6065–6074. [CrossRef]

35. Wang, J.J.; Jing, Y.Y.; Zhang, C.F.; Zhao, J.H. Review on multi-criteria decision analysis aid in sustainable energy decision-making. *Renew. Sustain. Energy Rev.* **2009**, *13*, 2263–2278. [CrossRef]

36. Karahalios, H. The application of the AHP-TOPSIS for evaluating ballast water treatment systems by ship operators. *Transp. Res. Part D Transp. Environ.* **2017**, *52*, 172–184. [CrossRef]

37. Dinmohammadi, A.; Shafiee, M. Determination of the Most Suitable Technology Transfer Strategy for Wind Turbines Using an Integrated AHP-TOPSIS Decision Model. *Energies* **2017**, *10*, 642. [CrossRef]

38. Guili, Y.; Jianhua, Z.; Tianhong, W.; Juan, D. Comprehensive energy-saving evaluation of thermal power plants based on TOPSIS gray relational projection and the weight sensitivity analysis. *J. Chin. Soc. Power Eng.* **2015**, *35*, 404–411.

39. Saaty, T.L. How to Make a Decision: The Analytic Hierarchy Process. *Eur. J. Oper. Res.* **1994**, *24*, 19–43. [CrossRef]

40. Zeng, F.; Cheng, X.; Guo, J.; Tao, L.; Chen, Z. Hybridising Human Judgment, AHP, Grey Theory, and Fuzzy Expert Systems for Candidate Well Selection in Fractured Reservoirs. *Energies* **2017**, *10*, 447. [CrossRef]

41. Zhao, H.; Yao, L.; Mei, G.; Liu, T.; Ning, Y. A Fuzzy comprehensive evaluation method based on AHP and Entropy for landslide susceptibility map. *Entropy* **2017**. [CrossRef]

42. Ishizaka, A.; Labib, A. Review of the main developments in the analytic hierarchy process. *Expert Syst. Appl.* **2011**, *38*, 14336–14345. [CrossRef]

43. Liang, J.; Shi, Z.; Li, D.; Wierman, M.J. Information entropy, rough entropy and knowledge granulation in incomplete information systems. *Int. J. Gen. Syst.* **2006**, *35*, 641–654. [CrossRef]

44. Zhou, R.; Pan, Z.; Jin, J.; Li, C.; Ning, S. Forewarning Model of Regional Water Resources Carrying Capacity Based on Combination Weights and Entropy Principles. *Entropy* **2017**, *19*, 574. [CrossRef]

45. Deng, J.L. Introduction grey system theory. *J. Grey Syst.* **1989**, *1*, 191–243.

46. Chen, D.; He, Q.; Wang, X. FRSVMs: Fuzzy rough set based support vector machines. *Fuzzy Sets Syst.* **2010**, *161*, 596–607. [CrossRef]

47. Chen, D.; Zhao, S. Local reduction of decision system with fuzzy rough sets. *Fuzzy Sets Syst.* **2010**, *161*, 1871–1883.

48. Men, B.; Liu, H.; Tian, W.; Liu, H. Evaluation of Sustainable Use of Water Resources in Beijing Based on Rough Set and Fuzzy Theory. *Water* **2017**, *9*, 852. [CrossRef]

49. Caixin, S.; Jian, L.; Haiping, Z.; Ji, Y.; Xiong, F. A New Method of Faulty Insulation Diagnosis in Power Transformer Based On Degree of Area Incidence Analysis. *Power Syst. Technol.* **2002**, *26*, 24–29.

50. Mingjian, C.; Yuanzhang, S.; Jun, Y.; Yuanlin, L.; Weinan, W. Power Grid Security Comprehensive Assessment Based on Multi-Level Grey Area Relational Analysis. *Power Syst. Technol.* **2013**, *37*, 3453–3460.

51. Jiang, S.-Q.; Liu, S.; Liu, Z.-X.; Fang, Z.-G. Grey incidence decision making model based on area. *Control Decis.* **2015**, *30*, 685–690.

52. Minfang, Q.; Zhongguang, F.; Yuan, J.; Ya, M. Comprehensive Evaluation Method of Power Plant Unit Based on Information Entropy and Principal Component Analysis. *Proc. CSEE* **2013**, *33*, 58–64.

53. Diakoulaki, D.; Mavrotas, G.; Papayannakis, L. Determining objective weights in multiple criteria problems: The critic method. *Comput. Oper. Res.* **1995**, *22*, 763–770. [CrossRef]

54. Zhang, Z. Attributes reduction based on intuitionistic fuzzy rough sets. *J. Intell. Fuzzy Syst.* **2016**, *30*, 1127–1137. [CrossRef]

55. Lai, C.; Chen, X.; Chen, X.; Wang, Z.; Wu, X.; Zhao, S. A fuzzy comprehensive evaluation model for flood risk based on the combination weight of game theory. *Nat. Hazards* **2015**, *77*, 1243–1259. [CrossRef]

56. Sun, L.; Liu, Y.; Zhang, B.; Shang, Y.; Yuan, H. An Integrated Decision-Making Model for Transformer Condition Assessment Using Game Theory and Modified Evidence Combination Extended by D Numbers. *Energies* **2016**, *9*. [CrossRef]

57. Carmone, F.J., Jr.; Kara, A.; Zanakis, S.H. A Monte Carlo investigation of incomplete pairwise comparison matrices in AHP. *Eur. J. Oper. Res.* **1997**, *102*, 538–553. [CrossRef]

58. Hsu, T.; Pan, F.F.C. Application of Monte Carlo AHP in ranking dental quality attributes. *Expert Syst. Appl.* **2009**, *36*, 2310–2316. [CrossRef]

59. Yaraghi, N.; Tabesh, P.; Guan, P.; Zhuang, J. Comparison of AHP and Monte Carlo AHP Under Different Levels of Uncertainty. *IEEE Trans. Eng. Manag.* **2015**, *62*, 122–132. [CrossRef]

60. Mohammad, A.; Hashem, S.; Reza, M. Monte Carlo Analytic Hierarchy Process (MAHP) approach to selection of optimum mining method. *Int. J. Min. Sci. Technol.* **2013**, *23*, 573–578.

61. Zhao, H.; Li, N. Performance Evaluation for Sustainability of Strong Smart Grid by Using Stochastic AHP and Fuzzy TOPSIS Methods. *Sustainability* **2016**, *8*, 129. [CrossRef]

62. Scannapieco, D.; Naddeo, V.; Belgiorno, V. Sustainable power plants: A support tool for the analysis of alternatives. *Land Use Policy* **2014**, *36*, 478–484. [CrossRef]

63. Gang, X.; Yang, Y.P.; Lu, S.Y.; Le, L.; Song, X. Comprehensive evaluation of coal-fired power plants based on grey relational analysis and analytic hierarchy process. *Energy Policy* **2011**, *39*, 2343–2351.

64. Liu, G.; Baniyounes, A.M.; Rasul, M.G.; Amanullah, M.T.O.; Khan, M.M.K. General sustainability indicator of renewable energy system based on grey relational analysis. *Int. J. Energy Res.* **2013**, *37*, 1928–1936. [CrossRef]

65. Pawlak, Z. Rough sets. *Int. J. Comput. Inf. Sci.* **1995**, *38*, 88–95. [CrossRef]

66. Zhang, X.; Mei, C.; Chen, D.; Li, J. Feature selection in mixed data: A method using a novel fuzzy rough set-based information entropy. *Pattern Recognit.* **2016**, *56*, 1–15. [CrossRef]

67. Rosenbloom, E.S. A probabilistic interpretation of the final rankings in AHP. *Eur. J. Oper. Res.* **1997**, *96*, 371–378. [CrossRef]

68. Fu, Z.; Qi, M. Study on the evaluation method of energy-saving and emission reduction of coal-fired units based on projection pursuit method coupled with maximum entropy. *Proc. CSEE* **2014**, *34*, 4476–4482.

69. Zhang, X.; Shi, G.; Liu, S. Study on zero discharge technology of wastewater from power plant. In Proceedings of the 2010 the 5th IEEE Conference on Industrial Electronics and Applications (ICIEA), Taichung, Taiwan, 15–17 June 2010; pp. 911–914.

 *sustainability*

Article

The Facilitation of a Sustainable Power System: A Practice from Data-Driven Enhanced Boiler Control

Zhenlong Wu [1], Ting He [1], Li Sun [2], Donghai Li [1,*] and Yali Xue [1]

[1]   State Key Lab of Power Systems, Department of Energy and Power Engineering, Tsinghua University, Beijing 100084, China; WZLsongshanshan@163.com (Z.W.); he-t14@mails.tsinghua.edu.cn (T.H.); xueyali@tsinghua.edu.cn (Y.X.)

[2]   Key Lab of Energy Thermal Conversion and Control of Ministry of Education, Southeast University, Nanjing 210096, China; sunli12@seu.edu.cn

*   Correspondence: lidongh@tsinguha.edu.cn

Received: 10 March 2018; Accepted: 4 April 2018; Published: 8 April 2018

Abstract: An increasing penetration of renewable energy may bring significant challenges to a power system due to its inherent intermittency. To achieve a sustainable future for renewable energy, a conventional power plant is required to be able to change its power output rapidly for a grid balance purpose. However, the rapid power change may result in the boiler operating in a dangerous manner. To this end, this paper aims to improve boiler control performance via a data-driven control strategy, namely Active Disturbance Rejection Control (ADRC). For practical implementation, a tuning method is developed for ADRC controller parameters to maximize its potential in controlling a boiler operating in different conditions. Based on a Monte Carlo simulation, a Probabilistic Robustness (PR) index is subsequently formulated to represent the controller's sensitivity to the varying conditions. The stability region of the ADRC controller is depicted to provide the search space in which the optimal group of parameters is searched for based on the PR index. Illustrative simulations are performed to verify the efficacy of the proposed method. Finally, the proposed method is experimentally applied to a boiler's secondary air control system successfully. The results of the field application show that the proposed ADRC based on PR can ensure the expected control performance even though it works in a wider range of operating conditions. The field application depicts a promising future for the ADRC controller as an alternative solution in the power industry to integrate more renewable energy into the power grid.

Keywords: Active Disturbance Rejection Control; Probabilistic Robustness; Monte Carlo; secondary air regulation

1. Introduction

In order to achieve sustainable energy development better and make full use of new energy, the use of renewable energy, such as solar, wind, and tidal power generation is expected to increase by 2.8% annually until the year 2040, and total renewable energy power generation will possess a share of one quarter of worldwide power by 2040 [1]. To integrate more renewable energy into the power grid, a growing requirement of load regulation is posed on conventional coal-fired units because of the intermittency of renewable energy. Now, more and more coal-fired units have to run in the load range of 50–100% and lift load frequently according to Automatic Generation Control (AGC) commands, which can be optimized by a modified differential evolution algorithm [2], and so on. Consequently, the dynamic characteristics of the control loops vary greatly, especially for those loops which are impacted by the load change, such as the superheated steam temperature loop, the main steam pressure loop, and the air loop. However, Proportional-Integral (PI) or Proportional-Integral-Derivative (PID) controllers in coal-fired units cannot ensure a satisfactory control effect, which eliminates the error passively and is tuned in the nominal condition. The control design of coal-fired units is becoming a challenging task in view of the above-mentioned reasons.

To accommodate the uncertainties, two aspects in view of engineering practice should be considered. Firstly, the alternative solution should have a strong ability to deal with uncertainties, and should not rely on precise mathematical models. Some advanced control strategies, such as PI observer [3,4] and linear/nonlinear Internal Model Control (IMC) [5,6], have attracted many researchers' attention. The PI observer realizes disturbance estimations by the state observer, which has a similar structure to the Luenberger observer. IMC has a strong ability to deal with the model-plant mismatch and unmeasured disturbances [5]; however, it is hard to implement in a Distributed Control System (DCS) platform because of computation complexities. In this paper, a data-driven control strategy called Active Disturbance Rejection Control (ADRC) is utilized because of its ability to deal with uncertainties and its ease of use [7,8]. ADRC was originally proposed by Professor Han as a result of long-term thinking about model-based control theory and engineering control cybernetics [9]. Its core idea is that the internal uncertainties and external disturbances, called the total disturbance, are estimated in real time by an Extended State Observer (ESO) and compensated for by a feedback controller [10]. Linear ADRC is proposed to simplify parameter tuning and the design framework [11]. ADRC is distinctly different from other control strategies in the following respects: (i) it can estimate and compensate for total disturbance in real time and its structure is simple; (ii) ADRC can estimate the total disturbance by an ESO, which is unlike other observers, such as the Unknown Input Observer (UIO) and the Disturbance Observer (DOB) [12]; and (iii) ADRC is an energy-saving control and has less computation while the computational intensity of Model Predictive Control (MPC) is large and would result in additional hardware costs [13]. Because of the advantages mentioned above, ADRC has been applied in various areas, such as motion control systems [14], chemical process control systems [15], fuel cell systems [16], load frequency systems [17], Atomic Force Microscope (AFM) scanning systems [18], pressurized water reactor power [19] and fractional order systems [20]. Besides this, ADRC also has solved many benchmark problems successfully, such as the ALSTOM gasifier benchmark problem [21], the two-mass-spring benchmark problem [22], and the four tank benchmark problem [23]. Secondly, not only nominal plants but also plants which are far from being in a nominal condition should be considered in the process of parameter tuning. Most tuning methods of ADRC are proposed for the nominal model rather than those plants which are probably in a whole range of operating conditions [11,24]. Some other methods have been proposed only for particular plants, such as First Order Plus Time Delay plants (FOPTD) [25] and unstable processes [26]. Randomized algorithms are attracting more and more attention to analyze uncertain systems which could consider parameter uncertainties in the whole parameter space, including the "worst-case" condition [27–29]. Probabilistic Robustness (PR), as one of these randomized algorithms, is a practical and powerful tool to analyze uncertain systems and is utilized for controller tuning and robustness analysis [30]. So, PR is proposed to optimize the parameters of ADRC due to its practicability and the consideration of plants that are probably in a whole range of operating conditions.

The main contributions of this paper are as follows:

(1)   A data-driven boiler control method is proposed to increase the flexibility of the conventional power plant, thus being able to integrate more renewables into the grid.
(2)   An algorithm which is able to depict the stable region of ADRC is presented.
(3)   The proposed tuning method for ADRC is applied to the secondary air regulation of a boiler unit successfully.

The rest of the paper is organized as follows: the problem formulation, the calculation of the stability region, and PR-based ADRC tuning are depicted in Section 2. In Section 3, five simulations illustrate the effectiveness of the proposed method. Then, a field test of ADRC based on PR for the secondary air regulation of a boiler unit was carried out and the effectiveness is proved. Finally, Section 5 offers concluding remarks.

2. Tuning of ADRC Based on PR

2.1. Problem Formulation

Parameter uncertainties, nonlinearity, and variation of dynamic characteristics caused by a wide range of operating conditions could be considered as parameter perturbation in a large space, so the problem formulation could be depicted as follows.

Considering a transfer function with parameter uncertainties in a parameter space Q

$$G_p(s) = \frac{c_m s^m + c_{m-1} s^{m-1} + \cdots + c_0}{a_n s^n + a_{n-1} s^{n-1} + \cdots + a_0} e^{-\tau s} \tag{1}$$

where $a_i (i = 0, 1, 2, \cdots n)$ and $c_j (j = 0, 1, 2, \cdots m)$ are the coefficient of the denominator and the numerator, respectively. The order of the numerator is not greater than that of the denominator and the delay time τ is a nonnegative real number. Due to the existence of parameter uncertainties, define $q = \{a_i, c_j, \tau\}$ as the random vector of parameter uncertainties throughout the parameter space Q according to the probability density function of p_r. The parameter space Q can be defined as:

$$Q = \left\{ \left[a_{i\_}, a_i^+ \right] \cup \left[c_{j\_}, c_j^+ \right] \cup \left[\tau\_, \tau^+ \right] \right\} \tag{2}$$

and the plant becomes a group of transfer functions.

The goal of control design is to meet design requirements in the whole parameter space, including the "worst-case" condition, that the design requirements may be the settling time, the integral of time and absolute error (ITAE), and overshoot.

The traditional tuning methods based on a nominal plant often have good control performance for the nominal plant. However, the control performance would be worse when the plant varies far from the nominal condition even though the controller is designed with robustness constraints and the simulations in Section 3 would emphasize this point. A seemingly good way is that controller tuning can go through any possible point in the parameter space Q, but it is impossible to be ergodic throughout the parameter space. The tuning method for ADRC based on PR could be a useful and highly efficient method which can consider probable plants.

2.2. The Fundamentals of ADRC

For ADRC design, a general nonlinear time-varying dynamic plant is assumed to have the following format:

$$y^{(n)}(t) = bu(t) + g\left(y^{(n-1)}(t), y^{(n-2)}(t), \cdots, y(t), w(t) \right) \tag{3}$$

where $y(t)$, $u(t)$, $w(t)$, and b are the output, input, disturbance, and gain parameters of the plant, respectively. g is the synthesis of time-variance, disturbances, dynamic uncertainties, etc. of the plant. Define $f = g\left(y^{(n-1)}(t), y^{(n-2)}(t), \cdots, y(t), w(t) \right) + (b - b_0)u$ as a synthesis of the unknown dynamics, time-variant, nonlinear, and external disturbances of the plant, which is denoted the total disturbance and is assumed to be unknown in ADRC design. Plant (3) can be written as:

$$y^{(n)}(t) = b_0 u(t) + f\left(y^{(n-1)}(t), y^{(n-2)}(t), \cdots, y(t), w(t) \right) \tag{4}$$

where b_0 is the approximation of the gain parameter b.

The core idea of ADRC is to estimate the unknown total disturbance f by extending the total disturbance as an additional state. Assume that f is differentiable and $\dot{f} = h$. The plant in Equation (4) can be written as:

$$\begin{aligned} \dot{x} &= Ax + Bu + Eh \\ y &= C^T x \end{aligned} \tag{5}$$

where $x = [x_1, x_2, \cdots, x_n, x_{n+1}]^T = \left[y, \dot{y}, \cdots, y^{(n)}, f\right]^T$, $h = \dot{f}$, $C^T = [1 \; 0 \; \cdots \; 0]_{(n+1)\times(n+1)}$, and

$$
A = \begin{bmatrix} 0 & 1 & & \\ \vdots & & \ddots & \\ 0 & & & 1 \\ 0 & \cdots & \cdots & 0 \end{bmatrix}_{(n+1)\times(n+1)}, \quad B = \begin{bmatrix} 0 \\ \vdots \\ 0 \\ b \\ 0 \end{bmatrix}_{(n+1)\times1}, \quad E = \begin{bmatrix} 0 \\ \vdots \\ 0 \\ 1 \end{bmatrix}_{(n+1)\times1}.
$$

The ESO for the plant in Equation (4) with y and u as inputs can be depicted as:

$$
\begin{cases}
\dot{z}_1 = z_2 + \beta_1(y - z_1) \\
\vdots \\
\dot{z}_{n-1} = z_n + \beta_{n-1}(y - z_1) \\
\dot{z}_n = z_{n+1} + \beta_n(y - z_1) + b_0 u \\
\dot{z}_{n+1} = \beta_{n+1}(y - z_1)
\end{cases}
\tag{6}
$$

or

$$
\dot{z} = A_e z + B_e y + C_e u
\tag{7}
$$

where $z = \begin{bmatrix} z_1 \\ z_2 \\ \vdots \\ z_{n+1} \end{bmatrix}_{(n+1)\times1}$, $A_e = \begin{bmatrix} -\beta_1 & 1 & & & \\ -\beta_2 & 0 & 1 & & \\ \vdots & & & \ddots & \ddots \\ -\beta_n & 0 & \cdots & 0 & 1 \\ -\beta_{n+1} & 0 & \cdots & 0 & 0 \end{bmatrix}_{(n+1)\times(n+1)}$, $B_e = \begin{bmatrix} \beta_1 \\ \beta_2 \\ \vdots \\ \beta_{n+1} \end{bmatrix}_{(n+1)\times1}$ and

$C_e = \begin{bmatrix} 0 \\ \vdots \\ b_0 \\ 0 \end{bmatrix}_{(n+1)\times1}$. $z_1, z_2, \ldots z_n$ aim at approximating $y(t)$ and its derivatives (up to order $n - 1$), and

z_{n+1} can approximate the total disturbance $f(t)$ when the observer gain vector B_e is tuned appropriately. We can design the control law as follows:

$$
u(t) = \frac{-z_{n+1}(t) + u_0(t)}{b_0}
\tag{8}
$$

where $u_0(t)$ is to be determined afterward. The conventional plant (4) becomes:

$$
y^{(n)}(t) = f - z_{n+1}(t) + u_0(t) \approx u_0(t)
\tag{9}
$$

then, the plant in Equation (4) can be seen as a cascaded integrators plant.

The final plant can be effectively controlled by using the following state-feedback law:

$$
u_0(t) = k_1(r(t) - z_1(t)) + k_2(\dot{r}(t) - z_2(t)) + \cdots + k_n\left(r^{(n-1)}(t) - z_n(t)\right)
\tag{10}
$$

where $r(t)$ is the reference signal.

The final control law can be straightforwardly expressed as:

$$
u(t) = \frac{k_1(r(t) - z_1(t)) + k_2(\dot{r}(t) - z_2(t)) + \cdots + k_n\left(r^{(n-1)}(t) - z_n(t)\right)}{b_0} - \frac{z_{n+1}(t)}{b_0}
$$
$$
=: K(\bar{r}(t) - z(t))
\tag{11}
$$

where $\bar{r}(t) = \left[r(t)\dot{r}(t) \cdots r^{(n-1)}(t)\,0\right]^T_{1\times(n+1)}$ and $K = \frac{1}{b_0}[k_1 k_2 \cdots k_n\,1]_{1\times(n+1)}$, and K is the feedback gain vector.

As mentioned above, the structure of ADRC is shown in Figure 1 and ADRC can be expressed as the following state-space form:

$$\begin{cases} \dot{z}(t) = A_e z(t) + B_e y(t) + C_e u(t) \\ u(t) = K(\bar{r}(t) - z(t)) \end{cases} \tag{12}$$

where the parameters of ADRC are K, B_e, and b_0. For simplifying tuning, K and B_e can be tuned based on the bandwidth-parameterization method as suggested in [11] and they can be determined by the controller bandwidth, ω_c, and the observer bandwidth, ω_0. The relationship of K and ω_c, B_e, and ω_0 can be shown as:

$$\begin{cases} k_i = \frac{n!}{(i-1)!(n+1-i)!}\omega_c^{n+1-i} & i = 1,2,\cdots,n \\ \beta_j = \frac{(n+1)!}{j!(n+1-j)!}\omega_0^{j} & j = 1,2,\cdots,n+1 \end{cases} \tag{13}$$

When $n = 1$, we can obtain $k_1 = \omega_c$, $\beta_1 = 2\omega_0$, and $\beta_2 = \omega_0^2$ for first-order ADRC.

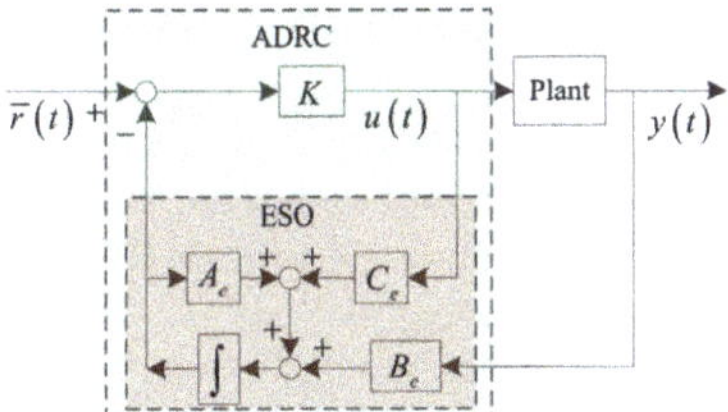

Figure 1. The structure of Active Disturbance Rejection Control (ADRC).

2.3. Stability Region of ADRC

It should be noted that ADRC has achieved a reasonable performance in spite of the order mismatch. First-order ADRC is widely used and is also the research focus in this paper.

In order to avoid the oscillation of the output, Equation (13) can be modified as follows:

$$\begin{cases} \beta_1 = 2\omega_0 \\ \beta_2 = \zeta\omega_0^2 \end{cases} \tag{14}$$

where ζ denotes a correction factor. The value of ζ is often set in the range from 0.01 to 10 appropriately by lots of simulations.

An ADRC controller is equivalent to a 2 degree-of-freedom (TDOF) PID controller [31], here is the method for a first-order ADRC controller.

The ESO in Equation (6) for first-order ADRC can be rewritten in the form of transfer functions:

$$\begin{cases} z_1(s) = \frac{b_0 s}{s^2+\beta_1 s+\beta_2}u(s) + \frac{\beta_1 s+\beta_2}{s^2+\beta_1 s+\beta_2}y(s) \\ z_2(s) = \frac{-\beta_2 b_0}{s^2+\beta_1 s+\beta_2}u(s) + \frac{\beta_2 s}{s^2+\beta_1 s+\beta_2}y(s) \end{cases} \tag{15}$$

By applying Mason's signal-flow gain formula, the structure of ADRC in Figure 1 can be transformed into an equivalent structure shown in Figure 2, and the specific derivations are omitted.

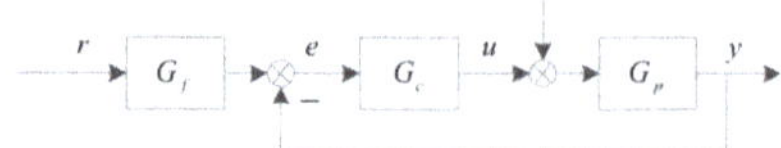

Figure 2. Equivalent structure of ADRC.

From Figure 2, the structure of ADRC is equivalent to a TDOF structure, where the feedback controller and the feedforward controller are:

$$G_c = \frac{(\xi\omega_0^2 + 2\omega_c\omega_0)s + \omega_c\xi\omega_0^2}{(s + 2\omega_0 + \omega_c)b_0 s} \tag{16}$$

$$G_f = \frac{\omega_c(s^2 + 2\omega_0 s + \xi\omega_0^2)}{(\xi\omega_0^2 + 2\omega_c\omega_0)s + \omega_c\xi\omega_0^2} \tag{17}$$

where the feedback controller G_c is equivalent to a PI controller plus a filter and the order of the numerator is bigger than that of the denominator for the feedforward controller.

It should be noted that ADRC can be formally equivalent to a TDOF structure, but it is still different from the traditional TDOF structure of a PID controller. Firstly, the feedback controller of the equivalent structure is a generalized PI controller plus a filter, which could avoid the occurrence of an 'impulse spike' in the controller's output [32]. Secondly, the feedforward controller is different from the typical set-point weighing feedforward controller in the TDOF structure because of the order of $G_f(s)$ [33]. Besides, to guarantee the stability of ESO, we should have $\beta_1 > 0$ and $\beta_2 > 0$, implying $\omega_o > 0$. Otherwise, the ESO would be unstable [15].

Now, the D-partition method can be applied to analyze the stability region of the ADRC controller. To simplify the stability region analysis, the plant transfer function is represented as:

$$G_p(i\omega) = r(\omega)e^{i\theta(\omega)} = a(\omega) + ib(\omega) \tag{18}$$

and the characteristic equation for the closed-loop system is represented as [33]:

$$1 + G_p(s)G_c(s) = 0 \tag{19}$$

where $G_l(s) = G_p(s)G_c(s)$ is called the open loop transfer function.

The boundaries of the stability region in the parameter plane consist of nonsingular boundaries for $\omega \in (0, -\infty) \cup (0, +\infty)$ and singular boundaries for $\omega = 0$ or $\omega = \pm\infty$ based on the D-partition method.

The boundaries of the stability region can be determined as follows:

(1) The singular boundary of ADRC for $\omega = 0$ is (called ∂D_0 surface)

$$\partial D_0 : \omega_c\xi\omega_0^2 c_0 = 0 \tag{20}$$

Because we have $\omega_o > 0$ and cannot always guarantee $c_o = 0$ for different plants, we can obtain that the singular boundary of ADRC is $\partial D_0 : \omega_c = 0$. Note that a pure time delay does not affect the singular boundary of ADRC ($e^0 = 1$).

(2) The singular boundaries of ADRC for $\omega = \pm\infty$ are (called ∂D_∞ surface)

$$\partial D_\infty : b_0 c_m = 0 \tag{21}$$

Considering that b_0 is the approximation of the real gain, we can obtain that $b_0 \neq 0$ and $c_m = 0$ do not contain any parameter of ADRC, so we can neglect this singular boundary of ADRC.

(3) The nonsingular boundaries of ADRC for $\omega \in (0, -\infty) \cup (0, +\infty)$ are (called ∂D_ω surface)

$$\partial D_\omega : \left(\xi \omega_0^2 + 2\omega_c \omega_0 \right) i\omega + \omega_c \xi \omega_0^2 (a(\omega) + ib(\omega)) + (i\omega + 2\omega_0 + \omega_c) b_0 i\omega = 0 \tag{22}$$

Separating the real and imaginary parts, we can obtain:
Above all, the stability region of first-order ADRC can be expressed as:

$$\partial D_\omega : \begin{cases} \omega_c \left(\xi \omega_0^2 - \omega^2 \right) a(\omega) - \left(\xi \omega_0^2 + 2\omega_c \omega_0 \right) b(\omega) \omega - \xi b_0 \omega^2 = 0 \\ \omega_c \left(\xi \omega_0^2 - \omega^2 \right) b(\omega) + \left(\xi \omega_0^2 + 2\omega_c \omega_0 \right) a(\omega) \omega + 2 b_0 \omega_0 \omega = 0 \end{cases} \tag{23}$$

$$\begin{cases} \partial D_0 : \omega_c = 0 \\ \partial D_\omega : \begin{cases} \omega_c \left(\xi \omega_0^2 - \omega^2 \right) a(\omega) - \left(\xi \omega_0^2 + 2\omega_c \omega_0 \right) b(\omega) \omega - \xi b_0 \omega^2 = 0 \\ \omega_c \left(\xi \omega_0^2 - \omega^2 \right) b(\omega) + \left(\xi \omega_0^2 + 2\omega_c \omega_0 \right) a(\omega) \omega + 2 b_0 \omega_0 \omega = 0 \end{cases} \\ \omega_o > 0 \end{cases} \tag{24}$$

By solving Equation (24) with ω varying from $-\infty$ to $+\infty$ when b_0 and ξ are fixed, we obtain three curves which constitute the stability region of first-order ADRC.

2.4. PR optimization

Define the parameters of an ADRC controller as $d = \{\omega_c, b_0, \omega_c, \xi\}$, which should be located in the stability region (22). When all of the uncertain plant and controller parameters are defined, the control system's performance can be evaluated by examining whether it satisfies the ith design requirements. For measuring design requirements on stability and performance quantitatively, a binary indicator function I is defined as follows

$$I_i = \begin{cases} 0 & \text{design requirements are not satisfied} \\ 1 & \text{design requirements are satisfied} \end{cases} \tag{25}$$

The probability P that the ADRC controller satisfies the design requirements can be described as the integral of the binary indicator function in the parameter space

$$P_i(d) = \int_Q I[G_p(q), G_c(d)] p_r(q) dq \tag{26}$$

By using this probabilistic framework, many control indices can be examined. Define the PR index as:

$$J(d) = fcn(P_1(d), P_2(d), \cdots) \tag{27}$$

where *fcn* is defined as the weights for each index.

The goal of ADRC controller design is to find the optimal controller parameters d^* that obtain the maximum value of the index $J(d^*)$ with the parameter uncertainties throughout the parameter space.

Considering that Equation (27) is difficult to integrate analytically in most cases, Monte Carlo simulation is a practical and useful tool to estimate the probability P [28]. The estimate of probability P and the PR index based on N samples

$$\hat{P}(d) = \frac{1}{N} \sum_{k=1}^{N} I[G_p(q), G_c(d)] \tag{28}$$

$$\hat{J}(d) = fcn(\hat{P}_1(d), \hat{P}_2(d), \cdots) \tag{29}$$

and the estimate $\hat{P}$ and $\hat{J}$ approach to the real probability P and J in the condition of $N \to \infty$, respectively. However, N cannot tend to infinity in practice and a finite N results in estimation errors. A minimum N which guarantees a certain confidence level to a risk parameter can be calculated based on the Massart Inequality [34,35]:

$$N > \frac{2\left(1 - \varepsilon + \frac{\alpha\varepsilon}{3}\right)\left(1 - \frac{\alpha}{3}\right)\ln\frac{2}{\delta}}{\alpha^2\varepsilon} \tag{30}$$

where ε denotes the given risk parameter, the confidence level is defined as $1 - \delta$, and $\alpha \in (0,1)$. Such a sample size ensures $P_r\{|P_x - K/N| < \alpha\varepsilon\} > 1 - \delta$, where P_x is the probability of the system to satisfy the design requirements, K/N is the estimated value of the probability, K is the number of design requirements satisfied in N samples, and the confidence interval is $[K/N - \alpha\varepsilon K/N + \alpha\varepsilon]$.

Considering the difficulty of non-convex problems for probability optimization [30], the genetic algorithm is proposed to optimize the controller parameters, which has a good global convergence ability and does not rely on any gradient or Hessian information [36].

Overall, the design procedure of an ADRC controller based on PR can be presented as:

(1) Define control indices. The settling time and overshoot are the selected indices in this paper, and the weights are defined simultaneously.

(2) The stability region D_c of the nominal plant is calculated according to Equation (24) as the search space of parameters.

(3) Randomly generate parameters of the ADRC controller in region D_c as the initial population for the genetic algorithm. Calculate the probability function $\hat{P}$ of the initial population as the object function for each set of parameters.

(4) The genetic algorithm is applied to optimize parameters for finding the largest value of the object function $\hat{P}$. Define the optimized parameters as the expected parameters of the ADRC controller d'.

(4) Test parameters d' by Monte Carlo simulation in the parameter space Q. If the result satisfies the requirement, the expected parameters are the optimal parameters d^*, otherwise return to step (3).

The estimated value of P can be fully close to the real value when N is big enough. To improve the calculation efficiency of genetic algorithms, a small value of N is set. A big enough value of N determined by the Massart Inequality according to risk parameter ε and the confidence level $1 - \delta$ in Equation (30) can be used to test the probability of design requirements. Set the risk parameter $\varepsilon = 0.1$, the confidence level $1 - \delta = 0.99$, and $\alpha = 0.2$ in this paper. Then, the minimum value of N for Monte Carlo simulation can be obtained as $N = 2442$ according to the Massart Inequality.

3. Simulations

The proposed method is applied to five typical industrial plants, the first-order plus dead-time plant, the second-order plus dead-time plant, the non-minimum phase plant, the integral plant, and the high-order plant. The control performance is compared with the traditional TDOF PI controller based on PR and the IMC controller and PID controller tuned by an optimization algorithm with the Integral of Time and Absolute Error (ITAE) index for a nominal plant. The mathematical form of the feedforward controller and the feedback controller of the TDOF PI controller can be depicted as follows:

$$G_{f\_TDOF} = \frac{bk_p s + k_i}{k_p s + k_i} \tag{31}$$

$$G_{c\_TDOF} = k_p + \frac{k_i}{s} \tag{32}$$

where b, k_p, and k_i are the parameters of the TDOF PI controller. The filter of the IMC controller is designed as:

$$G_{f\_IMC} = \frac{1}{T_f s + 1} \tag{33}$$

where T_f is the parameter of the filter and needs to be tuned. The mathematical form of the PID controller can be depicted as:

$$G_{PID} = k_{p1} + \frac{k_{i1}}{s} + k_{d1}s \tag{34}$$

where k_{p1}, k_{i1}, and k_{d1} are the parameters of the PID controller and they are tuned with the ITAE index.

It should be noted that Padé approximations are applied to the TDOF PI controller's design for $G_1(s)$ and $G_2(s)$ and the order of filters for the IMC controller is equal to the order of each plant. The necessary approximation is also applied to $G_4(s)$ based on suitable all-pass characteristics [5]. Besides this, the PI controller for $G_4(s)$ is a common PI controller because $G_4(s)$ is an integral plant. The following simulations for $G_4(s)$ are carried out with a PI controller rather than with a TDOF-PI controller.

The definitions of the PR index are varied based on design requirements, which can be a weight coefficient, a linear function, a nonlinear function, etc. Considering the conflicts between the settling time and overshoot, a linear function with a weighting coefficient is defined in this paper:

$$J(d) = 0.8P_{ts} + 0.2P_{\sigma} \tag{35}$$

where P_{ts}, P_{σ} are the binary indicator functions of the system satisfying the design requirements of the settling time T_s and overshoot σ, respectively.

Note that the settling time T_s is the time required for the response curve to reach and stay within a range about the final value of the size specified by an absolute percentage of the final value (usually 2% or 5%, the former is chosen in this paper) and overshoot σ is the relative proportion between the difference and the steady state value where the difference is between the maximum peak value and the steady-state value. The number of individuals in the initial population and Monte Carlo simulation and the maximum number of evolutionary iterations are set to 200, 500, and 20, respectively.

The typical nominal plants and the parameter perturbation range of nominal plants are shown in Table 1. Besides this, the design requirements and parameters of the Massart Inequality for each plant are shown in Table 2.

Table 1. The typical plants and the parameter perturbation range.

Plant	The Nominal Model	Parameter Perturbation Range
$G_1(s) = \dfrac{1}{Ts + 1}e^{-\tau s}$	$T = 20, \tau = 200$	$T \in [10, 30], \tau \in [180, 220]$
$G_2(s) = \dfrac{1}{(T_1s + 1)(T_2s + 1)}e^{-\tau s}$	$T_1 = 20, T_2 = 20, \tau = 90$	$T_1 \in [16, 24], T_2 \in [16, 24]$ $\tau \in [80, 100]$
$G_3(s) = \dfrac{k(a - s)}{(T_1s + 1)(T_2s + 1)}$	$T_1 = 5, T_2 = 0.4, a = 1.25, k = 4$	$T_1 \in [4.5, 5.5], T_2 \in [0.36, 0.44]$ $a \in [1, 1.5], k \in [4.8, 3.2]$
$G_4(s) = \dfrac{k}{s(Ts + 1)}$	$T = 11, k = 0.2$	$T \in [7, 15], k \in [0.1, 0.3]$
$G_5(s) = \dfrac{k}{(Ts + 1)^3}$	$T = 5, k = 1.3$	$T \in [3, 7], k \in [1, 1.6]$

Table 2. Design requirements and parameters of Massart Inequality.

Plant	Design Requirements	Parameters of Massart Inequality
$G_1(s)$	$T_s < 1000s, \delta < 5\%$	$\varepsilon = 0.01, \delta = 0.01, \alpha = 0.2, N = 24{,}495$
$G_2(s)$	$T_s < 700s, \delta < 5\%$	$\varepsilon = 0.01, \delta = 0.01, \alpha = 0.2, N = 24{,}495$
$G_3(s)$	$T_s < 10s, \delta < 5\%$	$\varepsilon = 0.1, \delta = 0.01, \alpha = 0.2, N = 2442$
$G_4(s)$	$T_s < 300s, \delta < 5\%$	$\varepsilon = 0.01, \delta = 0.01, \alpha = 0.2, N = 24{,}495$
$G_5(s)$	$T_s < 240s, \delta < 5\%$	$\varepsilon = 0.01, \delta = 0.01, \alpha = 0.2, N = 24{,}495$

Consider that b_0 is the approximation of b, b_0 is fixed as 1 for plant 1, plant 2, plant 4, plant 5, and as 5 for plant 3, and there are three parameters k_p, ω_o, and ξ to be tuned. Applying the proposed method, we can obtain the parameters of an ADRC controller based on PR. The parameters of the ADRC controller and other controllers are listed in Table 3.

Based on the parameters of the ADRC controller and contrasting controllers listed in Table 3, we can obtain step responses of these nominal plants in Table 1 with different controllers shown in Figure 3. A Monte Carlo simulation is also carried out N times, and the results are shown in Figures 4–8 with ITAE, T_s, and σ. Besides this, K/N, which is the estimated value of the probability, is listed in Table 4.

Table 3. The parameters of different controllers.

Plant	ADRC Controller	PID Controller	IMC Controller	TDOF-PI Controller
$G_1(s)$	$\omega_c = 0.2256$, $\omega_o = 5.3506$, $\xi = 0.010$	$k_{p1} = 0.38$, $k_{i1} = 0.0037$, $k_{d1} = 12.88$	$T_f = 112.3$	$b = 0.6031$, $k_p = 0.2509$, $k_i = 0.0030$
$G_2(s)$	$\omega_c = 0.0314$, $\omega_o = 2.9323$, $\xi = 0.0954$	$k_{p1} = 0.54$, $k_{i1} = 0.0072$, $k_{d1} = 14.22$	$T_f = 57.50$	$b = 0.4240$, $k_p = 0.2661$, $k_i = 0.0054$
$G_3(s)$	$\omega_c = 2.6275$, $\omega_o = 3.3546$, $\xi = 0.1464$	$k_{p1} = 0.85$, $k_{i1} = 0.11$, $k_{d1} = 0.24$	$T_f = 0.965$	$b = 1.000$, $k_p = 0.4245$, $k_i = 0.0853$
$G_4(s)$	$\omega_c = 0.0192$, $\omega_o = 0.1986$, $\xi = 2.6760$	$k_{p1} = 2.164$, $k_{i1} = 0, k_{d1} = 9.233$	$T_f = 4.01$	$k_p = 0.1186, k_i = 0$
$G_5(s)$	$\omega_c = 0.3002$, $\omega_o = 0.9273$, $\xi = 0.1529$	$k_{p1} = 0.346$, $k_{i1} = 0.0384$, $k_{d1} = 2.6507$	$T_f = 17.15$	$b = 0.9300$, $k_p = 0.1877$, $k_i = 0.0191$

PID = Proportional-Integral-Derivative; IMC = internal model control; TDOF-PI = 2 degrees of freedom proportional-integral.

Table 4. The estimated value of the probability.

Plant	ADRC Controller	PID Controller	IMC	TDOF-PI Controller
$G_1(s)$	0.9991	0.6487	0.9990	0.9989
$G_2(s)$	0.9991	0.8972	0.9992	0.9999
$G_3(s)$	0.9630	0.7732	0.9390	0.9523
$G_4(s)$	0.9999	0.8773	0.8385	0.9984
$G_5(s)$	1	0.8355	0.9844	0.9920

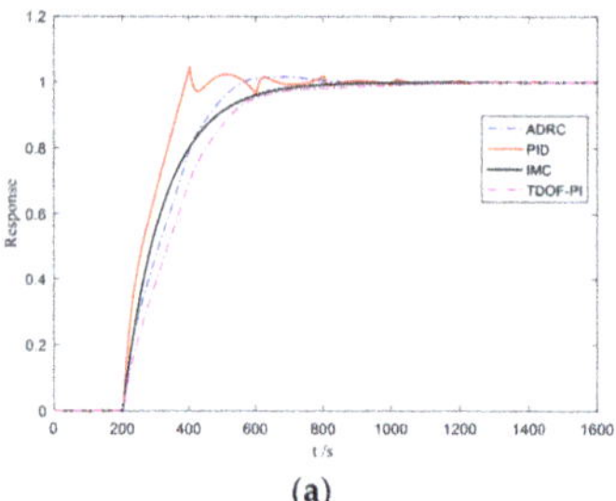
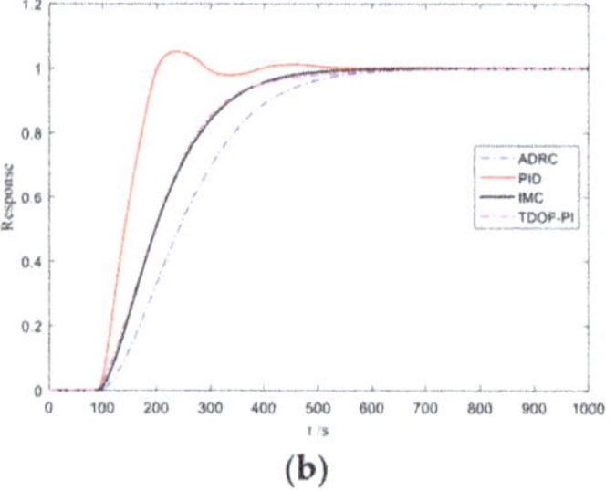

(a) (b)

Figure 3. *Cont.*

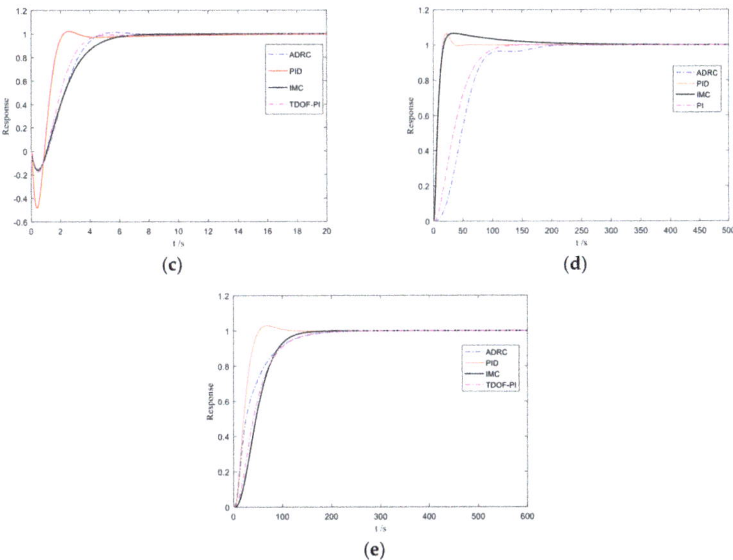

Figure 3. The step responses of different nominal plants with different controllers ((**a**) plant 1; (**b**) plant 2; (**c**) plant 3; (**d**) plant 4; (**e**) plant 5).

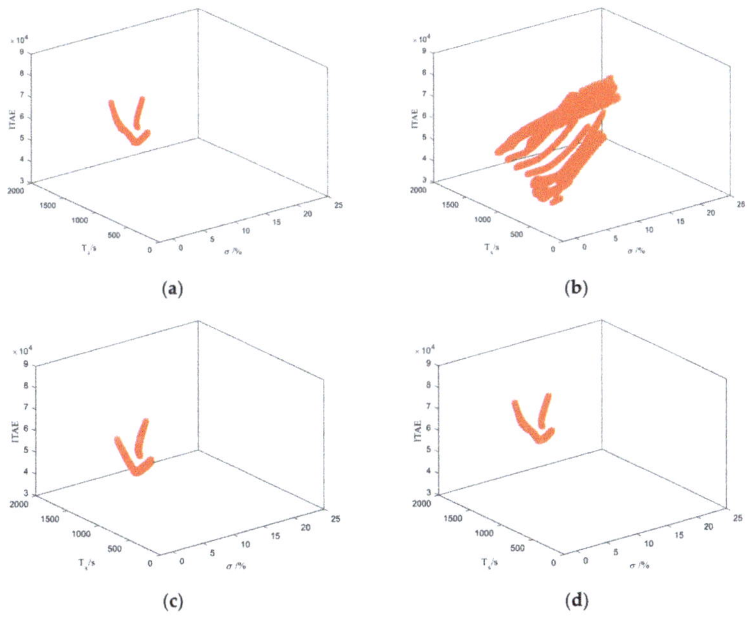

Figure 4. Results of Monte Carlo simulation for plant 1 ((**a**) ADRC; (**b**) PID; (**c**) IMC; (**d**) TDOF-PI). ITAE = Integral of Time and Absolute Error.

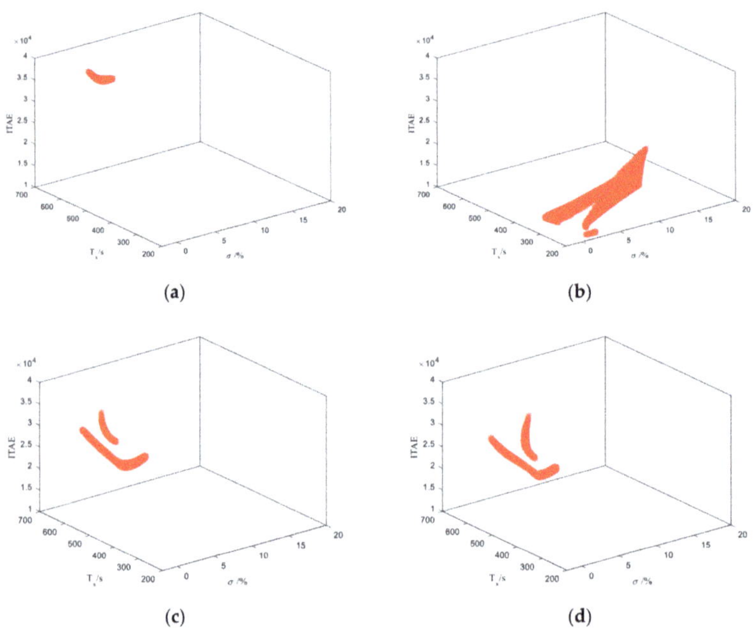

Figure 5. Results of Monte Carlo simulation for plant 2 ((**a**) ADRC; (**b**) PID; (**c**) IMC; (**d**) TDOF-PI).

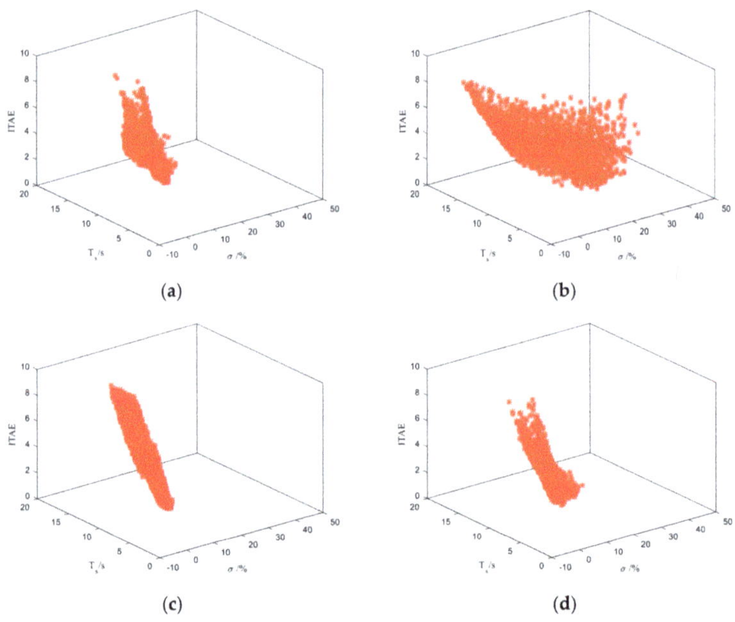

Figure 6. Results of Monte Carlo simulation for plant 3 ((**a**) ADRC; (**b**) PID; (**c**) IMC; (**d**) TDOF-PI).

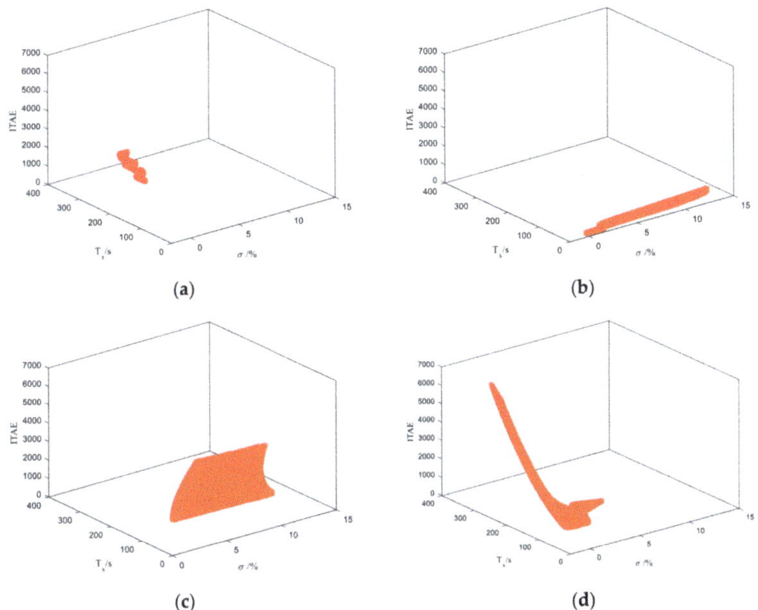

Figure 7. Results of Monte Carlo simulation for plant 4 ((**a**) ADRC; (**b**) PID; (**c**) IMC; (**d**) PI).

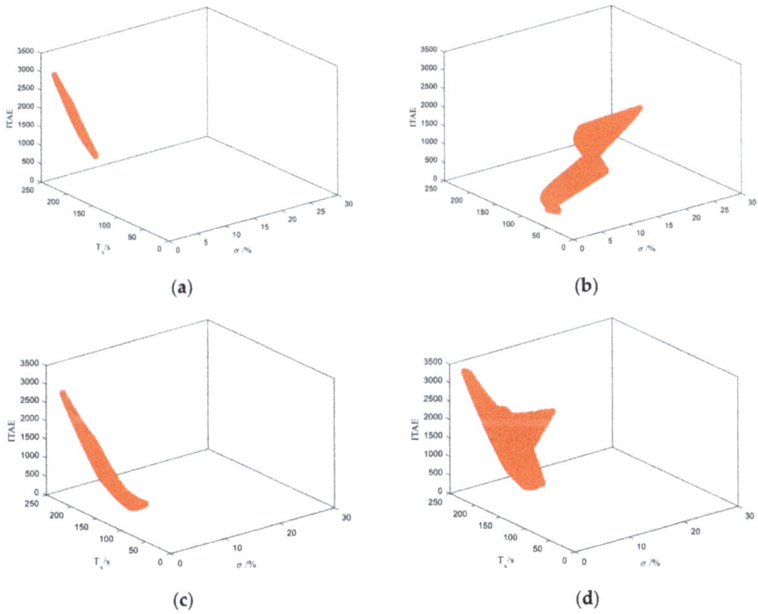

Figure 8. Results of Monte Carlo simulation for plant 5 ((**a**) ADRC; (**b**) PID; (**c**) IMC; (**d**) TDOF-PI).

From Figure 3, the PID controller has the fastest tracking speed, but it still has the largest overshoot for all plants except plant 4. The ADRC controller, TDOF-PI controller, and IMC controller have similar responses for plant 1, plant 2, plant 3, and plant 5, and they all have small overshoot (less than 2%)

and a similar settling time. Besides this, the IMC controller has the largest overshoot (6.6%) and the overshoot of the PID controller is 6.4% for plant 4, which means that the IMC controller and the PID controller cannot meet the design requirement of the overshoot. What needs to be stressed is that the ADRC controller and the TDOF-PI controller both meet the design requirements for all plants even though the settling time is larger than than that of the PID controller. Note that the responses shown in Figure 3 only reflect the control performance of the nominal plants and not all of the plants in the parameter space Q.

A denser distribution denotes stronger robustness and a smaller value denotes better control performance in Figures 4–8. The value of K/N listed in Table 4 is the specific value of the estimated value of the probability in Monte Carlo simulation, which means that a larger value denotes stronger robustness and a larger probability to meet the design requirements in the whole parameter space Q.

From Figure 4, we can know that the ADRC controller in Figure 4a, the TDOF-PI controller in Figure 4d, and the IMC controller in Figure 4c have a denser distribution than the PID controller in Figure 4b for plant 1, which means that the PID controller has the worst robustness, and the value of the PID controller in Table 4 also certifies this. Besides this, the ADRC controller, the TDOF-PI controller, and the IMC controller have a similar distribution, which means that they have a similar robustness, while the value in Table 4 tells that the ADRC controller has the largest probability to meet the design requirements in the whole parameter space Q.

The ADRC controller in Figure 5a has the densest distribution compared to the PID controller, the TDOF-PI controller, and the IMC controller even though the value of the ITAE index of the ADRC controller is the largest. The value of K/N of the ADRC controller for plant 2 is larger than 0.999 and it can ensure a large probability to meet the design requirements in the whole parameter space.

The value of K/N of the ADRC controller for plant 3 is the largest from Table 4, and this means that the ADRC controller has the strongest robustness. Besides this, the distribution of the PID controller is the sparsest in Figure 6b. Note that plant 3 has Right Half Plane (RHP) zero and this is why the value of K/N of the ADRC controller is smaller than that for other plants.

The ADRC controller in Figure 7a has the densest distribution compared to the PID controller in Figure 7b, the TDOF-PI controller in Figure 7c, and the IMC controller in Figure 7d, which means that the ADRC controller for plant 4 has the the largest probability to meet the design requirements in the whole parameter space Q, and the value of K/N of the ADRC controller for plant 4 in Table 4 is also certifies it.

From Figure 8, we can know that the ADRC controller in Figure 8a has the densest distribution compared to the PID controller, the TDOF-PI controller, and the IMC controller, and the value of K/N of the ADRC controller in Table 4 is 1, which means that ADRC can ensure that uncertain plants in the whole parameter space Q for plant 5 all meet the design requirements according to the Massart Inequality.

Summarizing the different controllers for the above five examples, ADRC based on the PR index has the largest estimated value of the probability for all plants except plant 2 and has the most possibility to obtain the design requirements throughout the whole parameter space. Note that ADRC based on the PR index still ensures a high probability (0.991) to meet the design requirements throughout the whole parameter space for plant 2. These illustrative simulations demonstrate the effectiveness of the proposed tuning method to enhance the ability of handling the uncertainties for a controller and they offer good support for the application of the proposed method to a boiler unit. A field application to the secondary air regulation of a boiler unit is described below.

4. A Field Application to the Secondary Air Regulation of a Boiler Unit

4.1. The Process Description

An air control system is an important subsystem for a boiler unit. An air control system can be depicted as in Figure 9. It mainly contains a primary air system, a secondary air system, a fluidizing air system, and an induced draft system. Secondary air enters into the furnace after passing through the air preheater under the force of secondary air fans as shown in the green part of Figure 9. The function of the secondary air system is to adjust the combustion and the temperature distribution by adjusting the frequency of secondary air fans. Secondary air flow is important for the economy and environmental protection of a boiler unit, which influences the combustion efficiency and generation of nitrogen oxides (NO_x) in the combustion [37,38]. The change of the amount of primary air and the bed temperature leads to a wide variety of dynamic characteristics of a secondary air system. What is worse, to integrate more renewable energy into the power grid, the frequent change of load output of a boiler unit results in this being worse.

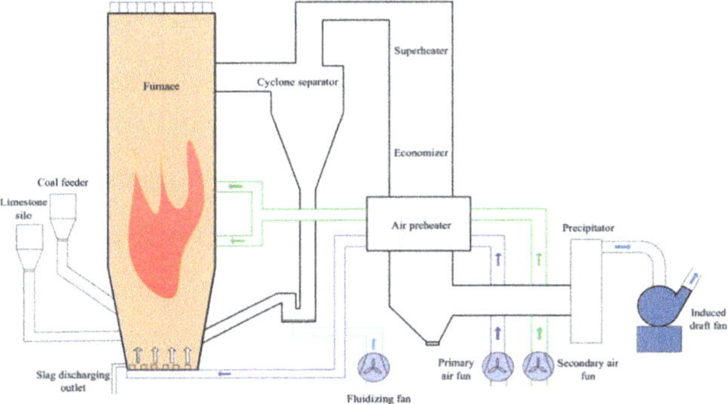

Figure 9. Air control system for a boiler unit.

To obtain the expected control performance in a wide range of operating conditions, ADRC controller tuning based on PR is applied to the secondary air regulation shown in Figure 9. This comparative field application was done in Tongda Power Plant, which is a 300 MW Circulating Fluidized Bed (CFB) subcritical unit located in Shanxi Province in China. Some key parameters of the unit in design condition are listed as follows: the output power is 300 MW, the bed temperature is 900 °C, the superheated steam temperature is 541 °C, and the pressure of the main steam is 17.5 MPa. Control logics of ADRC were configured in a DCS platform, which was produced by Zheda Zhongkong Co., Ltd., Hangzhou, China. Besides this, the total amount of the secondary air is about 84.6 m^3/s in Boiler Maximum Continue Rate (BMCR) condition and the rate of the fans is 1242.5 r/min in BMCR condition.

To obtain a simple model of the secondary air system, the opening loop test was done and the rough model is identified as Equation (36) with the boiler unit working in the condition of 200 MW, which is the common output for the unit. The control variable is the frequency of secondary air fans, there are two of the same fans in the secondary air system (their frequency donates MV1 and MV2, and the average frequency donates MV), and its unit is a percentage of the maximum (%). The output is the total amount of secondary air, which is expressed as a percentage (PV in Figure 10). The result comparison of the test data and the plant data is shown in Figure 10. It shows that the established model can reflect the real dynamic characteristics well. So, the model in Equation (36) is set as the nominal model to design an ADRC controller in the next subsection.

$$G = \frac{k_1}{(T_1 s + 1)^2} e^{-\tau_1 s} = \frac{1.967}{(14s + 1)^2} e^{-6s} \tag{36}$$

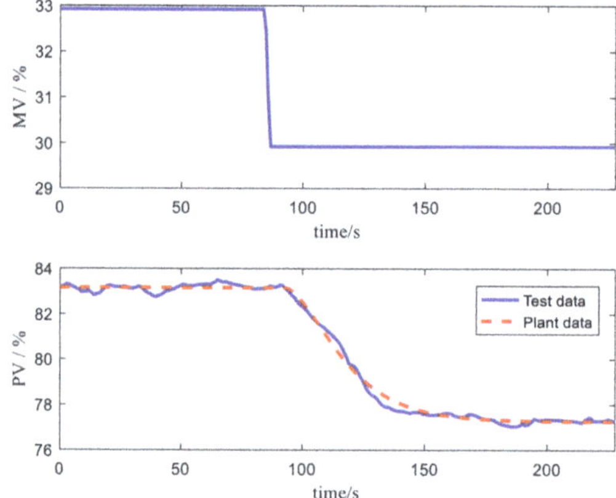

Figure 10. Comparison of test data and the model data.

4.2. ADRC Controller Design Based on PR

The ADRC controller based on PR is designed for the nominal model (36), and the parameter perturbation ranges are set as $k_1 \in [1.5, 2.5]$, $T_1 \in [11, 17]$, and $\tau_1 \in [3, 9]$ considering the load range of 50–100%. Design requirements and parameters of Massart Inequality are set as $T_s < 160s$, $\delta < 5\%$, and $\varepsilon = 0.01$, $\sigma = 0.01$, $\alpha = 0.2$, and $N = 24494$, respectively. Parameters of genetic algorithms are the same as those of the simulations in Section 3. b_0 is also fixed as 1 for the nominal model, and the other parameters are obtained as $\omega_c = 0.2504$, $\omega_o = 0.3458$, and $\xi = 0.3596$ by genetic algorithms.

The step responses of the nominal plant are shown as the black line in Figure 11 and varying plants in the parameter space are shown in Figure 11, and the result of the Monte Carlo simulation is shown in Figure 12. From Figures 11 and 12, we can know that ADRC based on PR can ensure the required control performance and also has a good robustness for the secondary air system working in a wide range. Besides, the value of the PR index J is 0.9912, which means that the ADRC controller based on the PR index has a large enough probability to meet the design requirements for the secondary air system whose parameters perturb in the perturbation ranges. The simulation based on the identified plant certifies the effectiveness of the proposed method and it gives us great confidence for the following field application.

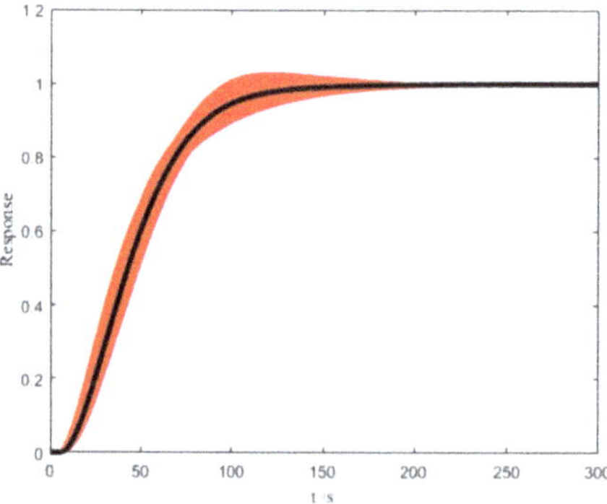

Figure 11. The step responses with the nominal plant and varying plants (Black line: the nominal plant, red line: varying plants).

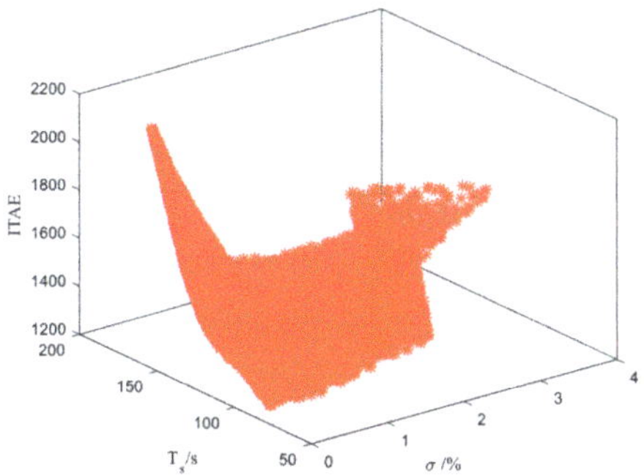

Figure 12. The robustness test by the Monte Carlo simulation.

4.3. The Field Application

Based on the simulation for the identified plant and some other protective measures done in the DCS platform, the field application of the ADRC controller with the parameters tuned by the proposed method was carried out. Besides this, a similar experiment of a PI controller whose parameters ($k_p = 1/3$, $k_i = 1/120$) are tuned by experienced engineers was also carried out.

The comparison results of the ADRC and PI controllers are shown in Figures 13 and 14, and the time spans are both 16 min. Figures 13a and 14a show the fluctuation of the unit load during the experiment. Note that the fluctuation range of the ADRC controller is from 228 MW to 245 MW, which is much larger than that of the PI controller from 230 MW to 240 MW. Besides, the settling time and the change amplitude of the secondary air set-point shown in Figures 13b and 14b, respectively, are both listed in Table 5. Obviously, the ADRC controller can track the set-point (SP in Figure 13b) with no static deviation in a short time while the PI controller will need more time to track the set-point (SP in Figure 14b) and sometimes even cannot reach the steady state in the last two set-point changes as shown in Figure 14b. Besides this, the average settling time of the ADRC controller is about 91.2 s, which is much smaller than that of the PI controller (no less than 151 s). Note that the reference signals of the secondary air system in Figures 13b and 14b are different because the load was always changing according to the AGC commands. The control variable of the ADRC controller shown in Figure 13c is smoother than that of the PI controller in Figure 14c, and this is of benefit for the long-time running of the unit.

The ability of the ADRC controller based on PR is verified by the comparative experiments, which can obtain expected control performance for the secondary air system working in a wider range of operating conditions. Besides, the load during the experiment is far from 200 MW, which is the load of the opening loop test, and this means that the secondary air system with the ADRC controller can obtain the expected control performance even though the working condition is far from the design condition. Generally, an ADRC controller can play an important role in frequent load regulation, which can facilitate the wide application of sustainable power systems.

Table 5. The settling time of ADRC and PI.

ADRC		PI	
The Change Amplitude	The Settling Time	The Change Amplitude	The Settling Time
$\Delta r = 2$	77 s	$\Delta r = 3$	161 s
$\Delta r = 5$	87 s	$\Delta r = 2$	130 s
$\Delta r = 3$	103 s	$\Delta r = 4$	>164 s
$\Delta r = 4$	95 s	$\Delta r = 5$	>149 s
$\Delta r = 8$	94 s		

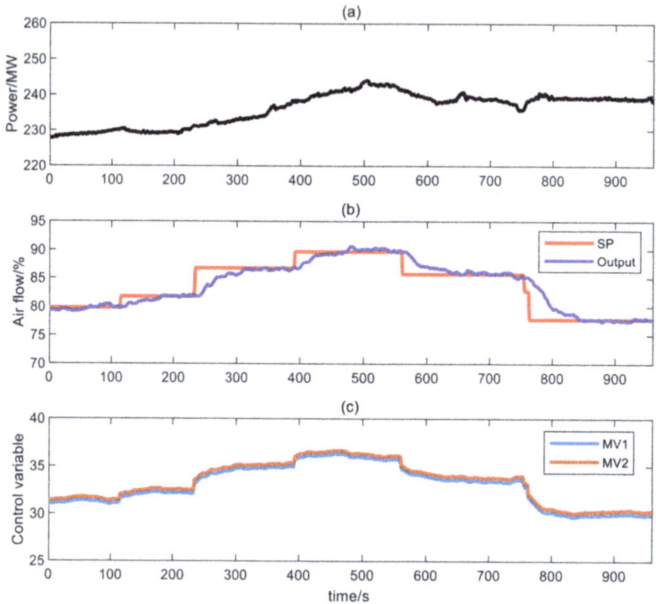

Figure 13. The field test result of the ADRC controller. SP = set-point. ((**a**) the power output; (**b**) the secondary air set-point and output; (**c**) control variable)

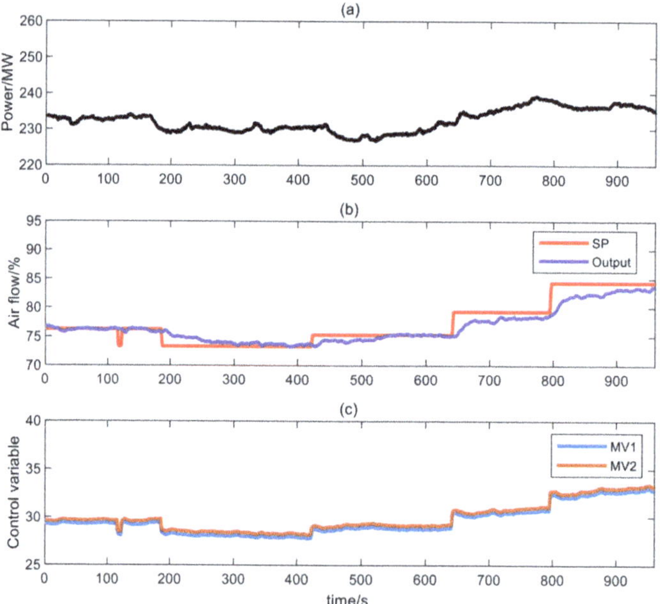

Figure 14. The field test result of the PI controller ((**a**) the power output; (**b**) the secondary air set-point and output; (**c**) control variable).

5. Conclusions

To achieve a sustainable future for renewable energy and integrate more renewable energy into the power grid, the conventional power plant has to lift its load frequently for a grid balance purpose. This inevitably brings some uncertainties to the power generation side and imposes new challenges for controller design. To improve the boiler control performance in conventional power generation, a data-driven control strategy, namely, ADRC based on PR, is proposed to handle the challenge in this paper. The stability region of the ADRC controller is analysed to provide the search space for the optimization of the PR index, and the design procedure is analysed and summarized. Five illustrative simulations are carried out to demonstrate the effectiveness of the proposed tuning method. With the confidence from the preliminary works, an ADRC controller was applied to the secondary air control of a boiler successfully. The running data shows that the proposed ADRC based on PR can ensure the expected control performance even though it works in a wider range of operating conditions and this depicts a promising future for ADRC controllers in the facilitation of sustainable power systems.

Acknowledgments: This work was supported by the National Key Research and Development Program of China under Grant No. 2016YFB0901405, the Natural Science Foundation of Jiangsu Province, China under Grant BK20170686, open funding of the state key lab for power systems, Tsinghua University, the Open Fund of the State Key Laboratory of Clean and Efficient Coal-fired Power Generation and Pollution Control and Coal-based Key Scientific and Technological Projects of Shanxi Province under Grant MD2014-07.

Author Contributions: Zhenlong Wu and Donghai Li designed the tuning procedure for ADRC based on probabilistic robustness, the simulations, and wrote this paper. Zhenlong Wu and Ting He designed and performed the field application to the boiler secondary air regulation. Li Sun and Yali Xue has enriched the manuscript in terms of scientific and technical expertise with improving the paper's quality. All authors contributed to bringing the manuscript into its current state.

Conflicts of Interest: The authors declare no conflict of interest. The founding sponsors had no role in the design of the study; in the collection, analyses, or interpretation of data; in the writing of the manuscript, and in the decision to publish the results.

References

1. *International Energy Outlook 2013 Report*; U.S. Energy Information Administration: Washington, DC, USA, 2013.
2. Nguyen, T.T.; Quynh, N.V.; Duong, M.Q.; Van Dai, L. Modified Differential Evolution Algorithm: A Novel Approach to Optimize the Operation of Hydrothermal Power Systems while Considering the Different Constraints and Valve Point Loading Effects. *Energies* **2018**, *11*, 540. [CrossRef]
3. Son, Y.I.; Kim, I.H.; Choi, D.S.; Shim, H. Robust cascade control of electric motor drives using dual reduced-order PI observer. *IEEE Trans. Ind. Electron.* **2015**, *62*, 3672–3682. [CrossRef]
4. Hussein, A.A.; Salih, S.S.; Ghasm, Y.G. Implementation of Proportional-Integral-Observer Techniques for Load Frequency Control of Power System. *Procedia Comput. Sci.* **2017**, *109*, 754–762. [CrossRef]
5. Kobaku, T.; Patwardhan, S.; Agarwal, V. Experimental Evaluation of Internal Model Control Scheme on a DC-DC Boost Converter Exhibiting Non-minimum Phase Behavior. *IEEE Trans. Power Electron.* **2017**, *32*, 8880–8891. [CrossRef]
6. Zumoffen, D.A.; Braccia, L.; Marchetti, A.G. Economic plant-wide control design with backoff estimations using internal model control. *J. Process Control* **2016**, *40*, 93–105. [CrossRef]
7. Han, J.Q. From PID to active disturbance rejection control. *IEEE Trans. Ind. Electron.* **2009**, *56*, 900–906. [CrossRef]
8. Madoński, R.; Herman, P. Survey on methods of increasing the efficiency of extended state disturbance observers. *ISA Trans.* **2015**, *56*, 18–27. [CrossRef] [PubMed]
9. Han, J.Q. Active disturbance rejection controller and its applications. *Control Decis.* **1998**, *13*, 19–23.
10. Fu, C.F.; Tan, W. Tuning of linear ADRC with known plant information. *ISA Trans.* **2016**, *65*, 384–393. [CrossRef] [PubMed]
11. Gao, Z.Q. Scaling and bandwidth-parameterization based controller tuning. In Proceedings of the American Control Conference, Minneapolis, MN, USA, 14–16 June 2006; pp. 4989–4996.
12. Sariyildiz, E.; Ohnishi, K. Stability and robustness of disturbance-observer-based motion control systems. *IEEE Trans. Ind. Electron.* **2015**, *62*, 414–422. [CrossRef]

13. Wu, Z.L.; Li, D.H.; Xue, Y.L.; Wang, L.; Wang, J. Active disturbance rejection control for fluidized bed combustor. In Proceedings of the 2016 16th International Conference on Control, Automation and Systems (ICCAS), Gyeongju, Korea, 16–19 October 2016; pp. 1286–1291.

14. Ye, Y.; Yue, Z.; Gu, B. ADRC control of a 6-DOF parallel manipulator for telescope secondary mirror. *J. Instrum.* **2017**, *12*, T03006. [CrossRef]

15. Sun, L.; Li, D.; Hu, K.; Lee, K.Y.; Pan, F. On tuning and practical implementation of active disturbance rejection controller: A case study from a regenerative heater in a 1000 MW power plant. *Ind. Eng. Chem. Res.* **2016**, *55*, 6686–6695. [CrossRef]

16. Sun, L.; Hua, Q.; Shen, J.; Xue, Y.; Li, D.; Lee, K.Y. A Combined Voltage Control Strategy for Fuel Cell. *Sustainability* **2017**, *9*, 1517. [CrossRef]

17. Rahman, M.M.; Chowdhury, A.H.; Hossain, M.A. Improved Load Frequency Control Using a Fast Acting Active Disturbance Rejection Controller. *Energies* **2017**, *10*, 1718. [CrossRef]

18. Tang, H.; Li, Y. Development and active disturbance rejection control of a compliant micro-/nano-positioning piezo-stage with dual mode. *IEEE Trans. Ind. Electron.* **2014**, *61*, 1475–1492. [CrossRef]

19. Liu, Y.; Liu, J.; Zhou, S. Linear active disturbance rejection control for pressurized water reactor power. *Ann. Nucl. Energy* **2018**, *111*, 22–30. [CrossRef]

20. Liu, R.J.; Nie, Z.Y.; Wu, M.; She, J. Robust disturbance rejection for uncertain fractional-order systems. *Appl. Math. Comput.* **2018**, *322*, 79–88. [CrossRef]

21. Huang, C.E.; Li, D.H.; Xue, Y.L. Active disturbance rejection control for the ALSTOM gasifier benchmark problem. *Control Eng. Pract.* **2013**, *21*, 556–564. [CrossRef]

22. Zhang, H.; Zhao, S.; Gao, Z.Q. An active disturbance rejection control solution for the two-mass-spring benchmark problem. In Proceedings of the American Control Conference (ACC), Boston, MA, USA, 6–8 July 2016; pp. 1566–1571.

23. Huang, C.; Sira-Ramírez, H. A flatness based active disturbance rejection controller for the four tank benchmark problem. In Proceedings of the American Control Conference (ACC), Chicago, IL, USA, 1–3 July 2015; pp. 4628–4633.

24. Zhao, C.Z.; Li, D.H. Control design for the SISO system with the unknown order and the unknown relative degree. *ISA Trans.* **2014**, *53*, 858–872. [CrossRef] [PubMed]

25. Wang, L.J.; Li, Q.; Tong, C.; Yin, Y.; Gao, Z.; Zheng, Q.; Zhang, W. On control design and tuning for first order plus time delay plants with significant uncertainties. In Proceedings of the American Control Conference (ACC), Chicago, IL, USA, 1–3 July 2015; pp. 5276–5281.

26. Tan, W.; Fu, C.F. Linear active disturbance-rejection control: analysis and tuning via IMC. *IEEE Trans. Ind. Electron.* **2016**, *63*, 2350–2359. [CrossRef]

27. Chen, X.; Zhou, K.; Aravena, J.L. Fast universal algorithms for robustness analysis. In Proceedings of the Decision and Control, Maui, HI, USA, 9–12 December 2003; pp. 1926–1931.

28. Calafiore, G.C.; Dabbene, F.; Tempo, R. Research on probabilistic methods for control system design. *Automatica* **2011**, *47*, 1279–1293. [CrossRef]

29. Calafiore, G.C. Repetitive scenario design. *IEEE Trans. Autom. Control* **2017**, *62*, 1125–1137. [CrossRef]

30. Wang, C.F.; Li, D.H.; Li, Z.; Jiang, X. Optimization of controllers for gas turbine based on probabilistic robustness. *J. Eng. Gas Turbines Power* **2009**, *131*, 054502. [CrossRef]

31. Wu, Z.L.; Xue, Y.L.; Pan, L.; Li, D.; He, T.; Sun, L.; Yang, Y. Active disturbance rejection control based simplified decoupling for two-input-two-output processes. In Proceedings of the Chinese Control Conference (CCC), Dalian, China, 26–28 July 2017.

32. Nandong, J. Heuristic-based multi-scale control procedure of simultaneous multi-loop PID tuning for multivariable processes. *J. Process Control* **2015**, *35*, 101–112. [CrossRef]

33. Åström, K.J.; Panagopoulos, H.; Hägglund, T. Design of PI controllers based on non-convex optimization. *Automatica* **1998**, *34*, 585–601. [CrossRef]

34. Wang, C.F.; Li, D.H. Decentralized PID controllers based on probabilistic robustness. *J. Dyn. Syst. Meas. Control* **2011**, *133*, 061015. [CrossRef]

35. Wang, C.F.; Li, D.H.; Jiang, X.Z. A PID Controller Design Method Based on Probablistic Robustness. *Proc. CSEE* **2007**, *32*, 020.

36. Ahn, C.W. *Practical Genetic Algorithms*; John Wiley & Sons: Hoboken, NJ, USA, 2006.

37. Kuang, M.; Li, Z.Q.; Ling, Z.Q.; Zeng, X. Improving flow and combustion performance of a large-scale down-fired furnace by shortening secondary-airport area. *Fuel* **2014**, *121*, 232–239. [CrossRef]

38. Luo, R.; Fu, J.P.; Li, N.; Zhang, Y.; Zhou, Q. Combined control of secondary air flaring angle of burner and air distribution for opposed-firing coal combustion. *Appl. Therm. Eng.* **2015**, *79*, 44–53. [CrossRef]

 *sustainability*

Article

Thermodynamic Cycle Concepts for High-Efficiency Power Plans. Part A: Public Power Plants 60+

Krzysztof Kosowski [1], Karol Tucki [2,*], Marian Piwowarski [1], Robert Stępień [1], Olga Orynycz [3,*], Wojciech Włodarski [1] and Anna Bączyk [4]

[1]   Faculty of Mechanical Engineering, Gdansk University of Technology, Gabriela Narutowicza Street 11/12, 80-233 Gdansk, Poland; kosowski@pg.gda.pl (K.K.); marian.piwowarski@pg.edu.pl (M.P.); rstepien@pg.edu.pl (R.S.); wwlodar@pg.edu.pl (W.W.)

[2]   Department of Organization and Production Engineering, Warsaw University of Life Sciences, Nowoursynowska Street 164, 02-787 Warsaw, Poland

[3]   Department of Production Management, Bialystok University of Technology, Wiejska Street 45A, 15-351 Bialystok, Poland

[4]   Department of Hydraulic Engineering, Warsaw University of Life Sciences, Nowoursynowska Street 159, 02-776 Warsaw, Poland; a.baczyk@levis.sggw.pl

*   Correspondence: karol_tucki@sggw.pl (K.T.); o.orynycz@pb.edu.pl (O.O.)

Received: 24 December 2018; Accepted: 17 January 2019; Published: 21 January 2019

Abstract: An analysis was carried out for different thermodynamic cycles of power plants with air turbines. Variants with regeneration and different cogeneration systems were considered. In the paper, we propose a new modification of a gas turbine cycle with the combustion chamber at the turbine outlet. A special air by-pass system of the combustor was applied and, in this way, the efficiency of the turbine cycle was increased by a few points. The proposed cycle equipped with a regenerator can provide higher efficiency than a classical gas turbine cycle with a regenerator. The best arrangements of combined air–steam cycles achieved very high values for overall cycle efficiency—that is, higher than 60%. An increase in efficiency to such degree would decrease fuel consumption, contribute to the mitigation of carbon dioxide emissions, and strengthen the sustainability of the region served by the power plant. This increase in efficiency might also contribute to the economic resilience of the area.

Keywords: thermodynamic cycle concepts; sustainability; modified cycle concepts; efficiency; energy systems

---

1. Introduction

Social and economic activity should aim to mitigate natural devastation. In recent years, the idea that improvements in quality of life should take place in greater harmony with the natural environment has gained popularity. Consequently, ecological problems have become the subject of reflection in various manufacturing sectors.

The development of production in various ecosystems should not only aim to keep up with demographic growth, but also to conserve equilibrium in the natural environment, so that it will be maintained in the best shape for the further development of future societies [1].

Mainly in reference to ecological politics, the definition of what is considered sustainable and durable development is essential. Ethical and economic values should be synchronized, aiming to match the premises of sustainable development. The term "sustainable development" is associated with many meanings. The multiplicity of this notion is reflected in various aspects and trends [2]. The concept of sustainable development focusses attention on environmental, social, and economic factors, and clearly underlines the interdisciplinary character of the energy sector [3].

Sustainable energetics should take into account: (1) consumption and how to supply energy without exposing society to danger and (2) how to assure economic development whilst caring for the natural environment [4]. The energetic system is based on the notion of equalization, which should, among other things, consider the production of energy with regard to long-lasting economic and environmental goals [5].

Disseminating sustainable development on a global scale [6] is the focus of activities aiming to conceptualize how energetics could be developed through skillful conservation of energy on the local level. The sustainable management of these resources has to be supported by sustainable development of energy [7], and should deliver information that enables the evaluation of the energetics in ecological, economic, and social dimensions.

The formation of a green economy requires the union of transformation processes across the various regions of the country, along with the active participation of enterprises. One of the factors assuring the development of the modern economy is the uninterrupted delivery of energy. The energy resources trade, as well as industrial activity, is bound to the pollution of the atmosphere. This often leads to regional conflicts and ecological threats, even on a global scale. Knowledge of the suitability of fuels and the their suitable selection is indispensable to the safe and effective design as well as exploitation of technical objects [8].

Energetic systems require a deep understanding, taking into account the changes in production technologies [9]. Therefore energetics must occupy an important place amongst the different areas of sustainable development. The aim of the present paper is to highlight the possibility of utilizing circulation power stations in the context of sustainable development. Such a development should be implemented universally, with regard to the politics of individual states.

A problem of the national power generation sector is the relatively low efficiency of energy generation from coal, which is additionally accompanied by high emissions of carbon dioxide [10–12]. The average efficiency of Polish power plants is lower than elsewhere in the EU. The most efficient power plants in Poland are the newest units: Łagisza II (41% efficiency), Pątnów II (41% efficiency), Bełchatów II (42% efficiency), and Opole II (45% efficiency). Note that although these values can differ depending on the quoted source, the differences never exceed two percentage points. The majority of Polish power plants have an efficiency level below 36%, whilst for the oldest ones it can even be below 30%. For the present efficiency level, the assessed CO_2 emissions from steam power plants is approximately equal to 1100–1200 kgCO$_2$/MWh (for lignite-fired power plants). Modern power units operating at supercritical steam parameters ensure higher efficiency (48%–50% for hard coal-fired power plants), which, however, is decreased by more than ten percentage points by the unavoidable use of CO_2 sequestration systems. It is also noteworthy that several Polish power plants, with a total installed power capacity of about 6 GW, are likely to be closed during the next few years, due to poor technical condition [13–16].

Problem Posing

In the case of distributed power generation, the efficiency of small electric power plants is even lower. Organic Rankine Cycle (ORC)-based thermal power plants have an efficiency as low as approximately 12%. At present, one of the most advanced and efficient electric power plants is a power unit planned for construction in the USA, which forms part of the framework of a program financed by the US Department of Energy (US DOE). This unit will operate at ultra-supercritical steam parameters (with a pressure of about 35 MPa and temperature up to 760 °C), and its efficiency is expected to reach approximately 55% [17,18]. Such high steam pressures and temperatures require special materials and technologies, the use of which in Poland remains confined to a rather distant future [19]. At present, the highest efficiencies are reached in combined-cycle power plants (slightly over 60%, designed by Siemens), but these power plants are not coal-fired [20–22]. Hence, the question arises as to which direction attempts should be made to increase the efficiency of electric power generation in Poland. This can be reached by increasing the thermal efficiency of the applied cycle and/or by increasing the efficiency of its individual components. In the case of the most recent and most technologically advanced large steam power plants, the efficiencies of their machinery and equipment are very high and we cannot expect that they can be further significantly increased [23–28]. For instance, the maximal efficiency of boilers ranges within 92–95%, high-pressure turbines 88–94%, medium-pressure turbines 90–97%, low-pressure turbines 88–95%, electric current generators 98.5–99%, and water pumps—about 85%. This is complemented by low losses in external glands and low mechanical losses in turbine sets, which are clearly lower than one percentage point. Therefore, the other option (i.e. increasing the thermodynamic cycle efficiency) seems more promising.

2. Materials and Methods

The topic of the present study concerned the possibility of increasing the energy efficiency of small turbines working at various structural configurations.

Thermodynamic analysis was applied to each technical configuration and this formed the main methodology applied in the current research. Five configuration variants were analyzed. Figure 1 shows five different configurations of gas turbine sets:

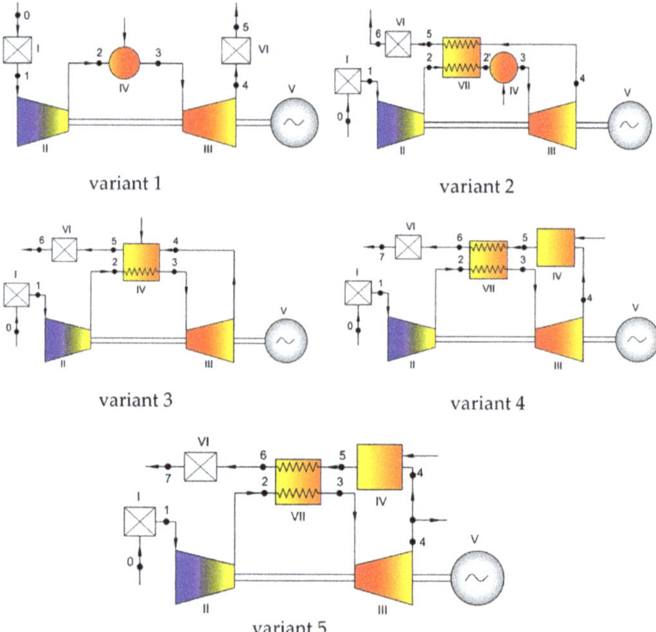

Figure 1. Analyzed turbine set arrangements. Variant 1: turbine set operating according to the simple open cycle; Variant 2: turbine set operating according to the open cycle with a regenerator; Variant 3: turbine set operating according to the open cycle with a combustion chamber at the turbine exit [29–31]; Variant 4: turbine set operating according to the open cycle with an external combustion chamber at the turbine exit and a high-temperature heat exchanger; Variant 5: turbine set operating according to the open cycle with partial bypassing of the external combustion chamber at the turbine exit and with a high-temperature heat exchanger.

In regenerator VII (variant 4), the exhaust gasses had a higher specific heat and a higher mass flow rate than the warmed up air. Thus, the temperature difference $T_5 - T_6$ was lower than $T_3 - T_2$. In variant 5, by drawing off some air, we decreased outlet temperature T_6 and reduced fuel consumption, as well as increasing cycle efficiency.

3. Results

Generally, the efficiency of the thermodynamic cycle is given by the well-known formula: $\eta = 1 - T_d/T_g$, where T_d and T_g represent the average temperatures of the hot reservoir (from which the heat is taken) and the cold reservoir (to which it is supplied), respectively. The highest possible temperature T_g and the lowest possible temperature T_d can be reached by the use of regeneration, as seen in Figure 2 (this figure shows the Brayton cycle [32], but the above is true for an arbitrary closed thermodynamic cycle).

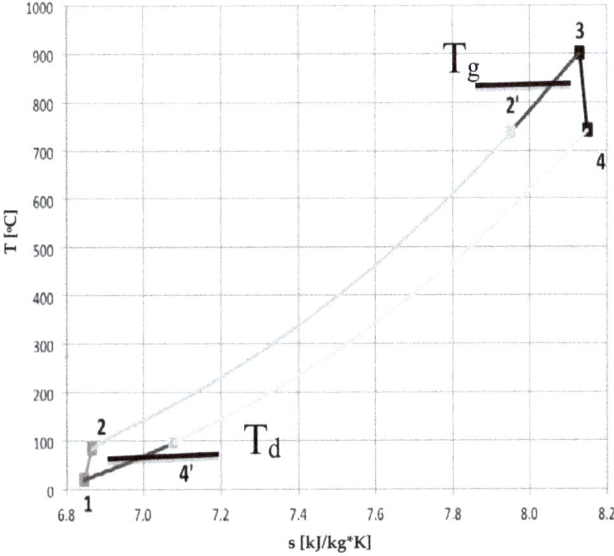

Figure 2. Cycle with regeneration (sample case).

The key role in regeneration is played by the so-called regeneration ratio ε (sometimes also referred to as the regenerator efficiency), which is defined as the ratio of the actual executed temperature increase to its theoretical maximal value for the heated medium, $\varepsilon = (T_{2'} - T_2)/(T_4 - T_2)$ (Figure 2). It is well-known that for the cycle of a gas turbine set with a regenerator, increasing the regeneration ratio ε leads to an increase in the cycle efficiency and to a decrease in the optimal compression value. Assuming that $\varepsilon = 1$, the efficiency of the ideal cycle of a gas turbine with a regenerator approaches the Carnot cycle efficiency when the compression approaches 1. However, in practice, the regeneration ratio has not exceeded $\varepsilon = 0.8$–0.85 for a long time, as increasing ε leads to an extreme increase of the regenerator heat transfer surface, which increases proportionally to $\varepsilon/(1 - \varepsilon)$. The next limitation was the maximal permissible temperature for regenerator material. The heat exchangers offered by the producers have a maximal temperature limit of approximately 700 °C. In recent years, there has been changes in this area, as a result of studying the use of CO_2 for cooling high-temperature gas reactors and developing the technology for production of so-called ceramic heat exchangers. At present, the production technology of heat exchangers operating at $\varepsilon = 0.98$ and at medium temperatures equal to 900–1000 °C, or even exceeding 1200 °C in the case of ceramic heat exchangers, is considered fully mature [33–38]. Adopting solutions of this type provides new opportunities for increasing the efficiencies of gas turbine sets and combined systems. The possibility of using highly efficient high-temperature heat exchangers has renewed the interest in closed gas cycles and gas turbine sets with a combustion chamber at the turbine exit.

4. Discussion

Within the analyzed variants, the variant that revealed the lowest efficiency (for the same assumed turbine inlet temperature and the same individual efficiencies of turbine set components) was Variant 1 (i.e., simple open cycle). The efficiency of Variant 2 (i.e., cycle with regenerator) was higher by a few percentage points. Even higher efficiency was reached by Variant 3, in which the air flowed to the combustion chamber directly from the turbine exit, which corresponded to a cycle with a regenerator of 100% efficiency, $\varepsilon = 1$. Variant 4 was intended for systems in which, in practice, arbitrary fuel can be combusted. However, due to the limited temperature difference $T_5 - T_3$, its efficiency was close to that in Variant 2. In Variant 5, part of the air leaving the turbine bypassed the combustion chamber, which improved the efficiency and provided wider opportunities for the use of a combined system. However, it should be noted that the heat exchange area of the regenerators in Variants 2, 4, and 5 were approximately 55–70% higher than in Variant 3.

The efficiency analysis was performed for large power turbine sets, at the assumed relatively high efficiencies of individual elements. In particular, the assumed efficiencies of the turbine and compressor were equal to 90%, while they were 98% for both the electric current generator and the combustion chamber. The assumed design values and working media parameters for particular points of the optimized cycles are shown in Tables 1 and 2, respectively.

Table 1. Assumed design parameters.

Parameter	Unit	Value	
$\eta_{compressor}$	[-]	0.900	
$\eta_{turbine}$	[-]	0.900	
η_{mech}	[-]	0.980	
$\eta_{leakage}$	[-]	0.980	
$\eta_{Generator}$	[-]	0.980	
$\eta_{comb.cham}$	[-]	0.980	
p_i/p_{i-1}	[-]	0.995	air inlet duct/filter
p_i/p_{i-1}	[-]	0.995	exhaust gases duct/filter/silencer
p_i/p_{i-1}	[-]	0.99	combustion chamber
p_i/p_{i-1}	[-]	0.99	regenerator
LHV	[MJ/kg]	24	

The relative efficiency values obtained for the analyzed variants after assuming the turbine inlet temperature $T_3 = 900\ ^\circ C$ are shown in Figure 3 with Variant 1 as the reference. They confirm the above observations. Figure 4 shows optimal compression values for the analyzed variants. The optimum values of compression ratio for each case were obtained as a result of the calculation of a large number of different variants. Note that at high efficiency (ε) of heat exchangers, the compression values in the analyzed systems were extremely low. On the one hand, this would make the turbine and compressor structures simpler and cheaper to produce, but on the other hand would decrease the specific power of the turbine set.

The efficiency values were heavily affected by the turbine inlet temperature and the efficiencies of heat exchangers. For instance, decreasing the final temperature difference $T_5 - T_2$ from 50 °C to 10 °C in Variant 3 increased the efficiency by about 9.5%. The highest efficiencies were obtained for combined gas–steam systems (see Figure 5). In those cases, exceeding 60% efficiency became possible even when the air turbine inlet temperature was as low as $T_3 = 900\ ^\circ C$. Figure 6 shows a sample case of a compressor and an air turbine cooperating in the combined power plant, with an output power of about

55 MW. Meanwhile, Figure 7 presents the flow part of the steam turbine for this power plant. Even higher efficiencies could be reached in combined cycles consisting of a closed gas system and a steam cycle.

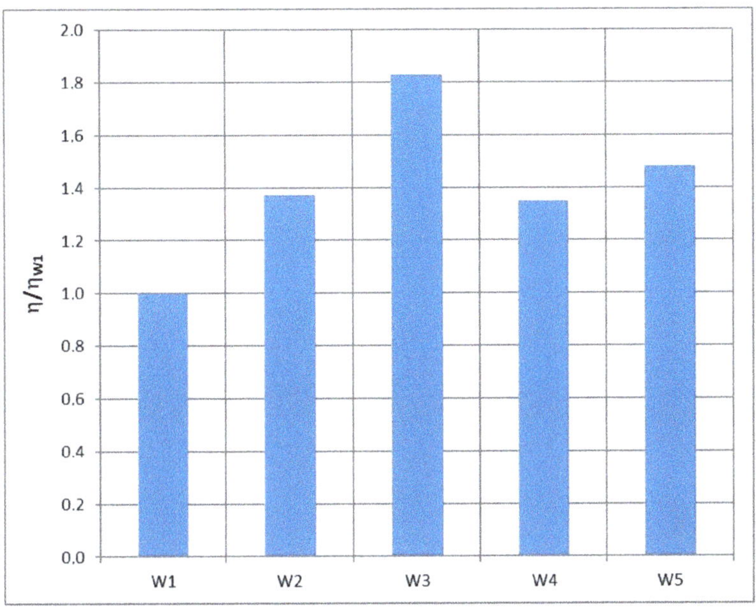

Figure 3. Relative frequencies of analyzed turbine sets.

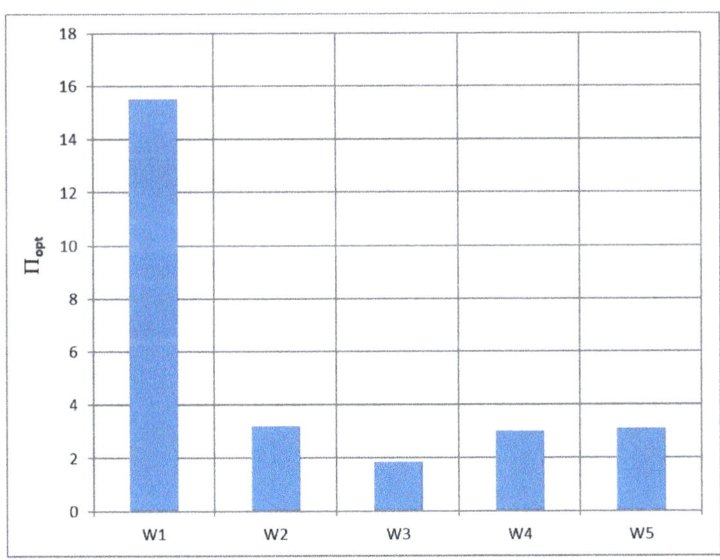

Figure 4. Optimal compression values of analyzed turbine sets.

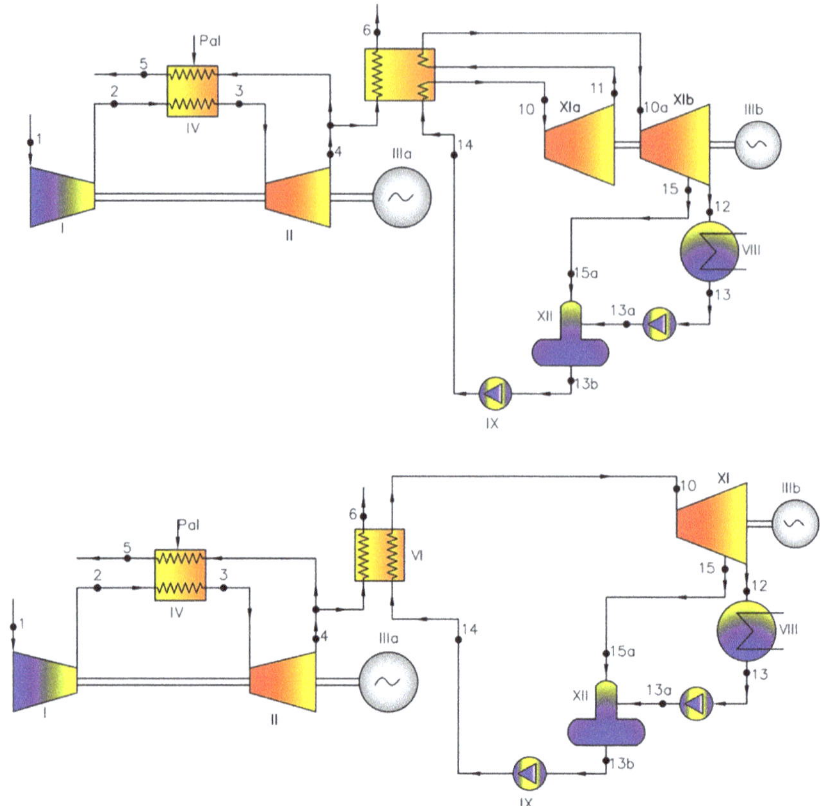

Figure 5. Combined steam–air system (sample solutions).

Figure 6. Compressor (**left**) and air turbine (**right**) for 55 MW combined steam–air system.

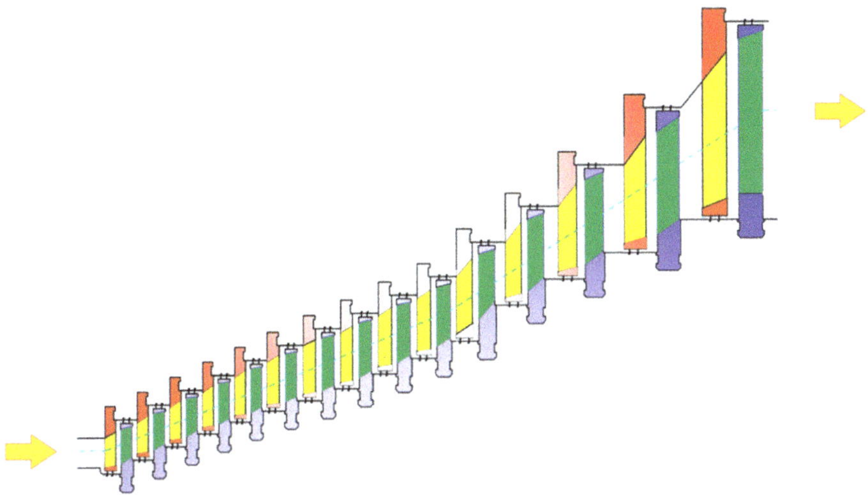

Figure 7. Schematic of steam turbine flow part for 55 MW combined steam–air system.

Table 2. Working media parameters in particular cycle points.

		W1	**W2**	**W3**	**W4**	**W5**
Π	[-]	15.50	3.20	1.85	3.00	3.10
p_0	[MPa]	0.1000	0.1000	0.1000	0.1000	0.1000
p_1	[MPa]	0.0995	0.0995	0.0995	0.0995	0.0995
p_2	[MPa]	1.5423	0.2985	0.1841	0.2985	0.3085
$p_{2'}$	[MPa]	-	0.2955	-	-	-
p_3	[MPa]	1.5268	0.2926	0.1822	0.2955	0.3054
p_4	[MPa]	0.1005	0.1015	0.1015	0.1025	0.1025
p_5	[MPa]	0.1000	0.1005	0.1005	0.1015	0.1015
p_6	[MPa]	-	0.1000	0.1000	0.1005	0.1005
p_7	[MPa]	-	-	-	0.1000	0.1000
T_0	[°C]	20	20	20	20	20
T_1	[°C]	20	20	20	20	20
T_2	[°C]	407.04	148.40	82.59	140.11	144.30
$T_{2'}$	[°C]	-	633.24	-	-	-
T_3	[°C]	900	900	900	900	900
T_4	[°C]	381.71	643.24	737.47	624.46	617.19
T_5	[°C]	381.71	199.89	92.59	910	910
T_6	[°C]	-	199.89	92.59	221.89	154.30
T_7	[°C]	-	-	-	221.89	154.30

In those solutions, the expected efficiencies could exceed 60% (60%–65%) or even 65% at higher air turbine inlet temperatures T_3 = 1000–1200 °C. Note that these temperatures are lower than those currently recorded in the most technologically advanced gas turbine sets.

5. Conclusions

Logging and husbanding energy constitute the essential elements of sustainable development. Laying the foundations for sustainable development relates to the energetics upon which many social and economic processes depend. The achievement of sustainable energetics should be the object of further investigations. As a future direction, Poland should aim to enlarge its ecological consciousness with the implementation of sustainable energetics. The proposed creation of circulation-combined power stations is one of the possible solutions to the problems of the Polish coal industry and the national energy production sector. It might also bring closer wide-ranging changes in the energetic politics.

The proposed combined power plant cycle is a solution variant for the problems faced by the Polish coal-based energy production sector. It would create opportunities for building highly efficient coal power plants (>60% efficiency). It would also respond to Poland's need to meet environmental protection requirements, as high efficiency will make CO_2 capturing and storage unnecessary, while simultaneously ensuring the high profitability of energy generation from coal.

The proposed technology is distinguished by:

- The original scheme of the thermodynamic cycle of power plant;
- The low parameters of the working medium;
- Its high efficiency, equal to about 60%–65% (or more). This means it would be possible to nearly double the amount of electric energy generated from national coal (for a given amount of coal, as compared to the present state of technology). In practice, this would be equivalent to decreasing fuel consumption by half. It would also be accompanied by a smaller consumption of cooling water (by at least 2–4 times) and smaller amounts of heat being released to the environment;
- The fuel used can be black or brown coal, resources which are plentiful in Poland. This solution will also allow for diversification, that is, the possible use of other fuels such as oil, natural gas, biofuels, biomass, bio gas, wooden pellets, and/or agricultural and municipal waste;
- Lower power plant construction costs, compared to, for instance, both subcritical and supercritical steam power plants, or modern combined gas-steam systems;
- Cheaper power plant operation and maintenance. Flow parts of the compressor and turbine would remain clean during the entire useful life of the power plant, as they would not be polluted with exhaust and would not require repairs and cleaning as in other highly efficient solutions;
- Possibility of application in thermal and electric power plants, in cogeneration and trigeneration systems (i.e., for the simultaneous production of electricity, heat, and useful cold);
- The absence of carbon dioxide capture and storage installation, which significantly decreases the power plant construction costs and avoids the considerable efficiency drop related to the operation of this installation;
- Halving of CO_2 emissions, down to below 600 $kgCO_2$/MWh (i.e., to a value which is comparable with the CO_2 emissions from present natural gas-fired power plants). Additionally, the emission of greenhouse gases would be reduced to half that of the emissions of present coal-fired devices. This solution would meet, with a surplus, all EU requirements (program 3×20), and would simultaneously be a rescue for Polish coal, which could then be considered as "clean fuel".

The proposed solutions would provide sales for Polish coal and would positively affect the energy security of the country. An optional solution to that described in the article is a combined system consisting of a closed CO_2 cycle and steam cycle. Design calculations performed for this variant and preliminary designs of gas and steam turbines for a 335 MW power plant confirmed the possibility of its implementation and of reaching the "60% plus" efficiency.

Author Contributions: Conceptualization, K.K. and M.P.; R.S.; W.W.; Methodology, K.T. and O.O.; Validation, K.K. and M.P.; A.B.; Investigation, R.S. and W.W.; Writing—Original Draft Preparation, K.T. and O.O.; A.B.; Funding Acquisition, K.K.

Funding: This research received no external funding.

Acknowledgments: The Authors wish to express their deep gratitude to Gdansk University of Technology for financial support given to the present publication (Krzysztof Kosowski) The research contributing to the present publication, conducted by Olga Orynycz, has been performed under the financial support No. S/WZ/1/2015 financed by the Polish Ministry of Science and Higher Education from the funds dedicated to science.

Conflicts of Interest: The authors declare no conflict of interest. The funders had no role in the design of the study; in the collection, analyses, or interpretation of data; in the writing of the manuscript, and in the decision to publish the results.

References

1. Orynycz, O.; Świć, A. The Effects of Material's Transport on Various Steps of Production System on Energetic Efficiency of Biodiesel Production. *Sustinability* **2018**, *10*, 2736. [CrossRef]
2. Ciegis, R.; Ramanauskiene, J.; Martinkus, B. The Concept of Sustainable Development and its Use for Sustainability Scenarios. *Econ. Cond. Enterp. Funct.* **2009**, *2*, 28–37.
3. Schirone, L.; Pellitteri, F. Energy Policies and Sustainable Management of Energy Sources. *Sustinability* **2017**, *9*, 2321. [CrossRef]
4. Patterson, W. *Keeping the Lights on: Towards Sustainable Electricity*; Earthscan: London, UK, 2009.
5. Mitchell, C. *The Political Economy of Sustainable Energy*; Palgrave Macmillan: Basingstoke, UK, 2008.
6. Kissinger, M.; Rees, W.E.; Timmer, V. Interregional sustainability: Governance and policy in an ecologically interdependent world. *Environ. Sci. Policy* **2011**, *14*, 965–976. [CrossRef]
7. Lior, N. Sustainable energy development (May 2011) with some game-changers. *Energy* **2012**, *40*, 3–18. [CrossRef]
8. Nunes, L.J.R.; Matias, J.C.O.; Catalão, J.P.S. Biomass combustion systems: A review on the physical and chemical properties of the ashes. *Renew. Sustain. Energy Rev.* **2016**, *53*, 235–242. [CrossRef]
9. Köppl, A.; Schleicher, S.P. What Will Make Energy Systems Sustainable? *Sustainability* **2018**, *10*, 2537. [CrossRef]
10. Pięta, P. Clean Coal Technologies in Every Stage of Coal Production in Polish Industry. *Min. Sci.* **2018**, *22*, 113–124.
11. Kuchler, M.; Bridge, G. Down the black hole: Sustaining national socio-technical imaginaries of coal in Poland. *Energy Res. Soc. Sci.* **2018**, *41*, 136–147. [CrossRef]
12. Jonek Kowalska, I. Challenges for long-term industry restructuring in the Upper Silesian Coal Basin: What has Polish coal mining achieved and failed from a twenty-year perspective? *Resour. Policy* **2015**, *44*, 135–149. [CrossRef]
13. Kamiński, J. A blocked takeover in the Polish power sector: A model-based analysis. *Energy Policy* **2014**, *66*, 42–52. [CrossRef]
14. Wierzbowski, M.; Filipiak, I.; Lyzwa, W. Polish energy policy 2050—An instrument to develop a diversified and sustainable electricity generation mix in coal-based energy system. *Renew. Sustain. Energy Rev.* **2017**, *74*, 51–70. [CrossRef]
15. Budzianowski, W.M. Target for national carbon intensity of energy by 2050: A case study of Poland's energy system. *Energy* **2012**, *46*, 575–581. [CrossRef]
16. Kamiński, J. Market power in a coal-based power generation sector: The case of Poland. *Energy* **2011**, *36*, 6634–6644. [CrossRef]
17. Chmielak, T.; Ziębik, A. *Obiegi Cieplne Nadkrytycznych Bloków Węglowych [in Polish]*; Wydawnictwo Politechniki Śląskiej: Gliwice, Poland, 2010; pp. 19–43.
18. Elsner, W.; Kowalczyk, Ł.; Niegodajew, P.; Drobniak, S. Thermodynamic Analysis of a Thermal Cycle of Supercritical Power Plant. *Mech. Mech. Eng.* **2011**, *15*, 217–225.
19. Kotowicz, J.; Brzęczek, M. Analysis of increasing efficiency of modern combined cycle power plant: A case study. *Energy* **2018**, *153*, 90–99. [CrossRef]
20. Bejan, A. *Advanced Engineering Thermodynamics*; John Wiley & Sons: Hoboken, NJ, USA, 2016.
21. Bugge, J.; Kjaer, S.; Blum, R. High–efficiency coal–fired power plants development and perspectives. *Energy* **2006**, *31*, 1437–1445. [CrossRef]
22. Tumanovskii, A.G. Prospects for the development of coal-steam plants in Russia. *Therm. Eng.* **2017**, *64*, 399–407. [CrossRef]

23. Nieć, M.; Sermet, E.; Checko, J.; Górecki, J. Evaluation of coal resources for underground gasification in Poland. Selection of possible UCG sites. *Fuel* **2017**, *208*, 193–202. [CrossRef]
24. Miller, B.G. Clean Coal Technologies for Advanced Power Generation. In *Clean Coal Engineering Technology*, 2nd ed.; Butterworth-Heinemann: Oxford, UK, 2017; pp. 261–308.
25. Ibrahim, T.K.; Mohammed, M.K.; Awad, O.I.; Abdalla, A.N.; Basrawi, F.; Mohammed, M.N.; Najafi, G.; Mamat, R. A comprehensive review on the exergy analysis of combined cycle power plants. *Renew. Sustain. Energy Rev.* **2018**, *90*, 835–850. [CrossRef]
26. Naserabad, S.N.; Mehrpanafi, A.; Ahmadi, G. Multi-objective optimization of HRSG configurations on the steam power plan repowering specifications. *Energy* **2018**, *159*, 277–293. [CrossRef]
27. Blumberg, T.; Assar, M.; Morosuk, T.; Tsatsaronis, G. Comparative exergoeconomic evaluation of the latest generation of combined-cycle power plants. *Energy Convers. Manag.* **2017**, *153*, 616–626. [CrossRef]
28. Yue, T.; Lior, N. Thermodynamic analysis of hybrid Rankine cycles using multiple heat sources of different temperatures. *Appl. Energy* **2018**, *222*, 564–583. [CrossRef]
29. Kosowski, K. *Steam and Gas Turbines with the Examples of Alstom Technology*; Alstom: Saint-Ouen, France, 2007; ISBN 978-83-925959-3-9.
30. Masłow, L.A. *Ship Gas Turbines*; Sudostroene: Leningrad Region, Russia, 1973. (In Russian)
31. Kostiuk, A.G.; Serstiuk, A.N. *Gas Turbines Units*; Wyschaya Skola: Moscow, Russia, 1979.
32. Zhang, Y.; Li, H.; Han, W.; Bai, W.; Yang, Y.; Yao, M. Improved design of supercritical CO_2 Brayton cycle for coal-fired power plant. *Energy* **2018**, *155*, 1–14. [CrossRef]
33. Zhang, X. Recent Developments in High Temperature Heat Exchangers. *Rev. Front. Heat Mass Transf.* **2018**, *11*, 18.
34. HTI. Available online: www.heatxfer.com (accessed on 19 January 2019).
35. BOSAL. Available online: www.bosal.com (accessed on 19 January 2019).
36. Brayton Energy. Available online: www.braytonenergy.net (accessed on 19 January 2019).
37. National Energy Technology Laboratory. Available online: www.netl.doe.gov (accessed on 19 January 2019).
38. VED Delmas. Available online: www.vdldelmas.com (accessed on 19 January 2019).

MDPI
St. Alban-Anlage 66
4052 Basel
Switzerland
Tel. +41 61 683 77 34
Fax +41 61 302 89 18
www.mdpi.com

Sustainability Editorial Office
E-mail: sustainability@mdpi.com
www.mdpi.com/journal/sustainability